Logic, Methodology and Philosophy of Science and Technology

Bridging Across Academic Cultures

Proceedings of the Sixteenth International Congress in Prague

Logic, Methodology and Philosophy of Science and Technology

Bridging Across Academic Cultures

Proceedings of the Sixteenth International Congress in Prague

Edited by

Tomáš Marvan

Hanne Andersen

Hasok Chang

Benedikt Löwe

Ivo Pezlar

© Individual author and College Publications, 2022
All rights reserved.

ISBN 978-1-84890-368-5

College Publications
Scientific Director: Dov Gabbay
Managing Director: Jane Spurr

http://www.collegepublications.co.uk

All rights reserved. No part of this publication may be reproduced, stored in a retrieval system or transmitted in any form, or by any means, electronic, mechanical, photocopying, recording or otherwise without prior permission, in writing, from the publisher.

Contents

Editors' Preface	iii
Words of Welcome from the DLMPST/IUHPST	v
Words of Welcome from the Local Organizing Committee	vii

INVITED PAPERS

Maryanthe Malliaris
Notes on Model Theory and Complexity   3
(Congress Section A.1)

Dunja Šešelja
What Kind of Explanations Do We Get from Agent-Based Models of Scientific Inquiry?   17
(Congress Section B.3)

Jonathan Okeke Chimakonam
Decolonising Scientific Knowledge: Morality, Politics and a New Logic   47
(Congress Section B.5)

Michael Matthews
Philosophy in Science Teacher Education   69
(Congress Section B.7)

Gerhard Heinzmann
Mathematical Understanding by Thought Experiments   85
(Congress Section C.1)

Hans Halvorson
Objective Description in Physics 111
(Congress Section C.2)

Ray Turner
Representation and Abstraction in Theories of Operations and Classes 133
(Congress Section C.6)

SYMPOSIA AND THEMATIC PANELS

Symposia and Thematic Panels Organized at the Congress 167
(a) Summaries of Symposia and Panels
    (i) DLMPST/IUHPST Commissions' Symposia 167
        Panels 207
    (ii) Contributed Symposia 209
(b) Other Contributed Symposia 283

APPENDICES

Appendix A 287
Congress Sections

Appendix B 288
List of Plenary, Invited and Contributed Talks

Appendix C 317
Mario Bunge (1919–2020)

Appendix D 319
Division of Logic, Methodology and Philosophy of Science and Technology of the International Union of History and Philosophy of Science and Technology Bulletin No. 23

Editors' Preface

This volume contains papers based on some of the invited lectures from the 16th International Congress of Logic, Methodology and Philosophy of Science and Technology. Apart from these, the opening statements by the representatives of the DLMPST and of the Local Organizing Committee are also printed, together with short descriptions of congress symposia written by their organisers. The texts are further supplemented with additional materials relating to the congress and DLMPST, printed as Appendices A–D.

*

The congress was held in Prague, Czechia, on August 5–10, 2019, under the auspices of the Division of Logic, Methodology, and Philosophy of Science and Technology (DLMPST) of the International Union of History and Philosophy of Science and Technology (IUHPST). The congress was held by the invitation of the Czech National Committee of Logic, Methodology and Philosophy of Science, and hosted by the Institute of Philosophy of the Czech Academy of Sciences. The overarching theme of the congress was 'Bridging across Academic Cultures'. During the congress week, four plenary lectures, twenty-one invited lectures and nearly seven hundred regular and symposia talks were presented. Abstracts of all the talks presented at the congress were published in the Book of Abstracts (available at the DLMPST website).

As a world congress that brings together philosophers of all career stages from all over the world, the CLMPST offers special opportunities for showcasing the full scope of our profession and for providing scholarly exchange on the many different facets of professional life. At the same time, such a huge event also creates special challenges for the participants in getting an overview of who is there and of what goes on, and that can make it difficult to navigate in the myriad of events that the congress offers. Some participants enjoy the large selection that a world congress offers. They take pleasure in walking around between sessions and talks, whether to pursue a very specific interest or to be surprised and enlightened by the many different topics and approaches represented. Others find a huge selection of parallel sessions confusing. They may feel lost among the many participants, or they may find it distracting that others walk in and out during sessions.

The CLMPST 2019 aimed at optimizing the conference experience for different types of participants by creating thematic tracks in the program that would remain physically in the same rooms over a longer period of time. That enabled those participants who were particularly interested in a specific topic to stay in same room together with others who shared their interests. In this way, micro-communities focused on particular topics could emerge within the big event. That also meant that participants who found the totality of the world congress overwhelming had the opportunity to go to such a track, and then stay in the same room and meet many of the same faces during the duration of the track. At the same time, other participants who enjoyed the huge selection could move freely between tracks, between sessions and between talks. However, to minimize irritation from clashes in preferences between those who want to move around between individual talks, and those who dislike the disturbance created by people moving in and out of a session between talks, sessions were organized in one-hour slots with only two talks, and then short breaks between these slots would enable participants to smoothly move between sessions.

With its division into 20 thematic sections, the 16th CLMPST offered a very broad coverage of the many topics discussed in logic, general philosophy of science and philosophy of the individual scientific disciplines. Each thematic section included contributed symposia with a number of papers focused on a specific topic; sessions with contributed individual papers scheduled together by the program committee; and invited lectures by scholars who had been selected by the program committee to present recent advancements from within their specialty. To enable participants, especially junior scholars and other newcomers to the profession, to get an overview of current discussions across the profession, the special invited talks for each thematic session were planned as an easily identifiable track of non-overlapping talks. In this way, the invited lectures together worked as an overview track that enabled participants to get introduced to exemplary work from across logic, methodology and philosophy of science. A similar overview track was created from the sessions organized by the regional and thematic societies.

*

The texts printed in this volume were prepared into a camera ready version in Prague by Tomáš Marvan and Ivo Pezlar. We thank Anna Pilátová, Anna Bryson Gustová, Eliška Končelová and Allen Wertheim for their meticulous work on the proofs in various stages of preparation of the volume, and to Vít Gvoždiak for technical assistance with recovering the congress data. Finally, we express our gratitude to College Publications for their willingness to continue the tradition of publishing the proceedings of the CLMPST congresses in their dedicated series.

Prague, Copenhagen, Cambridge, November 2022
Tomáš Marvan, Hanne Andersen, Hasok Chang, Benedikt Löwe, and Ivo Pezlar

Words of Welcome from the DLMPST/IUHPST

> *Editorial Note.* Traditionally, the websites of our congresses have words of welcome from the officers of DLMPST situating the congress in the tradition of congresses and explaining the theme of the congress. These texts are then reprinted in the conference booklets given to the participants. The audience of these texts were therefore both researchers considering to attend the congress, researchers who had already decided to go and were preparing for the congress participation, and the participants of the congress. In this proceedings volume, we reprint these texts for documentation purposes.

Dear logicians and philosophers of science,

on behalf of the Division of Logic, Methodology and Philosophy of Science and Technology of the International Union for History and Philosophy of Science and Technology (DLMPST/IUHPST), we should like to welcome all researchers and scholars in our fields to the website of the XVIth Congress on Logic, Methodology, and Philosophy of Science and Technology (CLMPST) and invite them cordially to join us in Prague in August 2019.

Our fields, logic and philosophy of science and technology, are traditionally at the nexus between disciplines, equally at home in the sciences, the humanities, and the social sciences. Consequently, many logicians and philosophers of science frequently act as mediators between disciplinary traditions and serve as translators between mutually unintelligible academic vernaculars.

Thus, the theme of the 2019 Congress, 'Bridging across Academic Cultures', is central to the experience of many among us.

The theme refers to the aforementioned academic cultures, so forcefully evoked in C. P. Snow's 1959 Rede lecture, 'The Two Cultures' (celebrating its 60th anniversary in 2019), but also to the pertinent cultural differences between the various countries and regions around the globe. As the global representative of our fields, the DLMPST/IUHPST desires to enrich our scientific dialogue by including the voices of those traditionally under-represented in our academic fields, be it due to geography, culture, ethnicity, or gender.

We are looking forward to having papers and symposia on all topics pertaining to our fields, from mathematical logic to metaphilosophy, from the foundations of the exact sciences to the philosophy of the social sciences; presented by researchers from all regions and cultures of the world. This will make the Prague Congress a scientifically inspiring, inclusive, and exciting event that truly bridges across academic cultures.

Menachem Magidor
President

Benedikt Löwe
Secretary General

Words of Welcome from the Local Organizing Committee

Dear colleagues,

on behalf of the local organising team of the 16th CLMPST, let me welcome you all to Prague—a lively, safe and eye-catching city. In the past, Prague has been an important place in the development of both logic and the philosophy of science. It was the home of great logicians and philosophers such as Bernard Bolzano (1781–1848), Ernst Mach (1838–1916), Philipp Frank (1884–1966) and Rudolf Carnap (1891–1970). In 1929, Prague hosted the first preparatory congress on the epistemology of the exact sciences, organised jointly by the Ernst Mach Society and the Berlin-based Society for Empirical Philosophy. It was on this occasion that the famous Vienna Circle manifesto of the scientific conception of the world was first made public, 2019 marking the ninetieth anniversary of this event. Prague hosted the Congress again in 1934. During the communist régime, the activity of Czech philosophers and logicians was subdued. Nonetheless, important works in mathematical logic and other fields of study were produced. For example, Petr Vopěnka (1935–2015) developed alternative set theory and worked in the philosophy of geometry, while Petr Hájek (1940–) has been working on the mathematical foundations of fuzzy logic. Since the revolution in 1989, Prague and Czechia have once again become a focal point for research and a busy meeting place of logicians and philosophers of science.

The host of the Congress, the Institute of Philosophy of the Czech Academy of Sciences (est. 1990), is a leading Czech research institution within the fields of philosophy and the humanities. It participates in a wide array of international research projects and regularly hosts guest lectures, conferences and workshops, such as the internationally renowned LOGICA Conferences and the Ernst Mach Workshops which present cutting-edge research in contemporary philosophy of science. Moreover, the Institute contains no fewer than three departments that focus particularly on logic, methodology and the philosophy of science.

The sixteenth edition of the Congress brings together logicians and philosophers of science from quite literally all over the world. The academic programme of the Congress covers all major areas of logic and the philosophy of science and is spread across 20 thematic sections. Some of these sections make their debut here in Prague: the empirical and experimental philosophy of science, educational aspects of the philosophy of science, philosophy of the biomedical and health sciences, the philosophy of computing and computation, and the philosophy of emerging sciences. The Congress will host four plenary lectures, twenty-two invited lectures, forty thematic symposia and hundreds of contributed talks. If that is not enough, you are welcome to stay in Prague for a further week to enjoy dozens and dozens of talks on mathematical logic during the Logic Colloquium, our partner conference, at the same venue.

Enjoy the best of logic and the philosophy of science in the heart of Europe!

Tomáš Marvan
Chair of the Local Organising Committee

Invited Papers

Notes on Model Theory and Complexity

MARYANTHE MALLIARIS[1]

Abstract. The aim is to provide some background and intuition for Keisler's order on theories, complementing a talk of the author at the Congress in 2019 which discussed the recent theorem of Malliaris and Shelah that Keisler's order has the maximum number of classes.

Keywords: saturation of ultrapowers, classification theory, Keisler's order.

1 Preamble

An unapparent connection is stronger than one which is obvious.
Heraclitus (fragment 54)

What does an algebraically closed field, a dense linear order, or a random graph look like "on average"? How does changing the average change the answer? What does the range of possibility tell us about the original object and its simplicity or complexity compared to other such objects? One interesting approach to this question is via the ultrapower construction. Take many copies of the model in question, "average" the sequence using an ultrafilter, and ask how the choice of ultrafilter affects the outcome. Even better, take many copies of *each* model in question, "average" *each* sequence separately using the same ultrafilter, and look for characteristic differences in the ultrafilter's effect on each sequence, as a way of "comparing complexity."

Keisler's suggestion was that the *level of saturation* of the resulting ultrapower was a useful distinguishing question to ask. Some (sequences of) models are easy to saturate—they are relatively simple. Some require the ultrafilter to have a certain kind of strength. Some are so complex that any ultrafilter which can saturate them must saturate any other model. Focusing on so-called *regular* ultrafilters, we will see this becomes a question about theories, independent of the choice of model. Then Keisler's order compares complete, countable theories T_1, T_2 by putting $T_1 \trianglelefteq T_2$ if for any infinite set I, any regular

---

[1] Department of Mathematics, University of Chicago, 5734 S. University, Chicago, IL 60637, USA.

ultrafilter $\mathcal{D}$ on I, any choice of models $M_1 \vDash T_1$, $M_2 \vDash T_2$, we have that if $(M_2)^I/\mathcal{D}$ is reasonably (i.e., $|I|^+$-) saturated, so is $(M_1)^I/\mathcal{D}$.

The exploration of the structure of this order over the last fifty years is changing our understanding of model-theoretic complexity. We will discuss some of its basic aspects below, with pointers to open problems.

All theories are complete and countable, unless otherwise stated.

2 Preliminaries

These notes may be read with a minimum of preliminaries, which more or less amount to understanding "saturation" and "regular ultrapowers." That is, it would be useful to know three main kinds of things. First is an understanding of what model theorists mean by types, along with the definition of "the model M is κ-saturated," the fact that it is sufficient to realize types in one free variable, and the idea that models are saturated precisely when they are homogeneous and universal (Chang & Keisler, 2012, 5.1.14). Second is an understanding of the ultrapower construction, as in Chang & Keisler (2012, §4.1). Third would be a few facts on regular ultrapowers: a definition of "the ultrafilter $\mathcal{D}$ is regular" (repeated below), the fact that regular ultrafilters exist on any infinite cardinal (Chang & Keisler, 2012, 4.3.5), the fact that the size of a regular ultrapower of any infinite model is the full size of the underlying Cartesian power (Chang & Keisler, 2012, 4.3.7), and the fact that any regular ultrapower of a model in a countable language is $\aleph_1$-saturated, (Chang & Keisler, 2012, 6.1.1). (Note that we call $M^I/\mathcal{D}$ a regular ultrapower simply to mean that the ultrafilter $\mathcal{D}$ is regular.) For the interested reader, the section on saturated ultrapowers in Chang & Keisler (2012, §6.1) will reinforce much that is said below and is worth reading in full. That section also contains a very informative and relevant earlier proof, Keisler's proof under GCH that elementarily equivalent models have isomorphic ultrapowers.

The reader is also encouraged to look at Keisler's paper (2009), in which the order is first defined.

3 Definition of Keisler's order

To start with the definition:

Definition 3.1 (Keisler, 1967)**.** *Let T_1, T_2 be complete countable theories. We say that*

$$T_1 \trianglelefteq T_2$$

when for any infinite cardinal λ, any regular ultrafilter $\mathcal{D}$ on λ, any $M_1 \vDash T_1$, and any $M_2 \vDash T_2$, we have that if $(M_2)^\lambda/\mathcal{D}$ is λ^+-saturated, then $(M_1)^\lambda/\mathcal{D}$ is λ^+-saturated.

Ultrapowers are constructed from only two ingredients, the model M and the ultrafilter $\mathcal{D}$, and a priori we would expect the saturation of the ultrapower to reflect something about both ingredients and their interaction. As we will see below, Keisler proved a key lemma about regular ultrafilters: if $M \equiv N$ in a countable language, and $\mathcal{D}$ is a regular ultrafilter on λ, then $M^\lambda/\mathcal{D}$ is λ^+-saturated if and only if $N^\lambda/\mathcal{D}$ is λ^+-saturated. Informally, the saturation of a regular ultrapower depends only on the elementary class of the model we begin with, and the ultrafilter. In other words, when $\mathcal{D}$ is regular and T is a (complete, countable) theory, we may simply say "$\mathcal{D}$ saturates T" when for some, equivalently every, $M \vDash T$ we have that $M^\lambda/\mathcal{D}$ is λ^+-saturated. Then Keisler's order says that $T_1 \trianglelefteq T_2$ iff every regular ultrafilter which saturates T_2 also saturates T_1. Keisler observed:

Observation 3.2. *There is a minimum class in Keisler's order.*

Proof. It suffices to show that there are theories which are saturated by any regular ultrafilter.[2] Let T be the theory of algebraically closed fields of some fixed characteristic, and let M be any countable (and necessarily infinite) model of T. Let D be any regular ultrafilter on λ, and let $N = M^\lambda/\mathcal{D}$. Since $\mathcal{D}$ is regular, $|M^\lambda/\mathcal{D}| = |M^\lambda| = 2^\lambda$. Since uncountable algebraically closed fields have transcendence degree equal to their cardinality, and since an algebraically closed field is λ^+-saturated iff it has transcendence degree at least λ^+, N is necessarily λ^+-saturated.

□

In order to see more subtle interactions of types and ultrafilters, let's look at a basic picture of how they interact.

---

[2] A priori there need not have been theories like this, but since there are, they are minimal.

4 Distributions

In this section we follow Malliaris (2009, Chapter 1) (with the change of adding "accurate" to the definition of distribution; there we had distinguished between distributions and accurate distributions).

Recall that the elements of an ultrapower, $M^I/\mathcal{D}$, are equivalence classes of elements of M^I which are equal on a set belonging to $\mathcal{D}$. Before working with any ultrapower, we will fix a representative of each equivalence class, so that expressions like "for $\alpha \in M^I/\mathcal{D}$ and for $t \in I$, consider $\alpha[t]$" make sense, by always evaluating such a projection in terms of the fixed representative. When $\bar{a} = \langle a_0, \ldots, a_{l-1} \rangle$ is a tuple of elements of $M^I/\mathcal{D}$, write $\bar{a}[t]$ to mean the tuple $\langle a_0[t], \ldots, a_{l-1}[t] \rangle$.

Convention 4.1. *For this section, fix a regular ultrapower $N = M^I/\mathcal{D}$, $|I| = \lambda$, and a type or partial type p over some set $A \subseteq N$, $|A| \leq \lambda$. We will use $N, M, I, \mathcal{D}, \lambda, p, A$ in these roles unless otherwise said, and use "small" to mean of size $\leq \lambda$ (e.g., "all types over small sets").*

We start with the idea that types in the ultrapower are only really types "on average": in other words, sets of formulas with the property that the projections of any finite subset are consistent on a set in $\mathcal{D}$, but for each finite subset, it may be a different set in $\mathcal{D}$. The question of saturation of ultrapowers draws our attention to the fact that large sets of formulas (the type in question) can appear consistent in an ultrapower even though the projections to the index models may be quite inconsistent (but for constantly different reasons). Ultimately, the interest of the problem of Keisler's order seems to relate to the idea that the relevant properties of the boolean algebra of definable sets—those which affect how an ultrafilter may use local inconsistency to allow a type to be on average consistent but not on average realized—are model-theoretically significant. In this section we'll be interested in how to start to isolate such properties, by discussing how to best represent the relationship between types and their various projections in the factor models.

Definition 4.2. *Let p be a type or partial type over a small set A in the ultrapower N. Let $\langle \varphi_\alpha(x, \bar{a}_\alpha) : \alpha < \lambda \rangle$ be an enumeration of p. Define the Łos map to be the map*

$$Ł : [\lambda]^{<\aleph_0} \to \mathcal{D}$$

given by: for each finite $u \subseteq \lambda$,

$$u \mapsto \{t \in I : M \vDash \exists x \bigwedge_{\alpha \in u} \varphi_\alpha(x, \bar{a}_\alpha[t])\}$$

Observation 4.3. *The Łos map really does send each finite u to an element of $\mathcal{D}$, by Łos theorem.*

Remark 4.4. *The Łos map is monotonic[3], meaning that $u \subseteq v$ implies $f(v) \subseteq f(u)$. As a result, for each $t \in I$, we may ask how large is the set $\{\alpha : t \in Ł(\{\alpha\})\}$. A priori, it need not be finite.*

Observation 4.5. *Suppose that for each $t \in I$, the set of formulas $\{\varphi_\alpha(x, \bar{a}_\alpha[t]) : t \in Ł(\{\alpha\})\}$ is consistent and has a realization b_t in M. Then the element $b := \langle b_t : t \in I \rangle / \mathcal{D}$ realizes p in N.*

Proof. For every $\alpha < \lambda$, $\{t \in I : M \vDash \varphi_\alpha[b[t], \bar{a}_\alpha[t]]\} \in \mathcal{D}$, so $N \vDash \varphi_\alpha(b, \bar{a}_\alpha)$ by Łos theorem. □

Example 4.6. Suppose we consider $(\mathbb{Q}, <)^I / \mathcal{D}$ and let A be the diagonal embedding of $\mathbb{Q}$ in the ultrapower (i.e., the canonical copy of the original model as an elementary submodel of the ultrapower, given by sending $q \in \mathbb{Q}$ to the equivalence class of the constant sequence $(q, q, \ldots)$; in slight abuse of notation, identify $\mathbb{Q}$ with its diagonal embedding). Consider a type describing the cut at π, e.g., $\{x > q : q \in \mathbb{Q}, q < \pi\} \cup \{x < q : q \in \mathbb{Q}, q > \pi\}$. Then the Łos map will send every formula to I. So, although this type is realized in the ultrapower, e.g., by $\aleph_1$-saturation, the Łos map doesn't make this obvious: the "consistent and realized" conditions in 4.5 are not both met.

Regularity is a kind of strong incompleteness. As we will see, the *regularizing families* in regular ultrafilters will allow us to refine the Łos map of our type p to a map which only involves a finite amount of information in each index model.[4]

Definition 4.7. *The ultrafilter $\mathcal{D}$ on I, $|I| = \lambda$, is regular if there is $X = \{X_\alpha : \alpha < \lambda\} \subseteq \mathcal{D}$, called a* regularizing family, *such that the intersection of any infinitely many elements of X is empty.*

---

[3] This is the terminology, though "anti-monotonic" would also have been appropriate.
[4] One could define κ-regular to mean existence of a regularizing family of size κ, and say *regular* means $|I|$-regular.

Note that since $\mathcal{D}$ is a ultrafilter, the intersection of any finitely many elements of X must belong to $\mathcal{D}$; the condition asks that any larger intersection must be empty. Note also that regularity is an existential condition. For $\mathcal{D}$ to be regular, it must contain at least one regularizing family.

Example 4.8. *Let $\mathcal{D}$ be a nonprincipal ultrafilter on $\mathbb{N}$. The family $\{\mathbb{N}\setminus\{n\} : n \in \mathbb{N}\}$ is not a regularizing family, since it contains an infinite subfamily with nonempty intersection, but the family $\{\mathbb{N}\setminus\{0,\ldots,n\} : n \in \mathbb{N}\}$ is.*

From this example, it is easy to see that any nonprincipal ultrafilter on a countable set is regular.

Observation 4.9. *If X is a regularizing family for $\mathcal{D}$, this is equivalent to saying that every $t \in I$ belongs to only finitely many elements of X.*

Let's see how a regularizing family can help us to distribute the information in a type among the factor models in a more delicate way.

Definition 4.10 (Building a distribution). *Let p be any type over a small set A in the regular ultrapower N. Let $\langle \varphi_\alpha(x, \bar{a}_\alpha) : \alpha < \lambda \rangle$ be an enumeration of p. Let $\{X_\alpha : \alpha < \lambda\} \subseteq \mathcal{D}$ be a regularizing family. Consider the map:*

$$f : [\lambda]^{<\aleph_0} \to \mathcal{D}$$

given by: for each finite $u \subseteq \lambda$,

$$u \mapsto \{t \in I : M \models \exists x \bigwedge_{\alpha \in u} \varphi_\alpha(x, \bar{a}_\alpha[t])\} \cap \bigcap_{\alpha \in u} X_\alpha$$

The key properties of the function in 4.10—monotonic, refines Łos map, assigns finitely many formulas to each t, and is in a natural sense accurate—are listed in the following definition.

Definition 4.11. *Call f a distribution of p when, fixing an enumeration of p,*
(1)   *f is monotonic and refines the Łos map,*
(2)   *the image of f is a regularizing family,*
(3)   *for each $t \in I$, and any subset $v \subseteq \{\alpha < \lambda : t \in f(\{\alpha\})\}$, $t \in f(v)$ if and only if*
(4)

$$M \vDash \exists x \bigwedge_{\alpha \in v} \varphi_\alpha(x, \overline{a}_\alpha[t])$$

Observation 4.12. *Let f be a distribution of p and suppose that for each $t \in I$,*

$$M \vDash \exists x \bigwedge_{t \in f(\{\alpha\})} \varphi_\alpha(x, \overline{a}_\alpha[t])$$

(informally, the projections of the finitely many formulas assigned to index t have a common witness in M). Then p is realized in N.

Proof. Define $b \in N$ to be the $\mathcal{D}$-equivalence class of $\langle b_t : t \in I \rangle$, where each b_t is a common witness at index t, and notice by Łos theorem $\{t \in I : M \vDash \varphi_\alpha[b[t], a_\alpha[t]]\} \in \mathcal{D}$ for all $\alpha < \lambda$.

□

Example 4.13. Use a distribution to give another proof that the theory of algebraically closed fields of some fixed characteristic is minimal. (Over any small set A in $N = M^\lambda / \mathcal{D}$, it suffices to realize the type describing a transcendental element, say $\langle \varphi_\alpha(x, \overline{a}_\alpha) : \alpha < \lambda \rangle$ where each φ_α asserts that x is not a root of some polynomial over A. Let f be a distribution of p, so by definition, it assigns finitely many conditions to each index t. In any algebraically closed, hence infinite, field, one can always find an element which is not a root of any of a given finite set of polynomials.)

Claim 4.14. *The following are equivalent:*
(a)   *p is realized in N.*
(b)   *there is a distribution of p such that for each $t \in I$,*

$$M \vDash \exists x \bigwedge_{t \in f(\{\alpha\})} \varphi_\alpha(x, \overline{a}_\alpha[t])$$

Proof. One direction is 4.12, and for the other, suppose b is a realization of p in N. Fix an enumeration of p and let $\{X_\alpha : \alpha < \lambda\}$ be a regularizing family. Consider the map g given by

$$u \mapsto \{t \in I : M \vDash \bigwedge_{\alpha \in u} \varphi_\alpha(b[t], \bar{a}_\alpha[t])\} \cap \bigcap_{\alpha \in u} X_\alpha$$

for every finite $u \subseteq \lambda$. □

More fundamentally, Keisler's idea of a good ultrafilter gives a beautiful and simple connection between realizing types and refinements of certain maps.[5] Recall that $f : [\lambda]^{<\aleph_0} \to \mathcal{D}$ is *monotonic* if $u \subseteq v$ implies $f(u) \supseteq f(v)$. Call $g : [\lambda]^{<\aleph_0} \to \mathcal{D}$ *multiplicative* if $g(u) \cap g(v) = g(u \cup v)$, and say that g *refines* f if $g(u) \subseteq f(u)$, for all finite u, v subsets of λ. (In the next definition, one could omit monotonic, and then show it suffices to check this for the monotonic functions.)

Definition 4.15. *The ultrafilter $\mathcal{D}$ on λ is λ^+-good, or just* good, *if for every monotonic $f : [\lambda]^{<\aleph_0} \to \mathcal{D}$ there is $g : [\lambda]^{<\aleph_0}$ which is multiplicative and refines f.*

Claim 4.16. *Let f be a distribution of p. Then the following are equivalent:*
(a)   *p is realized in N.*
(b)   *there is a map $g : [\lambda]^{<\aleph_0} \to \mathcal{D}$ which is multiplicative and refines f.*

Proof. For (b) implies (a), notice that g is a distribution: conditions (1) and (2) are immediate because it refines f, and condition (3) follows by multiplicativity, which tells us that at each t, the set of formulas assigned to index t all have a common realization.

For (a) implies (b), notice that the map g from the proof of Claim 4.14 can easily be modified to refine any given distribution f. (If g is defined as there, define h by $h(u) = g(u) \cap \bigcap_{\alpha \in u} f(\{\alpha\})$. This is the only place we use condition 4.11(3), the accuracy. If we hadn't asked for accuracy, there could a priori be indices t where α, β are both assigned by f to t, and even though $\{\varphi_\alpha(x, \bar{a}_\alpha[t]), \varphi_\beta(x, \bar{a}_\beta[t])\}$ are consistent, f nonetheless did not assign $\{\alpha, \beta\}$ to t. If so, $h(\{\alpha, \beta\})$ would not refine $f(\{\alpha, \beta\})$.)
□

We arrive to a theorem of Keisler:

Conclusion 4.17. *Let $\mathcal{D}$ be a good regular ultrafilter on λ and let M be a model in a countable language. Then $M^\lambda/\mathcal{D}$ is λ^+-saturated.*

---

[5] Good ultrafilters do not need to be regular a priori, though we will consider good regular ultrafilters.

Proof. By definition of good, every monotonic $f: [\lambda]^{<\aleph_0} \to \mathcal{D}$, and in particular every distribution of a type over a small set in the ultrapower has a multiplicative refinement. □

5 A lemma on regularity

Next, we discuss Keisler's lemma (Keisler, 1967, 2.1a), and since that is a well-written paper, let us use somewhat different language (which will motivate later discussions of the characteristic sequence, although when those arise, we will be working with φ-types for some fixed φ).

Lemma 5.1. *If $M \equiv N$ in a countable language, and $\mathcal{D}$ is a regular ultrafilter on λ, then $M^\lambda/\mathcal{D}$ is λ^+-saturated iff $N^\lambda/\mathcal{D}$ is λ^+-saturated.*

Proof. Suppose $p \in S(A)$ is a type over a small set in $M^\lambda/\mathcal{D}$, enumerated as $\langle \varphi_\alpha(x, \overline{a}_\alpha) : \alpha < \lambda \rangle$. It will suffice to show there is a type $q \in S(B)$ over a small set in $N^\lambda/\mathcal{D}$, enumerated as $\langle \varphi_\alpha(x, \overline{b}_\alpha) : \alpha < \lambda \rangle$, so that p is realized in $M^\lambda/\mathcal{D}$ if and only if q is realized in $N^\lambda/\mathcal{D}$.

The informal idea is that we simply use a distribution to see that p comes from a finite pattern in each index model, and then we use elementary equivalence of M and N to transfer these finite patterns over at each index t and then reassemble them in the second ultrapower.

More precisely, let $d: [\lambda]^{<\aleph_0} \to \mathcal{D}$ be a distribution of p. Fix an index $t \in I$. At t we have the set of formulas $\{\varphi_\alpha(x, \overline{a}_\alpha[t]): t \in d(\{\alpha\})\}$, and we may think of the distribution d as giving rise to the following auxiliary finite structure at index t. Let $\mathcal{L} = \{P_n: 1 \leq n < \omega\}$ where each P_n is an n-ary relation (not in our original language). The domain is the set of "vertices" $V(t) := \{\alpha : t \in d(\{\alpha\})\}$, finite by definition of distribution. On this domain, interpret the P_n's to be symmetric n-ary relations which record mutual consistency, that is, for each finite $1 \leq n < \omega$ and each $u \subseteq V(t)$, $|u| = n$, say that P_n holds on u if and only if $M \models \exists x \bigwedge_{\alpha \in u} \varphi_\alpha(x, \overline{a}_\alpha[t])$ (note by the accuracy condition on distributions, we could equivalently say, P_n holds on u if and only if $t \in d(u)$). Of course, the predicates P_n will only really be relevant for $n \leq |V(t)|$. They simply summarize the pattern of consistency at this point in the distribution.

Notice that the statement expressing that there exist parameter sequences for the formulas $\varphi_\alpha (\alpha \in V(t)$ with the precise pattern of consistency and inconsistency among these formulas recorded by the P_n's, is a first-order statement of $T = Th(M) = Th(N)$ (of course this statement will not mention the P_n's but rather the consistency or inconsisteny of the instances). Since $M \equiv N$, this statement also holds in N.

So in N we may find parameters $\overline{b}_{\alpha,t}$ for each $\alpha \in V(t)$ so that $\{\varphi_\alpha(x, \overline{b}_\alpha[t]): \alpha \in V(t)\}$ gives rise in N to precisely the same auxiliary $\mathscr{L}$-structure as did $\{\varphi_\alpha(x, \overline{a}_\alpha[t]): \alpha \in V(t)\}$ in M. For each $\alpha < \lambda$, let $\overline{b}_\alpha$ be the sequence in N so that $\overline{b}_\alpha[t] = \overline{b}_{\alpha,t}$ for all $t \in I$.

Since we have simply copied the pattern of p via the $\mathscr{L}$-structures, and p is a type, it is immediate by Łos theorem that the type enumerated as $\langle \varphi_\alpha(x, \overline{b}_\alpha) : \alpha < \lambda \rangle$ is a type over a small set in $N^\lambda/\mathscr{D}$. Call it q. Moreover, the distribution d of p we began with is itself a distribution for q. This is the heart of the matter: if the two types share a distribution (an accurate distribution!) then 4.16 tells us one cannot be realized without the other. □

6 The maximum class

Keisler's proof of existence of a maximum class in his order has two parts. First, Conclusion 4.17 tells us that good ultrafilters can handle any theory. Second, one proves that there are *some* theories which require goodness, in other words, the set $\{T$: for any infinite I, and any regular ultrafilter $\mathscr{D}$ on I, and any model $M \vDash T$, we have that $M^I/\mathscr{D}$ is λ^+-saturated if and only if $\mathscr{D}$ is good$\}$ is nonempty.

To this day, we have only this set-theoretic characterization of maximality: T is maximal in Keisler's order if and only if the only regular ultrafilters $\mathscr{D}$ which saturate T are the good regular ultrafilters. Finding a model theoretic necessary and sufficient condition for maximality is a very interesting open question, and recently, moving the known boundary to SOP_2 in Malliaris & Shelah (2016a) already involved some very interesting mathematics.

Why must a maximal theory exist? Observe that it would suffice to show that there is a theory such that for some $M \vDash T$, for any infinite λ and regular ultrafilter $\mathscr{D}$ on λ, and for any monotonic $f: [\lambda]^{<\aleph_0} \to \mathscr{D}$ whose image is a regularizing family, there is a type $p = \langle \varphi_\alpha : \alpha < \lambda \rangle \in S(A)$, $A \subseteq M^\lambda/\mathscr{D}$, $|A| \leq \lambda$ such that f is a distribution of p. If this is so, then f has a multiplicative refinement if and only if p is realized. When we can do this for any relevant f, $M^\lambda/\mathscr{D}$ is λ^+-saturated if and only if $\mathscr{D}$ is good. Put otherwise, T is able to detect any failure of goodness as the omission of a type. See Keisler's definition of a *versatile* formula in Keisler (1967). Such a formula can be seen to occur in reasonably complicated theories, such as set theory or number theory. (The reader may also try the following. Let M be the following model in the language $\{P, Q, \in\}$, where P, Q are unary relations and $\in$ is a binary relation. P^M and Q^M are both infinite and partition M. P^M contains a copy of $\mathbb{N}$, with no additional structure. Q^M contains an element corresponding to each finite subset of $\mathbb{N}$. We interpret $\in^M$ as the usual set membership relation: for $m \in P$, $u \in Q$, "$m \in^M u$" just in case the element m belongs to the subset corresponding to u. Let $T = Th(M)$.) It would have been reasonable to conjecture from such examples (in 1967) that maximality had something to do with the complexity of coding or at least the complexity of reflecting patterns. Shelah in (Shelah, 1978, VI, 2.6) changed the picture of maximality by proving that any theory with the strict order property (e.g., any theory of linear order) is maximal. One reason this might be a surprise is that pure linear order can't code much at all. Where then does its strength lie? As one very informal statement, consider the following difference between a type $p(x) = \{R(x, a_\alpha) \wedge \neg R(x, b_\alpha) : \alpha < \lambda\}$ in the random graph[6] and $q(x) = \{a_\alpha < x < b_\alpha : \alpha < \lambda\}$ describing a concentric cut in a linear order. Suppose we "miss" a formula of p: it is possible to continue to realize other formulas in the type. However, if we "miss" a formula of q, we miss all later formulas in the enumeration.

7 Stability and beyond

What happened next?

One of the surprising early results on Keisler's order, due to Shelah (1978, Chapter VI), was that the union of the first two classes in Keisler's order is precisely the stable theories. In some sense, this tells us that Keisler's order independently detects the major dividing line at stability, and it suggests that further dividing lines isolated by the order might be of great interest. For more on this, see Malliaris (2018). Moreover, like stability, Keisler's order has a local character (Malliaris, 2009) and so admits a certain kind of combinatorial analysis (Malliaris, 2012a) and (Malliaris, 2012b).

Since all theories with SOP are maximal, recalling the previous section, we see that all NIP theories fall into three classes (note that the union of these three classes strictly contains the NIP theories). For a long time, it was thought that Keisler's order had few classes overall, perhaps five or six, linearly ordered.

---

[6] The minimum unstable theory in Keisler's order, see Malliaris (2012a).

In a recent series of papers, Malliaris and Shelah have shown there is very interesting complexity in the region of the independence property, in some sense "near" the random graph (that is, "near" the minimum unstable class). Keisler's order has infinitely many classes among the simple unstable theories with trivial forking (Malliaris & Shelah, 2018), and indeed has the maximum number, continuum many (Malliaris & Shelah, 2021), already in that region. This has substantially changed our picture of the order and of its operation, by opening up a much closer connection between ultrafilters and theories, and illuminating the richness of the independence proeprty even in the absence of dividing. Perhaps this may lay the foundation for a deeper understanding of simple theories, which for a long time seemed fairly close to stable ones. Some first remarks on this are in the introduction to the recent paper (Malliaris & Shelah, 2021). There are interesting emerging connections to finite combinatorics, as noted, e.g., in Malliaris & Shelah (2021b) and Malliaris & Shelah (2021, §3).

The interested reader might now look at Malliaris (2018), Casey & Malliaris (2017), Malliaris & Shelah (2021). For open problems, see the papers (Malliaris & Shelah, 2017), (Malliaris, 2017) as well as the problems outlined in the research papers (Malliaris & Shelah, 2016b), (Malliaris & Shelah, 2016c) and (Malliaris & Shelah, 2021).

Acknowledgements. Partially supported by NSF CAREER award 1553653.

Bibliography

Casey, D., & Malliaris, M. (2017). Notes on cofinality spectrum problems. *Appalachian Set Theory Lecture Notes.* arXiv Preprint arXiv:1709.02408. Retrieved from https://arxiv.org/abs/1709.02408.

Chang, C. C., & Keisler, H. J. (2012). *Model Theory* (3rd ed.). Mineola, NY: Dover.

Heraclitus, fragment 54. (1987) Translated in *Heraclitus: Fragments, A Text and Translation with a Commentary by T. M. Robinson* (p. 39). Toronto: University of Toronto Press.

Keisler, H. J. (1967). Ultraproducts which are not saturated. *Journal of Symbolic Logic*, *32*, 23–46.

Malliaris, M. (2009). *Persistence and Regularity in Unstable Model Theory* (Doctoral dissertation, University of California, Berkeley). Retrieved from https://math.uchicago.edu/~mem/Malliaris-Thesis.pdf.

Malliaris, M. (2009). Realization of φ-types and Keisler's order. *Annals of Pure and Applied Logic*, *157*, 220–224.

Malliaris, M. (2012a). Hypergraph sequences as a tool for saturation of ultrapowers. *Journal of Symbolic Logic*, *77*(1), 195–223.

Malliaris, M. (2012b). Independence, order, and the interaction of ultrafilters and theories. *Annals of Pure and Applied Logic*, *163*(11), 1580–1595.

Malliaris, M. (2017). The clique covering problem and other questions. *The IfCoLog Journal of Logics and their Applications*, *4*(10), 3407–3429.

Malliaris, M. (2018). Model theory and ultraproducts. In B. Sirakov, P. N. de Souza, & M. Viana (Eds.), *Proceedings of the International Congress of Mathematicians – Rio de Janeiro 2018, Vol. II* (pp. 83–97). Hackensack, NJ: World Scientific.

Malliaris, M., & Shelah, S. (2016a). Cofinality spectrum theorems in model theory, set theory and general topology. *Journal of the American Mathematical Society*, *29*, 237–297.

Malliaris, M., & Shelah, S. (2016b). Existence of optimal ultrafilters and the fundamental complexity of simple theories. *Advances in Mathematics*, *290*, 614–681.

Malliaris, M., & Shelah, S. (2018). Keisler's order has infinitely many classes. *Israel Journal of Mathematics*, *224*(1), 189–230.

Malliaris, M., & Shelah, S. (2017). Open problems on ultrafilters and some connections to the continuum. *Contemporary Mathematics*, *690*, 145–160.

Malliaris, M., & Shelah, S. (2016c). Cofinality spectrum problems: the axiomatic approach. *Topology and Its Applications*, *213*, 50–79.

Malliaris, M., & Shelah, S. (2021). Keisler's order is not simple (and simple theories may not be either). *Advances in Mathematics*, *392*, 108036.

Malliaris, M., & Shelah, S. (2021b). Notes on the stable regularity lemma. *Bulletin of Symbolic Logic*, *27*(4), 415–425.

Shelah, S. (1978). *Classification Theory*. Amsterdam: North-Holland.

Author biography. Maryanthe Malliaris received her Ph.D. from the University of California at Berkeley, and is currently Professor of Mathematics at the University of Chicago. She was a winner of the 2010 Kurt Gödel Research Prize Fellowship, and in 2017 was awarded the Hausdorff medal for joint work with Saharon Shelah. Malliaris was an invited speaker at the 2018 International Congress of Mathematicians.

What Kind of Explanations Do We Get from Agent-Based Models of Scientific Inquiry?

DUNJA ŠEŠELJA[1]

Abstract. Agent-based modelling has become a well-established method in social epistemology and philosophy of science but the question of what kind of explanations these models provide remains largely open. This paper is dedicated to this issue. It starts by distinguishing between real-world phenomena, real-world possibilities, and logical possibilities as different kinds of targets which agent-based models (ABMs) can represent. I argue that models representing the former two kinds provide how-actually explanations or causal how-possibly explanations. In contrast, models that represent logical possibilities provide epistemically opaque how-possibly explanations. While highly idealised ABMs in the form in which they are initially proposed typically fall into the last category, the epistemic opaqueness of explanations they provide can be reduced by validation procedures. To this purpose, an examination of results of simulations in terms of classes of models can be particularly helpful. I illustrate this point by discussing a class of ABMs of scientific interaction and the claim that a high degree of interaction can impede scientific inquiry.

Keywords: agent-based models, highly idealised models, epistemically opaque how-possibly explanation, robustness analysis, scientific interaction.

1 Introduction

Computer simulations in the form of agent-based models (ABMs) have become a well-established formal method in social epistemology and philosophy of science. Following a long tradition in biomedical and social sciences, this computational method had quickly proven itself useful in the study of social aspects of scientific inquiry in subjects ranging from the impact of different social networks on the efficiency of knowledge acquisition and the division of cognitive labour all the way to research of the efficiency of scientific collaboration and studies of the norms that guide scientists facing disagreements. The primary advantage of using ABMs to examine such issues is that they allow us to study, in a controlled environment, how the various properties of individual agents representing scientists—such as their reasoning, decision-making, actions, and relations—bring about

---

[1] Ruhr University Bochum/Eindhoven University of Technology, dunja.seselja@rub.de.

various phenomena on the level of the scientific community, such as the success or a failure of the community to acquire knowledge.

Despite their popularity, studies based on computer simulations often meet with sceptical reactions of researchers who use other approaches to the philosophy of science, such as for instance historical case studies. Their primary concern is that the proposed models are highly idealised, which raises the question of validity of any findings such models may deliver. In particular, the simplicity with which the ABMs tend to represent scientific inquiry commonly leads to doubts regarding their explanatory value, such as: 'Do these models explain anything, and if so, what exactly?' 'Surely, they cannot be taken as explanatory of complex scientific episodes, which include a myriad of epistemic and non-epistemic causal factors?'

In this paper, I want to address these concerns and explain the nature of explanations which ABMs provide. I start (in Section 2) by distinguishing three focal points in the research on ABMs of science: the development of highly idealised models, studies of their robustness, and discussions of the epistemology of agent-based modelling. This will allow me to situate the current contribution within the third of the above-mentioned points. To examine the explanatory properties of highly idealised ABMs, I distinguish the different possible targets which ABMs can adequately represent, and then proceed to relate this classification to the types of explanation that can be inferred from each class (Section 3). I argue that highly idealised models that have not been validated provide *epistemically opaque how-possibly explanations*, that is, claims that express possible causal relationships although the conditions under which such relationships should hold are unclear. Further, I suggest that by the means of different validation procedures, ABMs can move from providing epistemically opaque explanations to *causal how-possibly explanations* (Section 4). I illustrate this point with a class of ABMs of scientific interaction and with a claim inferred on their basis, namely that a high degree of information flow can be detrimental to the efficiency of a scientific inquiry (Section 5). Section 6 then concludes the paper.

2 Research on ABMs of science

We can roughly distinguish three main directions in the research on ABMs of scientific inquiry developed within the philosophy of science. To explain the main questions raised within each of these focal points, let us first look at how the philosophical study of ABMs developed from other scientific domains.

Simulations of scientific inquiry are rooted in several parallel lines of research.[2] On the one hand, formal modelling was introduced into the philosophical study of social processes underlying scientific inquiry with the aim of gaining more precise insight into the tensions pervading scientific research, such as the tension between individual and group rationality or between epistemic and non-epistemic values.[3] That resulted in a number of analytical models, such as the model proposed by Goldman & Shaked (1991), which examined the relationship between the goal of one's professional success and promotion of truth acquisition, or Kitcher's models (1990, 1993), which tackled the division of cognitive labour against the background of individual rationality. These were later followed by several other proposals (e.g., Strevens, 2003; Zamora Bonilla, 1999; Zamora Bonilla, 2002).

Around the same time, computational methods entered the philosophical study of rational deliberation and cooperation in the context of game theory (Skyrms, 1990, 1996; Grim et al., 1998) and the study of opinion dynamics in social epistemology (Hegselmann & Krause, 2002, 2005). Computational models introduced in this literature already included ABMs: for instance, a cellular automata model of the Prisoner's Dilemma, or models examining how opinions change within a group of agents.

In a parallel development, agent-based modelling entered also the social sciences. In sociology of science, ABMs offered a novel way of analysing and explaining causal mechanisms underlying scientific inquiry, an approach that complemented the more entrenched method of quantitative empirical studies. The pioneering work of Gilbert (1997), aimed at simulating the structure of academic science, was closely related to a quantitative analysis of citation networks. Using a small number of simple assumptions, Gilbert's ABM was designed to reproduce certain quantitative relationships previously identified in empirical research (such as Lotka's Law concerning the distribution of citations among authors).

In contrast to ABMs developed in the sociology of science, which tended towards an integration of simulations and empirical studies used for their validation (cf. Gilbert

---

[2] For a recent overview of formal models of scientific inquiry and their role in philosophical literature, see Šešelja et al. (2020); for an overview of ABMs of scientific interaction see Šešelja (2022); for an overview of computational methods employed in philosophy, see Grim & Singer (2020) and Mayo-Wilson & Zollman (2021). For an earlier overview of ABMs of science, including both work done in sociology and in philosophy of science, see Payette (2012); for an overview of agent-based modelling and its role in social sciences and philosophy, see Klein et al. (2018); for a discussion of the use of computer models in science in general, see Imbert (2017).
[3] For an overview of economic approaches to social epistemology of science, which inspired discussions on the tension between the individual and group rationality, see Mäki (2005).

& Troitzsch, 2005), a parallel trend of abstract and highly idealised ABMs emerged in other social sciences, such as economics and archaeology. Most prominently, Schelling–Sakoda models of social segregation (Sakoda, 1971; Schelling, 1971, 1978; see also Hegselmann, 2017) and Axelrod's models of cooperation (e.g., Axelrod, 1984, 1997; Axelrod & Hamilton, 1981) paved the ground for agent-based modelling in the study of various social phenomena. These two trends gave rise to two distinct methodological approaches to ABMs that came to be known as KIDS (*Keep it Descriptive, Stupid*) and KISS (*Keep it Simple, Stupid*) strategies. The KIDS approach aims at developing models which are descriptively adequate with respect to central features of the target phenomenon and at integrating ABMs and empirical studies. The KISS approach, on the other hand, aims at the development of simple, highly idealised models which are based on a minimal set of assumptions about agents and their environment but sufficient to capture certain regularities on the community level.[4]

The development of ABMs in the philosophy of science has largely followed the KISS approach. The influential works of Hegselmann & Krause (2006), Zollman (2007, 2010), Muldoon & Weisberg (2011), Weisberg & Muldoon (2009), Grim (2009), Grim et al. (2013), and Douven (2010), among others, kickstarted research into abstract ABMs of scientific inquiry. This marks the first focal point in the research on ABMs in the philosophy of science. The development of ABMs aimed at demonstrating the contribution of agent-based modelling to the study of questions posed by philosophers of science and social epistemologists, such as the impact of social networks or division of cognitive labour on the efficiency of inquiry. The emphasis was on exploratory insights rather than validity of the models or a detailed analysis of their explanatory features. For modellers endorsing the KISS approach, this aim continued to be central.

Others, however, recognised the limitations of this approach. On the one hand, it is generally acknowledged that highly idealised models are sufficient to provide a 'proof of the concept', for instance, to show that a certain causal relationship is in principle possible (Šešelja, 2021). Similarly, highly idealised models are capable of producing conjectures about causal mechanisms underlying real-world phenomena. On the other hand, abstract models typically lack validation procedures, such as robustness analysis or studies of their representational adequacy (Aydinonat et al., 2020). This makes it difficult to assess whether and to what extent findings from these models can be considered informative of real-world scientific inquiries.

---

[4] For the KISS strategy see, e.g., Epstein & Axtell (1996), Axelrod (1997), Hegselmann & Krause (2002), Epstein (2006); for the KIDS one, see Edmonds & Moss (2004).

Such concerns gave rise to the second focal point in the research on ABMs in the philosophy of science: the study of robustness of previously developed models. To this end, previous models were adjusted and enhanced, resulting in what Aydinonat et al. (2020) called 'second generation models'.[5] The robustness analysis includes an examination of results delivered by a model with respect to changes in parameter values (sensitivity analysis) and changes to the idealising assumptions of the model (derivational robustness analysis). For example, with respect to Zollman's models (2007, 2010), Rosenstock et al. (2017) showed that the previously obtained results hold only for a small part of the relevant parameter space, while Frey & Šešelja (2020) and Borg et al. (2019) showed that Zollman's results do not obtain when some of the idealising assumptions are changed. Similar studies were conducted for Weisberg & Muldoon's (2009) model: others identified an error in the code of the model and critically assessed the robustness of results under different modelling assumptions (Alexander et al., 2015; Thoma, 2015; Pöyhönen, 2017; Pinto & Pinto, 2018).

Besides studies of robustness, enhancements of previously proposed ABMs have also led to their application to new research questions. For instance, a number of ABMs studying scientific polarisation, biases, or the spread of deceptive information were built on Zollman's work (see works by Holman & Bruner, 2015, 2017; O'Connor & Weatherall, 2018, 2019; Weatherall et al., 2018). Similarly, Weisberg & Muldoon's epistemic landscape model served as a starting point for various further studies: for instance, Balietti et al. (2015) studied the relationship between disciplinary fragmentation and scientific progress, Currie & Avin (2018) examined different types of scientific methods, while Harnagel (2018) and Avin (2019) focused on the mechanisms of allocation of research funding.

Finally, the third focal point in research on ABMs of science concerns the epistemology of agent-based modelling. What can we learn from ABMs? What kind of epistemic functions do they have? What are their limitations and prospects for future improvement? These interrelated questions have been examined in a number of studies. On the one hand, some have argued that unless ABMs are empirically embedded and validated, we will have a hard time ensuring their empirical adequacy (e.g., Martini & Pinto, 2016; Thicke, 2020; Bedessem, 2019; Frey & Šešelja, 2018, Šešelja, 2021; Politi, 2021). For instance, by using empirical data as the input for ABMs we can calibrate parameters in the model

---

[5] Research on ABMs in empirical sciences has followed a similar course. For instance, Thiele et al. (2014) identify two phases in their development: The first focused on gaining generic insights via ABMs rather than on their in-depth analysis. In the second phase, previously developed models are subjected to various types of robustness analyses with the goal of 'better mechanistic understanding of the model and on relating the model to real-world phenomena and mechanisms'.

(for example, Harnagel, 2018, used bibliometric data to this purpose). On the other hand, Mayo-Wilson & Zollman (2021) have argued that for some modelling purposes, such as illustrating that certain events or situations are possible, validation need not be necessary. Models can instead be justified by 'plausibility arguments' and by recourse to stylised historical case studies.

Central to the above discussion is the question of the epistemic purpose of a model. Aydinonat et al. (2020) have argued that this may be difficult to assess when examining a model in isolation. According to them, we instead ought to take a 'family-of-models perspective' and determine the contribution of an ABM using subsequent models that enable a better understanding of results delivered by the previous ones.[6] More precisely, Aydinonat and colleagues argue that we should view the ABMs as argumentative devices whose purpose is determined by the argumentative context in which they are used. An argument supported by a particular model can be further strengthened by analyses based on subsequent models.

This paper belongs to the third focal point in research on ABMs of science. While previous discussions examined the conditions under which ABMs can be explanatory of real-world phenomena, the question what kind of explanations highly idealised ABMs of science provide remained open. Using the perspective of Aydinonat et al. (2020), we can say that this boils down to the following questions: Can we use highly idealised ABMs of science to construct explanatory arguments, and if so, of what kind? An attempt to answer this question is the subject of the following section.

3 ABMs and their explanatory power

What can we learn from highly idealised ABMs of science and what exactly do they represent? The answer is far from trivial and it is closely related to the ongoing philosophical debate about the epistemic function of highly idealised or 'toy' models in empirical sciences (e.g., Alexandrova, 2008; Fumagalli, 2016; Grüne-Yanoff, 2009; Hoyningen-Huene, 2020; Nguyen, 2019; Reiss, 2012; Reutlinger et al., 2018, cf. also references in Footnote 9). While my aim is to address this question by focusing on ABMs in the philosophy of science, the bulk of this section is sufficiently general to apply to toy models in other disciplines as well. I start by distinguishing between different possible targets which models can adequately represent and I relate them to the different types of explanations a particular representation licenses. Then I turn to validation strategies,

---

[6] Similar methodological approaches have been endorsed in the context of ABMs in the social sciences, see, e.g., Page (2018), Kuhlmann (2021).

which help us move a model from one explanatory category to another. Finally, I go back to the ABMs of science and examine how they are to be classified both before and after passing a certain validation procedure.

3.1 What do ABMs represent?

According to Bolinska (2013), 'A vehicle is an *epistemic representation* of a given target system if and only if it is a tool for gaining information about this system'. Here, information denotes those considerations which are not readily accessible by directly observing the target but can be understood via a particular vehicle, in this case by the means of a particular model. In the remainder of this article, whenever I speak of a model representing a target, I refer to an epistemic representation of the target. After distinguishing different types of targets, I will specify the types of explanations that their representation warrants.

One way of categorising the representational properties of ABMs is according to whether they represent actual or possible phenomena. On the one end of the spectrum, there are ABMs that represent real-world phenomena (Figure 1).[7] These models were developed most prominently in urban planning and epidemiology, where they have been used for policy guidance. For instance, the UrbanSim (Waddell, 2002) set of models of urban planning was developed to guide urban policy and transportation investments. While UrbanSim was built based on empirical data, it was designed as a virtual experimental lab where various counterfactual scenarios can be represented and analysed (Bruch & Atwell, 2015). In other words, these models were built not merely to represent actual empirical processes, but also—and crucially—to model 'real-world possibilities', that is, scenarios that could take place once some factors are altered. This is the second kind of targets ABMs can represent. The capacity of models to represent possibilities is essential for drawing normative and descriptive conclusions from them, because it allows us to draw inferences about counterfactual dependencies concerning the purported target. For instance, models of herd immunity and disease spread, which are used to examine different policies of epidemics management, enable the acquisition of precisely this sort of knowledge (Epstein, 2009).

---

[7] This classification should not be taken as exhaustive since some issues are either lacking or require further disambiguation. For example, non-existent targets, which can be part of hypothetical modelling, may be physically impossible and yet informative of real-world phenomena and their possibilities (Weisberg, 2013, pp. 121–122).

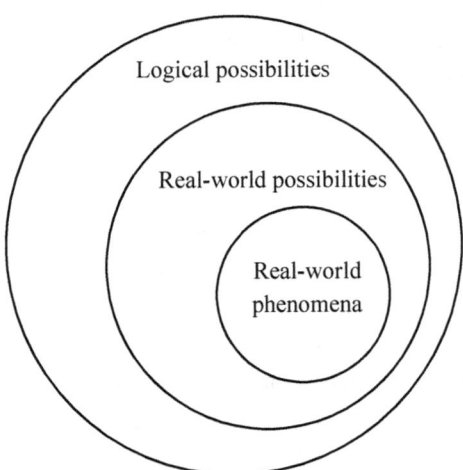

Figure 1. A simplified picture of different target phenomena represented by ABMs

Models mentioned in the previous paragraph increase our explanatory understanding of the phenomena they represent in the sense of expanding our ability to make reliable 'what-if' inferences about them (Ylikoski, 2014). In contrast to such models, other simulations represent only logical possibilities. These are scenarios that may but need not correspond to any interesting real-world possibilities. Importantly, determination of whether they correspond to real-world possibilities—and if so, which ones—is an open question. Hence, these are models from which we cannot draw reliable 'what-if' inferences. As I argue below, highly idealised models upon their initial development typically belong to this category.

To make this classification more precise, let us look into the kind of explanations that each class warrants.[8]

3.2 How-possibly and how-actually explanations

In view of the above classification of the modelled targets, it is helpful to make a related distinction between *how-actually* explanations (HAEs), and *how-possibly* explanations (HPEs). While the former notion concerns explanations simpliciter, that is, accounts of how phenomena actually occur, the latter was introduced to cover accounts of possible

---

[8] Explanation of phenomena is certainly not the only epistemic function of ABMs. For other epistemic functions of ABMs see, e.g., Edmonds et al. (2018), Epstein (2006), Frey & Šešelja (2018).

ways in which phenomena can occur.[9] Following Verreault-Julien (2019), we can characterise HAEs as expressing propositions of the form 'p because q (and initial conditions c)'. In contrast, HPEs express propositions of the form 'it is possible that: p because q (and initial conditions c)'. HPEs can express various types of modalities, such as mathematical or causal ones.

Based on the above, we can characterise the explanatory properties of ABMs as follows:

ABMs representing *real-world phenomena* and *real-world possibilities* provide one of the two following types of explanations:

- **HAEs** which express propositions of the form: 'p because q and initial conditions c', where we know which conditions these are and we know that they hold for a particular empirical target.

- **causal HPEs**,[10] which express propositions of the form: 'It is *causally* possible that: p because q and initial conditions c.', where we know which conditions these are although we may not know whether they hold for a particular empirical target, or we know that they do not hold for that particular target.[11]

ABMs representing *logical possibilities* provide:

- *epistemically opaque HPEs* (ep-op HPEs), which express propositions of the form:

> 'It is *logically* possible that: p because q', which is equivalent to 'It is *causally* possible that: p because q and initial conditions c', where we may not know which conditions these are, nor whether they hold for the given empirical target.

This classification is similar to Gräbner's (2018) proposal, where his 'full explanations' correspond to what I call HAEs, his 'partial explanations' to causal HPEs, and his

---

[9] The notion of HPE was introduced by Dray (1957) in the context of explanations in history. Subsequently, it became the subject of extensive debates in the literature on scientific modelling, especially in biology and social sciences (see, e.g., Bokulich, 2014, 2017; Forber, 2010, 2012; Hempel, 1965; Reydon, 2012; Ylikoski & Aydinonat, 2014). One can find different versions of this notion across literature. My approach here is in line with Verreault-Julien (2019) in terms of assigning a broad meaning to HPEs.
[10] I consider causal explanations because they are typically discussed in the context of the modelling in social sciences, see, e.g., Alexandrova (2008), Northcott & Alexandrova (2015), Reiss (2012).
[11] The latter case captures counterfactual scenarios, while the former one captures potential scenarios, which may be actual or counterfactual.

'potential explanations' include causal HPEs and ep-op HPEs.[12] The notion of ep-op HPEs is closely related to what Ylikoski & Aydinonat (2014) call 'causal mechanism schemes', which 'do not directly explain any particular empirical fact' but 'address only simplified theoretical explananda' (Ylikoski & Aydinonat, 2014, p. 27). By calling such HPEs epistemically opaque, we highlight the indeterminate nature of the represented target phenomenon.

The 'initial conditions' mentioned in the classification above stand for various contextual factors that must be satisfied for a particular regularity to hold. In case of ABMs of science, this may include for instance the size of the community, the nature of interaction among scientists, the nature of decision-making of scientists concerning theories they want to pursue, etc. Such factors are implicitly or explicitly assumed in the given model.

Note also that in the above, 'knowledge' is used in a colloquial rather than the strictly epistemological sense, and it could be replaced with 'having a justified belief'. The idea is that the conditions constraining a particular explanatory relationship are established via a suitable scientific method, in which case we have a good reason to believe which conditions these are or whether they hold for a given empirical target.

Most models fall somewhere in-between the above categories. Depending on the epistemic status of the initial conditions c (whether we are able to specify which ones they are and whether they hold for the empirical target in question), an ABM will be closer to one rather than another type. This is determined by the process of model validation, to which we shall turn now.

4 Verification and validation of ABMs

The main reason for running simulations of a scientific inquiry is to examine the impact of certain factors on the collective goals of research, such as efficiency, which would be difficult to estimate by analytical methods or by the means of qualitative analysis. This means that the results of an adequate ABM should not be merely obvious consequences of the underlying assumptions (Lazer & Friedman, 2007; Pöyhönen & Kuorikoski, 2016), because that would make the entire process of modelling superfluous. This, however, means that the link between the model and its purported target need not be obvious. In particular, when a highly idealised model is first proposed, the results it delivers may come with a degree of epistemic opacity in the sense that we do not understand the

---

[12] For a more general discussions on different types of explanations obtained by means of models see Bokulich (2017), Lawler & Sullivan (2020).

conditions under which the established causal dependency holds. To remove this veil of opacity, we need to turn to the validation procedures.[13]

Justification of models and their representational properties is conducted via two closely related processes: verification and validation. While verification is a method of evaluating the accuracy of the program of a given ABM based on its conceptual design, validation is the process of evaluation of links between the model and its purported target (e.g., Cooley & Solano, 2011; Gräbner, 2018). Irrespective of the purpose for which the model was built, it always requires some degree of verification to ensure that its simulation code does not suffer from bugs and other unintended issues. In short, that it corresponds to the modeller's conceptual idea. The type of required validation, however, directly depends on the purpose of the model and its intended target. In particular, examination of whether the model represents a logical possibility, a real-world possibility, or a real-world phenomenon requires different validation procedures.

Clearly, showing that a model represents a logical possibility will be the least demanding of the procedures alluded to above. All that needs to be shown is that there is a plausible interpretation of the model such that the inference 'it is logically possible that: p because q' is warranted. For instance, if an ABM is supposed to represent the impact of a certain division of cognitive labour among scientists on the success of their inquiry, we only need to show that we can plausibly interpret the model as representing scientific research and the division of cognitive labour among scientists. At the same time, we need not know under what particular conditions of inquiry (e.g., for how large a community, under what communication structure, under what research behaviour of scientists, etc.) the observed regularity (here between a specific division of labour and a particular measure of success) holds.

This need not, however, be the only goal we are interested in. Even in the case of abstract, highly idealised models, we are often after more than a mere logical possibility. For instance, we may be interested in showing that a typical case of scientific inquiry within a certain domain of study is at least 'susceptible' towards a particular regularity.[14] In other words, we may be interested in causal scenarios which are possible under a set of conditions typical of inquiries within a given scientific domain. To achieve this, we need

---

[13] This corresponds to what Bokulich (2011) calls a 'justificatory step' in establishing explanations obtained by a model, i.e., the domain of its applicability.
[14] For example, Nguyen (2019) takes the Schelling model as licensing the claim: 'A city whose residents have weak preferences regarding the skin colour of their neighbours has a susceptibility towards global segregation.' He does not tell us, however, in virtue of what exactly such susceptibility can be considered warranted.

a model that provides a causal HPE. To go back to the example above, it would mean showing the impact of a specific division of labour on the success of inquiry under a set of conditions typical for research in a particular scientific domain.

Since the difference between causal HPE and ep-op HPE rests in the epistemic status of the initial conditions under which the observed regularity holds, the better we can specify such conditions, the more we are able to move away from an ep-op HPE and towards a causal HPE. This is where various validation procedures enter the stage. On the one hand, their purpose is to help us determine the conditions in the model world under which the results of simulations remain stable. On the other hand, validation helps us to relate these conditions to empirical phenomena. The former is the task of robustness analysis and the latter of an empirical embedding of the model.

4.1 Robustness analysis

As the name suggests, robustness analysis is a method of examining the robustness, or stability, of results of a particular model under changes in its assumptions. Depending on the kind of assumptions we focus on, we can distinguish between two types of analyses:

a) *Sensitivity analysis* is a method of examining how sensitive the output of the model is to changes in parameters (Thiele et al., 2014).[15] This analysis is used to determine the scope of parameters within which the results of a simulation remain stable.

b) *Derivational robustness analysis* is a method of examining the robustness of results under changes in the (idealising) assumptions of the model.[16] This is especially important in the case of highly idealised models, where it is usually difficult to assess whether idealisations impact the results or not. One way of conducting a derivational robustness analysis is by using a family of ABMs to

---

[15] Gräbner (2018) considers sensitivity analysis a part of verification rather than validation, because its purpose is to explore the results, rather than link them to a specific target. This view, however, disregards the fact that sensitivity analysis can be informative in this sense as well. For example, if it turns out a particular result occurs only under a small portion of the parameter space, this would pose an additional requirement on examining whether these parameters correspond to any empirical circumstances.

[16] Derivational robustness construed this way includes both 'structural robustness' and 'representational robustness' as defined by Weisberg & Reisman (2008), where the former stands for stability of the results under changes in the causal structure of the modelled system, and the latter for stability of the results under changes in the representational framework of the model. For discussions on derivational and representational robustness, see Woodward (2006), Ylikoski & Aydinonat (2014), Lehtinen (2018), Railsback & Grimm (2011, pp. 302–306), Kuhlmann (2021).

gradually vary the assumptions of the initial model and examine how such changes impact the results (Aydinonat et al., 2020). Another option is to use structurally different models aimed at representing the same target phenomenon: this approach can help reveal the impact of implicit assumptions and idealisations.

While robustness analysis can help us to better understand the ABM in question, it is typically insufficient as a method of specifying the empirical conditions under which particular results hold (see, e.g., Houkes & Vaesen, 2012). For instance, if the analysis shows that the results are relatively stable, we may still have insufficient evidence to claim that they are representative of a given empirical target. Perhaps a specific assumption in the model whose impact has not yet been examined could be making all the difference. Or it could be the case that the empirical target is best represented in terms of very specific parameter values, which have not been carefully examined by robustness tests. To amend this problem, robustness analysis needs to be supplemented with, and guided by, an empirical embedding of the model.

4.2 Empirical embedding and model validation

As mentioned above, the robustness analysis can be guided towards an examination of those assumptions that correspond to the intended empirical target. This allows us to check whether the causal dependency inferred from the model holds under assumptions which are empirically relevant. But how does one make sure the relevant assumptions are well embedded and indeed correspond to the relevant empirical phenomena? This is done via different strategies jointly known as empirical validation of ABMs.[17] Following Gräbner (2018), I list some of the most relevant procedures.

a) *Process validation* concerns the question of how well mechanisms represented in the model reflect our empirical knowledge about them (Gräbner, 2018). To this end, the strategy of enhancing the theoretical realism of the model by information based on our knowledge from sociology and the philosophy of science can be helpful (Casini & Manzo, 2016; Šešelja, 2021). For instance, exchange of information among scientists has been typically represented as a simple sharing of results of scientific studies (e.g., Grim et al., 2013; Weisberg & Muldoon, 2009; Zollman, 2010), but qualitative philosophical accounts of

---

[17] Literature on this topic is plentiful, see, e.g., Arnold (2019), Beisbart & Saam (2018), Boero & Squazzoni (2005), Casini & Manzo (2016), Gräbner (2018), Guerini & Moneta (2017), Richiardi et al. (2006), Tesfatsion (2017), Thicke (2020).

scientific communication often emphasise critical interaction (e.g., Longino, 2002, Longino, 2022, Chang, 2012). For this reason, inclusion of this aspect in ABMs of science when examining the robustness of previously obtained results may be one way of conducting their process validation (e.g., Borg et al., 2018; Frey & Šešelja, 2020).

b) *Input validation* concerns the question of whether the exogenous inputs for the model are empirically meaningful and appropriate for the purpose at hand (Tesfatsion, 2017). This may include behavioural assumptions ascribed to the agents, the initial conditions, parameter values, etc. (Fagiolo et al., 2019). If parameters in the model are adjusted so as to reflect or include concrete numerical information, we say a model is 'empirically calibrated' (Boero & Squazzoni, 2005). In the case of ABMs of science, this would mean for example adjusting the number of agents in a model according to the size of a particular scientific community or representation of social networks in the model based on bibliometric data (Martini & Pinto, 2016; Perović et al., 2016; Thicke, 2019).

c) *Descriptive and predictive output validation* concern the question to what extent the output of the model replicates existing knowledge about the target and whether it can predict its future states (Gräbner, 2018; Tesfatsion, 2017; Thicke, 2019). For instance, if a model aims at representing a certain episode from the history of science, then under specific initial conditions the macrobehaviour of simulated agents should correspond to our historical knowledge of the case study in question.

All in all, validation of ABMs is essential for determining the details of targets they represent. In particular, validation supplements and guides the robustness analysis in determining the conditions under which the causal dependency identified via the model holds. By following the above validation strategies, we can move the explanation based on a particular model from 'epistemic opaqueness' to a causal HPE (or to a HAE). In the following section, I illustrate this point with a class of ABMs of scientific interaction.

5 ABMs of scientific interaction: zooming in on the target

In this section, I look into a class of ABMs which were developed to represent the effects of scientific interaction on the efficiency of inquiry. The main question these models aim to address is how different degrees of connectedness across a given scientific community impact the efficiency of knowledge acquisition. While at first sight, a high degree of interaction would seem purely beneficial, simulations have shown that this need not always be the case. For instance, if misleading information spreads quickly through the scientific community, scientists may collectively end up choosing a wrong theory.

To understand the root of this problem, it is useful to clarify the trade-off between 'exploration' and 'exploitation', to which it is closely related. The relationship between exploration (search for new possibilities) and exploitation (the use of existing options) has long been studied in theories of formal learning, organisational sciences, etc. (March, 1991). It is easy to see that a similar trade-off may take place in the context of scientific inquiry: given a particular scientific problem, one can either explore novel ideas and hope to find solutions which are better than the existing ones, or stick with the currently available hypotheses and use those instead. Depending on the difficulty of the problem, different strategies of balancing between exploration and exploitation are more suitable: for instance, if a solution to a problem is hard to find, scientists may need to invest their resources in exploration before focusing on exploiting existing ideas.

Simulations of scientific interaction were inspired by the idea that different communication networks among scientists, characterised by varying degrees of connectedness (see Figure 2), may have a different impact on the balance between exploration and exploitation. In particular, if an initially misleading idea is shared too quickly through the community, scientists may lock in on it and prematurely abandon their search for better solutions. Alternatively, if the information flow is slow and sparse, important insights gained by some scientists, which could lead to an optimal solution, may remain undetected by the rest of the community for a long time.

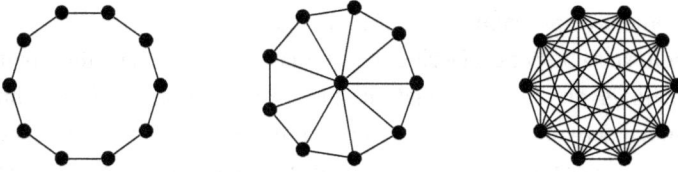

Figure 2. Three types of communication networks, representing an increasing degree of connectedness: a cycle, a wheel, and a complete graph. The nodes in each graph stand for scientists, while edges between the nodes stand for transmission of information between two scientists.

In what follows, I will look at a class of ABMs of scientific interaction starting with the pioneering work by Kevin Zollman. After suggesting an epistemically opaque how-possibly explanation (ep-op HPE) that can be drawn from his models, I proceed to examine how subsequent research allowed for specification of further conditions under which the observed regularity holds.

5.1 Scientific interaction and bandit problems

A set of ABMs developed by Zollman (2007, 2010, 2013) is based on the idea that scientific interaction can be studied in terms of 'bandit problems'. Bandit problems, well-known in economics and statistics, are a prime example of the exploration–exploitation trade-off. They concern a situation in which a gambler, or a group of gamblers, is confronted with multiple slot machines ('bandits'), which have different probabilities of success. While gamblers aim to maximise their overall reward, it is not immediately clear how long they should test each available machine and at which point they should stick with one that seems to give the highest payoff. If we further suppose that gamblers can share information among themselves and that each gambler sticks to the machine that seems to give the highest reward, we can ask: Which communication network will increase their chance to identify the machine with the highest payoff?

Zollman starts with the idea that this type of uncertainty is similar to one which scientists find themselves in when confronted with multiple rival hypotheses. Using a framework developed by Bala & Goyal (1998), he investigates which types of communication networks increase the chance that a scientific community, confronted with two rival hypotheses, successfully identifies the better of the two.

At the beginning of the simulation,[18] scientists—represented as Bayesian reasoners—are assigned random prior probabilities for two rival hypotheses, each of which has a designated objective probability of success. Agents always choose to pursue a theory which they believe to be better. During the simulation, they update their beliefs based on their own findings and the information they receive from their neighbours within a particular social network. Zollman examines three kinds of social networks from Figure 2. Scientists are successful if they manage to converge on the objectively better hypothesis (i.e., one that has a higher objective probability of success).

His results suggest that a high degree of interaction can be harmful. Because the initial findings about the hypotheses may be misleading, when scientists are linked via a complete graph the misleading information will spread quickly throughout the community. Consequently, the entire community may prematurely abandon the objectively better hypothesis.

---

[18] I am describing Zollman's (2010) model, which is a generalised version of his 2007 proposal.

Zollman also observes that if scientists start with extreme prior values, representing agents who stick to their hypotheses, the misleading information will not affect them early on. In fact, the complete graph is in such scenario more successful than the cycle.[19]

Altogether, the simulation results in the following ep-op HPE:

> **(High-inf)** It is logically possible that a scientific community prematurely abandons the better of two rival hypotheses because of a high degree of information flow among the scientists.

To turn a *High-inf* into a causal HPE from which we could make inferences about real-world possibilities, we need to specify the conditions under which this regularity holds. While Zollman provides one such condition, namely the absence of extreme priors, subsequent research has examined some additional factors.

5.2 The context of difficult inquiry

A number of related studies had shown that the main domain of application of Zollman's results is the context of a difficult inquiry. I take a brief look at these results and classify them according to the type of validation procedure they support.

Sensitivity analysis. Rosenstock et al. (2017) conducted a sensitivity analysis of Zollman's findings and showed that the 'Zollman effect'—the superior performance of the cycle versus the complete graph—holds only for a small part of the relevant parameter space. In particular, they show that the result obtains when the two relevant hypotheses are similar in terms of their objective probability of success, the population size is small, and the amount of data collected by scientists on each round is likewise small. The authors conclude that these factors are characteristic of *difficult learning*, because scientists either have a hard time distinguishing between the rival hypotheses or their data is sparse. Such conditions make it easier for misleading information to propagate through the community and sway it to the wrong hypothesis.

All in all, the results of sensitivity analysis restrict the application domain of *High-inf* to the context of difficult inquiry.

---

[19] This result is obtained by stopping the simulation after a certain number of rounds. Given sufficient time, agents in all networks end up on the correct hypothesis.

Derivational robustness. Restriction of the application domain to the context of difficult inquiry finds further support in results obtained by some structurally different ABMs. First, the ABM by Lazer & Friedman (2007), which was developed in organisational sciences, arrived at a similar conclusion. Their model is designed to study the problem-solving performance of agents linked via different social networks using a multidimensional epistemic landscape. The authors observe that in complex tasks that require a problem-solving capacity to extend over a longer period of time, highly connected networks perform worse than the less connected ones. Similar to what happens in Zollman's model, highly connected groups quickly converge on a single approach, thus failing to preserve the diversity of ideas needed to solve complex tasks.

Results supporting *High-inf* have also been obtained with subsequent ABMs based on epistemic landscapes (e.g., Grim, 2009; Grim et al., 2013; Derex et al., 2018), which suggests their derivational robustness (although see below).

Empirical output validation. Finally, the output of these models was reproduced by some empirical studies. For example, Mason et al. (2008) as well as Derex & Boyd (2016) conducted computer-based experiments in which participants linked via different communication networks were confronted with certain problem-solving situations. Both studies concluded that less interconnected groups outperform the more connected ones because they are able to preserve diversity and explore the space of possible solutions to a higher degree.

While all of these findings support *High-inf* under the conditions of difficult learning, we ought to be cautious with their extrapolation to actual scientific inquiries. One thing to note is that all of the above-mentioned studies are based on the assumption—integral to both the simulations and the experimental setup of empirical studies—that there is a trade-off between exploitation and exploration. But it should be noted that neither is actual scientific inquiry necessarily based on this trade-off, nor do results obtain once the trade-off assumption is relaxed.

5.3 Relaxing the exploration/exploitation trade-off

When scientists pursue a theory, it is not uncommon that along the way they acquire information relevant to the assessment of a rival theory. For example, scientists may detect some explanatory anomalies in their current theory (e.g., evidence that cannot be accounted for by that theory) that could be explained by the rival theory. As a result, research into the former (exploitation) could inspire and lead to research on the latter (exploration).

These considerations inspired ABMs and empirical studies that relaxed the exploration/exploitation trade-off. Here, I review some examples.

Derivational robustness in view of exploratory agents. Kummerfeld & Zollman (2016) developed an ABM of scientific interaction based on an analogy with bandit problems, but this time allowing agents who pursue one hypothesis to also occasionally acquire information about a rival hypothesis. Their results show that higher levels of exploration by agents go hand in hand with benefits of increased connectivity among them.

The positive impact of high levels of interaction has been observed also in a structurally different model: argumentation-based ABM (ArgABM) (Borg et al., 2019, 2017, 2018). ArgABM aims at capturing the argumentative dynamics underlying a scientific inquiry. The model employs an 'argumentative landscape' representing rival research programmes or theories in a given domain which scientists gradually explore. Each theory consists of 'arguments', which stand for studies supporting a particular theory. These arguments can be challenged ('attacked') by studies belonging to rival research programmes or defended by further arguments developed within the same programme. In this way, the argumentative landscape allows for the representation of both false positives (acceptance of a false hypothesis) and false negatives (rejection of a true hypothesis). The success of inquiry is measured in terms scientists converging on the theory that is predefined as fully defensible within the landscape (initially unknown to the agents).

The results of ArgABM indicate that a high degree of interaction among scientists is beneficial. The more connected agents are, the better their chances of converging on the best theory, and this holds under a variety of conditions of inquiry.

The main reason ArgABM delivers this result lies in the following two modelling assumptions. First, when agents explore a theory, they also gain information about rival theories in the form of argumentative attacks or defences of own theory. For instance, by finding an argument in my theory that attacks the rival theory, I identify a potential problem in the latter. Alternatively, if I encounter an attack on my own theory, I will learn the argument from the rival theory (this could represent a scenario in which proponents of the rival theory publish a study showing they are able to explain certain phenomena which our theory cannot explain that well). As a result, exploitation includes a degree of exploration.

Second, to accurately evaluate a theory (e.g., in terms of the number of 'anomalies' represented as attacked and undefended arguments in a theory, see Borg et al., 2019), agents need a sufficiently detailed knowledge of the argumentative landscape. If a scientist knows only a part of the landscape, she may assess a particular theory as unproblematic, while in fact she has not learned about its problematic parts. This corresponds to a scenario in which scientists, having read a few studies in favour of a particular research programme, conclude that the programme is feasible, but they failed to read other studies, which show that results presented in the former ones could not be replicated or are based on a methodological error. As a result, less connected groups will suffer from greater information losses, making it more likely that their assessment of a particular theory is inaccurate.

Empirical output validation. In contrast to the previously mentioned empirical studies, an experiment run by Mason & Watts (2012) resulted in the conclusion that a higher degree of connectivity is actually rewarding. Unlike the former experiments, this study is based on a relaxed assumption about the exploration/exploitation trade-off. Exploitation of existing ideas does not necessarily restrict participants to the local maxima. Instead, they have the option of going on to individually search for better solutions.

In sum, several studies that relaxed the assumption about the trade-off between exploration and exploitation failed to replicate *High-inf*, thus pointing to limitations of its application. Additionally, the ArgABM highlighted the negative aspect of information loss that can take place in loosely connected communities.

5.4 Alternative mechanisms of diversity

Derivational robustness under the assumption of cautious agents. Frey & Šešelja (2020) have conducted an additional derivational robustness analysis of Zollman's (2010) model by enhancing it with a number of assumptions characteristic of a difficult inquiry.[20] That study is therefore also a contribution to the *process validation* mentioned in Section 4, while more specifically, it focuses on the robustness of results once the process of difficult inquiry is captured in terms of empirically relevant assumptions.

The most important finding of those simulations is that even in the context of difficult inquiry, a high degree of information flow is not necessarily harmful. On the contrary, more connected networks may outperform the less connected ones. In particular, if diversity is generated in some other way than by the means of network structure, a high

---

[20] The code of their model, available at https://github.com/daimpi/SocNetABM/tree/RobIdeal, also includes Zollman's ABM as a nested variant and thus provides an easily accessible tool for its verification.

information flow will not have a negative impact on the efficiency of the group. For instance, if scientists are equipped with a dose of caution, or 'rational inertia', when deciding whether they should abandon their current theory and start pursuing a rival one, the cycle is no longer superior to the complete graph.

Moreover, addition of the assumption that agents interact critically does not on its own help the complete graph to catch up with the cycle: for that to happen, scientists must be cautious in their decision-making (for instance by displaying a degree of resistance against changing the theory they had endorsed).

All in all, these results further specify conditions under which the *High-inf* holds.

5.5 From ep-op HPE to causal HPE

To sum up, the studies reviewed above suggest that the explanation obtained from Zollman's original model can be expressed as follows:

(High-inf-causal) It is causally possible that a scientific community prematurely abandons the better of two rival hypotheses due to a high degree of information flow among scientists under the following conditions:

- that the inquiry is difficult
- theoretical diversity is not generated in some other way (e.g., by scientists having extreme priors or a tendency to stick to their hypotheses)
- that pursuit of one hypothesis does not allow for insights into its rivals (i.e., there is a strict exploration/exploitation trade-off)
- potentially some additional assumptions.

While in the original model, we could only draw an ep-op HPE without a clear application domain, subsequent studies allowed us to zoom in on the target that the model actually represents and for which the observed causal mechanism appears to hold. Of course, further studies may reveal that additional specifications are needed or that some of the existing ones ought to be revised.

The preceding discussion also illustrates that the difficulty of extrapolating findings from a model to an empirical application domain holds not only for ABMs but also for empirical experimental studies.

6 Conclusion

In this paper, I explored the epistemic benefits of running computer simulations in the philosophy of science and the kinds of inferences one can draw from them. I have argued that models can represent (i) logical possibilities, (ii) real-world possibilities, or (iii) real-world phenomena, where each category comes with specific explanatory features. By using strategies of verification and validation, we can identify the class to which a particular ABM belongs. While abstract, highly idealised models prima facie allow only for the inference of a causal possibility under unknown circumstances, the process of validation by the means of other ABMs as well as empirical studies can help reveal these conditions.

In conclusion, let me make a few general points. First, highly idealised ABMs of science should be appreciated even under conditions of a minimal degree of verification and validation required for obtaining ep-op HPE. In this form, they can assume a variety of epistemic functions, ranging from providing conjectures about scientific inquiry and starting a new family of models all the way to contributing to the validation of other ABMs. Second, the development of new ABMs and their subsequent validation is best considered in terms of broader inquiries consisting of classes of ABMs, but also empirical studies targeting the same phenomenon. Third, there is no reason to see the highly idealised nature of ABMs of science as their drawback. As long as the model is subjected to an adequate process of verification and validation with respect to its purported aim and target, it can be an important step forward in our understanding of scientific inquiry.

Acknowledgements. I am very grateful to Daniel Frey, Paul Hoyningen Huene, Christian Straßer, Emily Sullivan, Leonid Tiokhin, Philippe Verreault-Julien, and the Philosophy & Ethics Group at TU Eindhoven for fruitful discussions that helped me formulate a number of points in this paper. Research on this this paper was funded by Irène Curie Fellowship of TU Eindhoven and the Deutsche Forschungsgemeinschaft (DFG, German Research Foundation) – project number 426833574.

Bibliography

Alexander, J. McKenzie, Himmelreich, J., & Thompson, C. (2015). Epistemic landscapes, optimal search, and the division of cognitive labor. *Philosophy of Science, 82*(3), 424–453.

Alexandrova, A. (2008). Making models count. *Philosophy of Science, 75*(3), 383–404.

Arnold, E. (2019). Validation of computer simulations from a Kuhnian perspective. In C. Beisbart & N. J. Saam (Eds.), *Computer Simulation Validation – Fundamental Concepts, Methodological Frameworks, and Philosophical Perspectives* (pp. 203–224). Cham: Springer.

Avin, S. (2019). Centralized funding and epistemic exploration. *The British Journal for the Philosophy of Science, 70*(3), 629–656.

Axelrod, R. (1984). *The Evolution of Cooperation.* Basic Books.

Axelrod, R. (1997). *The Complexity of Cooperation: Agent-Based Models of Competition and Collaboration.* Vol. 3. Princeton: Princeton University Press.

Axelrod, R, & Hamilton, W. D. (1981). The evolution of cooperation. *Science, 211*(4489), 1390–1396.

Aydinonat, N. E., Reijula, S., & Ylikoski, P. (2020). Argumentative landscapes: The functions of models in social epistemology. *Synthese* (forthcoming).

Bala, V., & Goyal, S. (1998). Learning from neighbours. *The Review of Economic Studies, 65*(3), 595–621.

Balietti, S., Mäs, M., & Helbing, D. (2015). On disciplinary fragmentation and scientific progress. *PloS ONE, 10*(3), e0118747.

Bedessem, B. (2019). The division of cognitive labor: Two missing dimensions of the debate. *European Journal for Philosophy of Science, 9*(1), 1–16.

Beisbart, C., & Saam, N. J. (Eds.). (2019). *Computer Simulation Validation: Fundamental Concepts, Methodological Frameworks, and Philosophical Perspectives.* Cham: Springer.

Boero, R., & Squazzoni, F. (2005). Does empirical embeddedness matter? Methodological issues on agent-based models for analytical social science. *Journal of Artificial Societies and Social Simulation, 8*(4).

Bokulich, A. (2011). How scientific models can explain. *Synthese, 180*(1), 33–45.

Bokulich, A. (2014). How the tiger bush got its stripes: 'How possibly' vs. 'how actually' model explanations. *The Monist, 97*(3), 321–338.

Bokulich, A. (2017). Models and explanation. In L. Magnani & T. Bertolotti (Eds.), *Springer Handbook of Model-Based Science* (pp. 103–118). Springer: Cham.

Bolinska, A. (2013). Epistemic representation, informativeness and the aim of faithful representation. *Synthese, 190*(2), 219–234.

Borg, A., Frey, D., Šešelja, D., & Straßer, C. (2019). Theory-choice, transient diversity and the efficiency of scientific inquiry. *European Journal for Philosophy of Science,* 9, article number 26.

Borg, A., Frey, D., Šešelja, D., & Straßer, C. (2017). Examining network effects in an argumentative agent-based model of scientific inquiry. In A. Baltag, J. Seligman, & T. Yamada (Eds.), *Proceedings of Logic, Rationality, and Interaction: 6th International Workshop, LORI 2017* (pp. 391–406). Berlin: Springer.

Borg, A., Frey, D., Šešelja, D., & Straßer, C. (2018). Epistemic effects of scientific interaction: approaching the question with an argumentative agent-based model. *Historical Social Research*, *43*(1), 285–309.

Bruch, E., & Atwell, J. (2015). Agent-based models in empirical social research. *Sociological Methods & Research*, *44*(2), 186–221.

Casini, L., & Manzo, G. (2016). Agent-based models and causality: a methodological appraisal. *The IAS Working Paper Series*, 7.

Chang, H. (2012). *Is Water H2O? Evidence, Pluralism and Realism*. Cham: Springer.

Cooley, P., & Solano, E. (2011). Agent-based model (ABM) validation considerations. In *Proceedings of the Third International Conference on Advances in System Simulation, SIMUL 2011* (pp. 134–139).

Currie, A., & Avin, S. (2018). Method pluralism, method mismatch & method bias. *Philosopher's Imprint*, 19(13), 1–22.

Derex, M., & Boyd, R. (2016). Partial connectivity increases cultural accumulation within groups. *Proceedings of the National Academy of Sciences*, *113*(11), 2982–2987.

Derex, M., Perreault, C., & Boyd, R. (2018). Divide and conquer: Intermediate levels of population fragmentation maximize cultural accumulation. *Philosophical Transactions of the Royal Society B: Biological Sciences*, *373*(1743), 20170062.

Douven, I. (2010). Simulating peer disagreements. *Studies in History and Philosophy of Science Part A*, *41*(2), 148–157.

Dray, W. H. (1957). *Laws and Explanation in History*. 3rd edition. Oxford: Oxford University Press.

Edmonds, B., & Moss, S. (2004). From KISS to KIDS – an 'antisimplistic' modelling approach. In P. Davidsson, B. Logan, & K. Takadama (Eds.), *International Workshop on Multiagent Systems and Agent-Based Simulation* (pp. 130–144). Berlin: Springer.

Edmonds, B., Le Page, C., Grimm, V., Montanola, C., Ormerod, P., Root, H., & Flaminio, S. (2018). Different modelling purposes. *Journal of Artificial Societies and Social Simulation*, *22*(3), 1–30.

Epstein, J. M. (2006). *Generative Social Science: Studies in Agent-Based Computational Modelling*. Princeton: Princeton University Press.

Epstein, J. M. (2009). Modelling to contain pandemics. *Nature*, *460*(7256), 687–687.

Epstein, J. M., & Axtell, R. (1996). *Growing Artificial Societies: Social Science from the Bottom Up*. Washington, DC: Brookings Institution Press.

Fagiolo, G., Guerini, M., Lamperti, F., Moneta, A., & Roventini, A. (2019). Validation of agent-based models in economics and finance. In C. Beisbart & N. J. Saam (Eds.), *Computer Simulation Validation: Fundamental Concepts, Methodological Frameworks, and Philosophical Perspectives* (pp. 763–787). Cham: Springer.

Forber, P. (2010). Confirmation and explaining how possible. *Studies in History and Philosophy of Science Part C: Studies in History and Philosophy of Biological and Biomedical Sciences, 41*(1), 32–40.

Forber, P. (2012). Conjecture and explanation: A reply to Reydon. *Studies in History and Philosophy of Science Part C: Studies in History and Philosophy of Biological and Biomedical Sciences, 43*(1), 298–301.

Frey, D., & Šešelja, D. (2018). What is the epistemic function of highly idealized agent-based models of scientific inquiry? *Philosophy of the Social Sciences, 48*(4).

Frey, D., & Šešelja, D. (2020). Robustness and idealization in agent-based models of scientific interaction. *British Journal for the Philosophy of Science, 71*, 1411–1437.

Fumagalli, R. (2016). Why we cannot learn from minimal models. *Erkenntnis, 81*(3), 433–455.

Gilbert, N. (1997). A simulation of the structure of academic science. *Sociological Research Online, 2*(2), 1–15.

Gilbert, N., & Troitzsch, K. (2005). *Simulation for the Social Scientist*. London: McGraw-Hill Education.

Goldman, A. I., & Shaked, M. (1991). An economic model of scientific activity and truth acquisition. *Philosophical Studies, 63*(1), 31–55.

Gräbner, C. (2018). How to relate models to reality? An epistemological framework for the validation and verification of computational models. *Journal of Artificial Societies and Social Simulation, 21*(3), p. 8.

Grim, P. (2009). Threshold phenomena in epistemic networks In *AAAI Fall Symposium: Complex Adaptive Systems and the Threshold Effect* (pp. 53–60).

Grim, P., Mar, G., Paul St. Denis, P. (1998). *The Philosophical Computer: Exploratory Essays in Philosophical Computer Modelling*. Vol. 1. Cambridge, MA: MIT Press.

Grim, P., & Singer, D. (2020). Computational philosophy. In E. N. Zalta (Ed.), *The Stanford Encyclopedia of Philosophy* (*Spring 2020 Edition*).

Grim, P., Singer, D. J., Fisher, S., Bramson, A., Berger, W. J., Reade, C., Flocken, C., & Sales, A. (2013). Scientific networks on data landscapes: question difficulty, epistemic success, and convergence. *Episteme, 10*(4), 441–464.

Grüne-Yanoff, T. (2009). Learning from minimal economic models. *Erkenntnis, 70*(1), 81–99.

Guerini, M, & Moneta, A. (2017). A method for agent-based models validation. *Journal of Economic Dynamics and Control, 82*, 125–141.

Harnagel, A. (2018). A mid-level approach to modelling scientific communities. *Studies in History and Philosophy of Science Part A, 76*, 49–59.

Hegselmann, R. (2017). Thomas C. Schelling & James M. Sakoda: The intellectual, technical, and social history of a model. *Journal of Artificial Societies and Social Simulation, 20*(3).

Hegselmann, R., & Krause, U. (2002). Opinion dynamics and bounded confidence models, analysis, and simulation. *Journal of Artificial Societies and Social Simulation, 5*(3).

Hegselmann, R., & Krause, U. (2005). Opinion dynamics driven by various ways of averaging. *Computational Economics, 25*(4), 381–405.

Hegselmann, R., & Krause, U. (2006). Truth and cognitive division of labor: First steps towards a computer aided social epistemology. *Journal of Artificial Societies and Social Simulation, 9*(3), 10.

Hempel, C. (1965). *Aspects of Scientific Explanation and other Essays in the Philosophy of Science.* New York: Free Press.

Holman, B., & Bruner, J. (2015). The problem of intransigently biased agents. *Philosophy of Science, 82*(5), 956–968.

Holman, B., & Bruner, J. (2017). Experimentation by industrial selection. *Philosophy of Science, 84*(5), 1008–1019.

Houkes, W., & Vaesen, K. (2012). Robust! Handle with care. *Philosophy of Science, 79*(3), 345–364.

Hoyningen-Huene, P. (2020). The logic of explanation by abstract models (to appear).

Imbert, C. (2017). "Computer simulations and computational models in science". In: Springer handbook of model-based science. Cham: Springer, pp. 735–781.

Kitcher, P. (1990). The division of cognitive labour. *The Journal of Philosophy, 87*(1), 5–22.

Kitcher, P. (1993). *The Advancement of Science: Science without Legend, Objectivity without Illusions.* Oxford: Oxford University Press.

Klein, D., Marx, J., & Fischbach, K. (2018). Agent-based modelling in social science, history, and philosophy. An introduction. *Historical Social Research/Historische Sozialforschung, 43*(1), 7–27.

Kuhlmann, M. (2021). On the exploratory function of agent-based modelling. *Perspectives on Science, 29*(4), 510–536.

Kummerfeld, E, & Zollman, K. J. S. (2016). Conservatism and the scientific state of nature. *The British Journal for the Philosophy of Science, 67*(4), 1057–1076.

Lawler, I., & Sullivan, E. (2020). Model explanation versus model-induced explanation. *Foundations of Science, 26*, 1049–1074.

Lazer, D., & Friedman, A. (2007). The network structure of exploration and exploitation. *Administrative Science Quarterly, 52*(4), 667–694.

Lehtinen, A. (2018). Derivational robustness and indirect confirmation. *Erkenntnis, 83*(3), 539–576.

Longino, H. E. (2002). Science and the common good: Thoughts on Philip Kitcher's Science, Truth, and Democracy. *Philosophy of Science, 69*(4), 560–568.
Longino, H. E. (2022). What's social about social epistemology? *The Journal of Philosophy, 119*(4), 169–195.
Mäki, U. (2005). Economic epistemology: Hopes and horrors. *Episteme, 1*(3), 211–222.
March, J. G. (1991). Exploration and exploitation in organizational learning. *Organization Science,* 2(1), 71–87.
Martini, C., & Pinto, M. F. (2016). Modelling the social organization of science. *European Journal for Philosophy of Science, 7*, 221–238.
Mason, W., & Watts, D. J. (2012). Collaborative learning in networks. *Proceedings of the National Academy of Sciences, 109*(3), 764–769.
Mason, W. A., Jones, A., & Goldstone, R. L. (2008). Propagation of innovations in networked groups. *Journal of Experimental Psychology: General, 137*(3), 422.
Mayo-Wilson, C., & Zollman, K. J. S. (2021). The computational philosophy: simulation as a core philosophical method. *Synthese, 199*, 3647–3673.
Muldoon, R., & Weisberg, M. (2011). Robustness and idealization in models of cognitive labor. *Synthese, 183*(2), 161–174.
Nguyen, J. (2019). It's not a game: Accurate representation with toy models. *The British Journal for the Philosophy of Science, 71*(3).
Northcott, R., & Alexandrova, A. (2015). Prisoner's dilemma doesn't explain much. In M. Peterson (Ed.), *The Prisoner's Dilemma. Classic Philosophical Arguments* (pp. 64–84). Cambridge: Cambridge University Press.
O'Connor, C., & Weatherall, J. O. (2018). Scientific polarization. *European Journal for Philosophy of Science, 8*(3), 855–875.
O'Connor, C., & Weatherall, J. O. (2019). *The Misinformation Age: How False Beliefs Spread.* New Haven, CT: Yale University Press.
Page, S. E. (2018). *The Model Thinker: What You Need to Know to Make Data Work for You.* London: Hachette.
Payette, N. (2012). Agent-based models of science. In A. Scharnhorst, K. Börner, & P. van den Besselaar (Eds.), *Models of Science Dynamics* (pp. 127–157). Cham: Springer.
Perović, S., Radovanović, S., Sikimić, V., & Andrea Berber, A. (2016). Optimal research team composition: data envelopment analysis of Fermilab experiments. *Scientometrics, 108*, 83–111.
Pinto, M. F., & Pinto, D. F. (2018). Epistemic landscapes reloaded: An examination of agent-based models in social epistemology. *Historical Social Research/ Historische Sozialforschung, 43*(1), 48–71.
Politi, V. (2021). Formal models of the scientific community and the value-ladenness of science. *European Journal for Philosophy of Science, 11*, article number 97.

Pöyhönen, S. (2017). Value of cognitive diversity in science. *Synthese, 194*(11), 4519–4540.
Pöyhönen, S., & Kuorikoski, J. (2016). Modelling epistemic communities. In M. Fricker, P. J. Graham, D. Henderson, N. Pedersen, & J. Wyatt (Eds.), *The Routledge Handbook of Social Epistemology*. Abingdon: Routledge.
Railsback, S. F., & Grimm, V. (2011). *Agent-Based and Individual-Based Modelling: A Practical Introduction*. Princeton: Princeton University Press.
Reiss, J. (2012). The explanation paradox. *Journal of Economic Methodology, 19*(1), 43–62.
Reutlinger, A., Hangleiter, D., & Hartmann, S. (2018). Understanding (with) toy models. *The British Journal for the Philosophy of Science, 69*(4), 1069–1099.
Reydon, T. A. C. (2012). How-possibly explanations as genuine explanations and helpful heuristics: A comment on Forber. *Studies in History and Philosophy of Science Part C: Studies in History and Philosophy of Biological and Biomedical Sciences, 43*(1), 302–310.
Richiardi, M., Leombruni, R., Saam, N. J., & Sonnessa, M. (2006). "A common protocol for agent-based social simulation. *Journal of Artificial Societies and Social Simulation, 9*(1), 15.
Rosenstock, S., O'Connor, C., & Bruner, J. (2017). Epistemic networks, is less really more? *Philosophy of Science, 84*(2), 234–252.
Sakoda, J. M. (1971). The checkerboard model of social interaction. *The Journal of Mathematical Sociology, 1*(1), 119–132.
Schelling, T. C. (1971). Dynamic models of segregation. *Journal of Mathematical Sociology, 1*(2), 143–186.
Schelling, T. C. (1978). *Micromotives and Macrobehavior*. New York: W. W. Norton & Company.
Šešelja, D. (2021). Exploring scientific inquiry via agent-based modelling. *Perspectives on Science, 29*(4).
Šešelja, D. (2022). Agent-based models of scientific interaction. *Philosophy Compass* (forthcoming).
Šešelja, D., Straßer, C., & Borg, A. (2020). Formal models of scientific inquiry in a social context: An introduction. *Journal for General Philosophy of Science, 51*, 211–217
Skyrms, B. (1990). *The Dynamics of Rational Deliberation*. Cambridge, MA: Harvard University Press.
Skyrms, B. (1996). *Evolution of the Social Contract*. Cambridge: Cambridge University Press.
Strevens, M. (2003). The role of the priority rule in science. *The Journal of Philosophy, 100*(2), 55–79.

Tesfatsion, L. (2017). Modelling economic systems as locallyconstructive sequential games. *Journal of Economic Methodology*, *24*(4), 384–409.

Thicke, M. (2020). Evaluating formal models of science. *Journal for General Philosophy of Science*, *51*, 315–335.

Thiele, J. C., Kurth, W., & Grimm, V. (2014). Facilitating parameter estimation and sensitivity analysis of agent-based models: A cookbook using NetLogo and R. *Journal of Artificial Societies and Social Simulation*, *17*(3), 11.

Thoma, J. (2015). The epistemic division of labor revisited. *Philosophy of Science*, *82*(3), 454–472.

Verreault-Julien, P. (2019). How could models possibly provide howpossibly explanations? *Studies in History and Philosophy of Science Part A*, *73*, 23–33.

Waddell, P. (2002). UrbanSim: Modelling urban development for land use, transportation, and environmental planning. *Journal of the American Planning Association*, *68*(3), 297–314.

Weatherall, J. O., O'Connor, C., & Justin Bruner, J. (2018). How to beat science and influence people: Policy makers and propaganda in epistemic networks. *The British Journal for the Philosophy of Science*, *71*(4), 1157–1186.

Weisberg, M. (2013). *Simulation and Similarity: Using Models to Understand the World*. Oxford: Oxford University Press.

Weisberg, M., & Muldoon, R. (2009). Epistemic landscapes and the division of cognitive labor. *Philosophy of Science*, *76*(2), 225–252.

Weisberg, M., & Reisman, K. (2008). The robust Volterra principle. *Philosophy of Science*, *75*(1), 106–131.

Woodward, J. (2006). Some varieties of robustness. *Journal of Economic Methodology*, *13*(2), 219–240.

Ylikoski, P. (2014). Agent-based simulation and sociological understanding. *Perspectives on Science*, *22*(3), 318–335.

Ylikoski, P., & Aydinonat, N. E. (2014). Understanding with theoretical models. *Journal of Economic Methodology*, *21*(1), 19–36.

Zamora Bonilla, J. (1999). The elementary economics of scientific consensus. *Theoria: An International Journal for Theory, History and Foundations of Science*, *14*(3), 461–488.

Zamora Bonilla, J. P. (2002). Scientific inference and the pursuit of fame: A contractarian approach. *Philosophy of Science*, *69*(2), 300–323.

Zollman, K. J. S. (2007). The communication structure of epistemic communities. *Philosophy of Science*, *74*(5), 574–587.

Zollman, K. J. S. (2010). The epistemic benefit of transient diversity. *Erkenntnis*, *72*(1), 7–35.

Zollman, K. J. S. (2013). Network epistemology: Communication in epistemic communities. *Philosophy Compass*, *8*(1), 15–27.

Author biography. Dunja Šešelja is a Professor at the Institute for Philosophy II, Ruhr University Bochum and a head of the research group "Reasoning, Rationality and Science". She is also a member of the Philosophy & Ethics Group at TU Eindhoven. She serves as a Co-Editor-in-Chief of the *European Journal for Philosophy of Science*, and she is the project initiator of the DFG Research Network "Simulations of Scientific Inquiry". Previously, she held visiting professorships at the University of Vienna and Ghent University, and postdoctoral positions at Ghent University, Ruhr-University Bochum, and MCMP, LMU Munich. Her research aims at the integration of historically informed philosophy of science, social epistemology and formal models of scientific inquiry.

Decolonising Scientific Knowledge: Morality, Politics and a New Logic

JONATHAN OKEKE CHIMAKONAM[1]

Abstract. I argue that philosophers of science have been neglecting the coloniality of knowledge as an issue in the production of scientific knowledge. Coloniality of knowledge imposes the scientific protocol of the West on the rest of the world while gatekeeping the protocols of epistemologies of the South. This has led to what can be called the *bordering problem*—an intellectual segregation by the *norm* that discounts the epistemic vision of the *normalised*. But are procedures that can yield scientific knowledge exhausted in the testable, demonstrable, empirical protocol of modern (Western) science? My claim is that they are not. Coloniality of knowledge bases its logic on the pretension that the Western episteme is acontextual and therefore superior to the rest that is not. Here, I employ a decolonial strategy called *disbordering* to confront the bordering problem couched in the ethics that seeks to regulate what counts as scientific knowledge, and the politics that seeks to determine who counts in the production of scientific knowledge. My aim is to challenge the imperialisation of modern scientific protocol and question the residualisation of knowledge formations from the Global South.

Keywords: decoloniality, coloniality, scientific knowledge, ethics, politics, epistemologies of the South, logic.

1 Introduction

The goals of philosophers of science in the twenty-first century should, I believe, include the decolonisation of scientific knowledge by de-imperialising the modern scientific protocol in order to make way for "collective", "shared", epistemic "cultural identity" (Hall, 1990, 223). The cultural identities of different scientific traditions will lead to the emergence of what can be called a Collective Ecology of Knowledge (CEK). The CEK involves 1) the re-organisation of colonially created artificial borders to follow the line of cultural identities, 2) the recognition of the identities of different knowledge ecologies, 3) the recognition of the right and power of diverse knowledge ecologies to produce scientific knowledge through their own protocols, 4) and the horizontalisation of conversations between diverse knowledge ecologies through creative struggle. This

---

[1] University of Pretoria, Faculty of Humanities: Pretoria, Gauteng, ZA/Conversational School of Philosophy, University of Calabar, NGN.

horizontalisation of conversations involves 4i) a recognition that other knowledge ecologies can be sites for credible scientific knowledge even though they may follow different protocols; 4ii) and a recognition of the freedom and right of epistemic agents from other places to produce credible scientific knowledge.

This requirement to decolonise is imperative for every field of study. This is what some of the intercultural philosophers have been driving at. Mikhail Bakhtin, Paulo Freire, Martin Buber, David Bohm, Ramón Flecha and, in a more systematic way, Heinz Kimmerle (1994) and Ram Adhar Mall (2000) have favoured dialogue as the approach of choice for the realisation of a CEK. Franz Wimmer (1996, 2007) promotes polylogue and more recently, some elements of the Conversational School of Philosophy promote conversational thinking (Chimakonam, 2017a; Egbai & Chimakonam, 2019; Chimakonam & Chimakonam, 2022). It is beyond the scope of the present paper to dwell on these approaches, but it is sufficient to say that the idea of CEK can engender a liberal outlook where the artificial, repressive borders constructed by modernity for trapping reason, restricting interaction and preventing the formation of a critical mass of intelligences, will be rolled back using decolonial strategies.

What stands against the CEK is the tradition of modernity instituted by the dominant Western culture, which lays out rules for determining what counts as knowledge in different disciplines, establishing borders that draw a line between the zone of knowledge and non-knowledge. This border automatically defines who is qualified to produce, regulate and disseminate authentic knowledge. It enthrones a system of cultural segregation, racial subordination and epistemic repression. The artificially plotted cultural borders therefore translate to intellectual borders in which the positive side of the power relations becomes the zone of existence while the negative side becomes the zone of non-existence. At issue, therefore, is the epistemic or intellectual border between the West and the rest. It is a bold 90 degrees North-Westerly red line delineating one-quarter of the global hemisphere. While this one-quarter is the zone of existence, epistemic sagacity and authority, the three-quarters is the zone of non-existence and epistemic folly. For lack of a better description, I want to call this false ideology (that the right and authority to produce, regulate, operationalise and disseminate authentic knowledge is determined by one's geographical origin) the 'Bordering Problem,' and it exists in the sciences as in other fields.

Briefly, for more than a century, philosophers of science have been questioning the methods and conclusions reached by scientists. By doing so, they have drawn attention to ethical and methodological issues, opened new vistas for research, extended the frontiers of knowledge and even pushed the scientists closer to God at one end, and at the other end closer to playing God. But despite all these achievements, there are areas that have been largely ignored in the literature. So, if philosophers of science were challenged to mention one of the areas to which serious attention has not been paid by their discipline, what would be the response?

It is easy to imagine answers referring to new breakthrough areas of science, as in robotics and genetic enhancement or even some older areas such as nuclear energy, but all these would be better studied within the domains of ethics of knowledge production and use. Questions of ethics would normally concern which scientific procedure is right or wrong, which technology is right or wrong, and what is the right or wrong way to apply a technology... While not trivialising any of these concerns, we need to consider an area of great importance to the programme of CEK that has remained under-explored. This is the need to breach and transcend the artificial borders of science as plotted in European modernity.

Most philosophers of science are largely unaware of the pervasive politics of coloniality that marginalises the perspectives from other cultures—what Boaventura du Suosa Santos (2014) describes as the perspectives of epistemologies of the South. The dominant Western culture is imperialistic, galvanising its procedures as the norm in order to residualise the rest and force them to abandon their epistemic formation and subscribe to the imposed Western formation. In this essay, I want to problematise the question of bordering in knowledge by exploring its political and ethical twists. I contend that the bordering problem reflects the three phases of coloniality, namely, coloniality of power, knowledge, and being. Focusing on coloniality of knowledge, I will show that the presentation of modernity, and specifically its scientific protocol as acontextual and absolute is ideological and directed to the suppression of ideas in other cultures. As a result, I will argue that a specific type of logic that traces its origin back to Aristotle is at the foundation of the modern civilisation that has created the bordering problem, and that a shift in logic would be necessary for the decolonisation of scientific knowledge. I present the bordering problem as a new frontier deserving of the attention of philosophers of science.

2 The grand scheme of the global matrix or European modernity

When Europe left the Dark Ages and moved into the Renaissance, and from there to the Enlightenment, the stage was set for the emergence of a new civilisation. Francis Bacon (1620/1855) and René Descartes (1637/1968) tapped into the logical and intellectual legacies of the past, especially those of Aristotle, to systematise the empirical and rational approaches to nature that eventually crystallised in the protocol of modern science. The agrarian revolution that first ensued gave birth to the transatlantic slave trade, which saw more than 20 million people from Africa forced into slave labour in various farms in North and South America as well as in Europe. This was the first signal that the new civilisation would prioritise expedience over morality. Looking back to the way in which that history unfolded, from the vantage point of coloniality, one can argue that the builders of European modernity (some actors in the Western intelligentsia at the time) had one particular problem: *atychiphobia* or the fear of failure. The horrors that reason had suffered in the long night of the Dark Ages had probably left them terrified of the prospect of a future regression of the same kind, and so they set themselves the ambitious goal of building a civilisation that would never again collapse. In this way, atychiphobia became the single, salient determinant of the orientation of modern Western civilisation and the foreign policies of nations in the West.

Looking into history, these actors could see the previous civilisations that humankind built across the ages and how they all collapsed at some point, and the difficult interludes that set in before a new civilisation rose in another part of the world. They could see that the Songhai and Mali civilisations in Africa never rose again. They could see that the Chinese and the Mesopotamian civilisations never rose again. They could see that in Europe, the Greek and the Roman empires never rose again, and the effects of the long interlude of the Dark Ages after the collapse of the Roman empire were still fresh and harrowing. To build a civilisation that would never collapse, some of these actors most likely came up with two agenda. The first was to construct a modernity fitted with the mechanism of the 'global matrix'. The global matrix was to be a structure that would link four spheres of human existence (the economic, political, cultural and intellectual together) and would suck every region of the world into itself, making it possible for the builders of modern civilisation to 1) maintain control of other regions of the world in perpetuity, 2) use this power to exploit other regions to the advantage of the West, and 3) ensure that no new civilisation ever sprung again from the regions that lie outside the West.

The second was to develop a structure of colonialism with which to fragment the regions that lie outside the West using multiple artificial borders, and so control them. Where possible, these borders disregarded ethnic, cultural and linguistic lines so as to eliminate the possibility of cohesion of local intelligentsias and interaction between the best human minds. Furthermore, the borders were strictly policed in order to prevent the movement of local intelligentsias from one territory to the other, which might lead eventually to the formation of a critical mass in any location. With these colonial mechanisms in place, it was possible for colonialists to control the rest of the world, exploiting them through the matrices of European modernity and condemning the exploited regions to perpetual disadvantage.

Today economic, military and political sanctions, neo-colonialism and joint military actions are used as weapons to destroy and impoverish many countries in the South. The effectiveness of these weapons is secured by the system of global matrix. If we must mention a few instances, they include Zimbabwe, Libya, the Democratic Republic of Congo, Somalia and the Central African Republic in Africa, Afghanistan, Iraq, Syria, Yemen, Cambodia, and North Korea in Asia, and Venezuela, Peru, Haiti, Cuba and El Salvador in South America.

The visa system is another powerful weapon created by the global matrix and used against people in different places. Most people from the South require visas not just to travel to the West but to move around in the South, but most people in the West do not require visas to travel to the South, let alone move around in the West. In other words, the West, which invented the visa system, can travel freely to most parts of the world but people from the South cannot, not even within the South. One may wonder why governments in the South have not taken a decision against this level of control, but how many governments in the South actually make their own decisions? Through the system of the global matrix, the West deploys the coloniality of power to dictate to governments in the South. This is what some people have labelled neo-colonialism (Nkrumah, 1965). It paves the way for sustained control, repression and exploitation.

The idea of exploiting the regions of the global South to nourish the West appears to have been borrowed from the medieval belief in witches and vampires who farmed humans to provide the vampire clan with blood nourishment. It would then not be inappropriate to describe modern civilisation built by the West as a 'vampire civilisation'. Capitalism[2] was the instrument devised in this constructed modernity to simulate the vampire scheme. It is brutal, merciless, and purely expedient. In this light, it can be argued that the First and Second World Wars were not, as propaganda has it, between good and bad people, but were wars between villains who were vying for the leadership of the vampire civilisation. Evidence for this can be seen in what happened at the end of the Second World War. After the supposedly good side defeated the so-called bad side, the USSR and the USA, which were supposed to be on the same side, began their own war that lingered until the

---

[2] Socialism is not different in this regard except that while capitalism exploits by under remuneration of labour, where the exploited know that they are being exploited but are powerless to confront their exploiters, socialism exploits by deluding the exploited into thinking that they are willingly contributing what they do not need to the patriotic project of nation building.

collapse of the Soviet Union, and the United States emerged as the leader of the vampire civilisation. Francis Fukuyama (1992) paints an interesting picture of this historical moment, which for lack of space, I will not discuss here.

I will now briefly discuss the structure of European modernity. Based on Aristotle's two-valued logic, both Descartes and Bacon formulated methodologies that marginalise an aspect of reality. The Cartesian methodic doubt, for example, draws a line between reason and the senses, just as Bacon's inductive method separated science from faith. The two-valued logic is a logic of binary opposition in which the value True is superior to the value False. It presents a structure for a lopsided dichotomy in reality. This logic lay at the foundation of modernity by first shaping the enlightenment, the agrarian and scientific revolutions, and then crystallising into modernity. By its law of identity, it divests every variable of any connection with any other thing else and its ability to transform. The law of contradiction pits every variable against any other that is seemingly opposed to it. The law of excluded middle eliminates the possibility of seemingly opposed variables ever complementing each other. Thus, every aspect of modernity is caught up in the divisive structure of superior/inferior, self/other (leading to slave trade, colonialism, racism, religious bigotry, gender inequality, coloniality); rich/poor, insider/outsider (leading to classism); normal/abnormal (leading to sexism), etc.

On the one hand, then, we have the global matrix of modernity that imposed the Western particular on the rest of the world, thereby sucking in every culture, region and people, such that to do anything at all and do it well, one must follow the accepted methods of modernity. On the other hand, we have modernity that is structured to marginalise and segregate the rest of the cultures. It is a structure of power relations between the West and the rest. The West is the zone of being, power and knowledge, while the rest is the zone of nothingness—total oblivion. In this grand scheme, everything, including the power to control the production, regulation and dissemination of scientific knowledge, lies within the purview of the West.

3 The bordering problem in philosophy of science

It is now becoming clear that the main problem of the twenty-first century is the problem of border lines. Whether artificial or natural, borders are today dividing humanity into in-group and out-group, superior and inferior, norm and normalised. What is alarming is that such divisions are causes of famine, conflicts, hate crimes, xenophobia and wars. Various actors on the positive side of the power relationship are attempting to weaponise borders to trap the human spirit and prevent the free interaction and movement of intelligentsias, thus animating tropes such as epistemic injustice, epistemic marginalisation and epistemicide. This is the world that modernity as constructed has built, a world of slavery, racialism, colonialism and imperialism.

Colonialism is not an independent system, as most people erroneously suppose. It is a mechanism, among other mechanisms that make up the engine of constructed modernity. In constructed modernity, you have instruments or mechanisms like enslavement, capitalism, socialism, creed, the organised polity, the protocol of the sciences and colonialism. The latter replaced 'the slave trade', which gave way to coloniality. This was why the much-anticipated post-colonialism did not materialise at the end of colonialism, because colonialism was not an independent system and the system that controlled it simply used it to go on to deploy the tool of coloniality, which is expected to have a longer life span. At the expiration of coloniality, it can be predicted that *borderism*, a new type of ideology for epistemic affirmation/negation that will depend on one's or a group's geographical location, will emerge. This is, however, beyond the scope of the present paper.

Here I will focus on teasing out the implications of bordering for the intellectual sphere, specifically the philosophy of science. The problem of bordering for scientific knowledge has two components; right and power. Now, one easy mistake would be to assume that this problem is the same as the problem of demarcation. Bordering pre-dates demarcation in science, but curiously, it has largely been ignored by most philosophers of science. While demarcation distinguishes what qualifies as science from what does not in accordance with the criteria of science prescribed by the self-styled model or norm, bordering draws a line between that norm and the normalised cultures as a determination of which side has the right and the power to produce, regulate and disseminate what counts as scientific knowledge and which does not. In other words, bordering relates to the imperialisation of knowledge by the self-styled model culture and the residualisation of the normalised cultures, based on an imaginary map that charts intellectual borders.

Even though this intellectual border is informed by physical ones through the intervention of colonialism, they have since morphed into intellectual forms through coloniality. Thus, the bordering problem is an intellectual one. It is a line between the West and the rest that separates being from nothingness, power from powerlessness, and knowledge from ignorance.[3] This construal becomes a problem because it is mired in a lopsided ethics and

---

[3] Boaventura du Sousa Santos (2016, p. 20) has also discussed what he calls the 'abyssal line,' "A line that is so important that it has remained invisible. It makes an invisible distinction sustaining all the distinctions we make between legal and illegal, and between scientific, theological and philosophical knowledges. This invisible distinction operates between metropolitan societies and colonial societies". But the bordering problem, for me, is not just a line that separates; it is a weapon of mass segregation, suppression and repression.

politics that discount a section of humanity in the production, regulation and dissemination of knowledge.

The right to produce scientific knowledge: The tricky thing about this problem is that it is not about expertise but about racial chauvinism (See Hebga, 1954; Serequeberhan, 1991), i.e., which culture has the right to produce what counts as scientific knowledge. The assumption for the dominant Western culture is that there can be only one answer to that question, namely, the West. The West is the model while the rest are the segregated other. There is only one way of producing credible scientific knowledge and it is the proud possession of Western culture. It is the incomparable testable, empirical, demonstrable protocol. In this, there are two basic assumptions; first, anyone who does not abide by its prescription is not doing science and whatever is produced therefrom is not scientific knowledge. Second, and this is curious, anyone who abides by its criteria, insofar as they are not of the West, is a mere imitator and copycat. And no matter how one may look at it, whatever is produced, even if it is a work of genius, has been made possible only by the Western epistemic structure and is for that reason alone a legacy of Western culture. The case of Albert Einstein and his theory of relativity is a clear example. For example, some Western colleagues like the German, Philipp Lenard, racially abused and discounted him, but when his research findings became incontestable, they were a legacy of western intellectual tradition. Lenard once said of Einstein that he brings the foul "Jewish Spirit" into physics and that "just because a goat is born in a stable does not make it a noble thoroughbred" (Gunderman, 2015). This assumption points to coloniality of knowledge and being.

Some post-colonial scholars have observed quite correctly that colonialism did not end outright. Ramon Grosfoguel (2007) describes the assumption that the end of colonial administration ushered in a post-colonial era as one of the most powerful myths of the twentieth century. For Santos (2018), colonialism simply morphed into postcolonialism (without a hyphen), rather than post-colonialism, which indicates an era after colonialism. Postcolonialism, for him, is an imperialism of a sort or a subtle continuation of colonialism in ways that make it less incompatible with nationalism (Kennedy, 2016). For some decolonial scholars, colonialism did not exactly end, but mutated and morphed into another monster that they describe and study using the concept of coloniality which has three strands namely, coloniality of knowledge, power and being (Quijano, 2000; Grosfoguel, 2007; Ndlovu-Gatsheni, 2015; Seroto, 2018). According to Maldonado-Torres:

> Coloniality is different from colonialism. Colonialism denotes a political and economic relation in which the sovereignty of a nation

or a people rests on the power of another nation, which makes such a nation an empire. Coloniality, instead, refers to long-standing patterns of power that emerged as a result of colonialism, but that define culture, labour, intersubjectivity relations, and knowledge production well beyond the strict limits of colonial administrations. Thus, coloniality survives colonialism. It is maintained alive in books, in the criteria for academic performance, in cultural patterns, in common sense, in the self-image of peoples, in aspirations of self, and so many other aspects of our modern experience. In a way, as modern subjects we breathe coloniality all the time and every day. (Maldonado-Torres, 2007, p. 243.)

Maldonado-Torres' description of the difference between colonialism and coloniality reveals a deep frustration with the failure of a post-colonial world to result from the end of colonialism. Rather, what ensued was coloniality which, like an octopus, seeks to spread and maintain the West's preferred structure of the world and prejudiced structuring of various subjectivities. This point is made clearer by Anibal Quijano (2007, p. 170), who maintains that despite the end of political colonialism, coloniality has emerged as "the most general form of domination in the world today […]". In his discussion, he outlines four main areas in which the Euro-North-American world order controls the world. These include the economy, authority/power, gender/sexuality and knowledge. In these spheres, Quijano argues that the dominant Western culture imposes its structures on the rest and denies the subjugated other, cultural, epistemic and ontological legitimacy. Robert Young (2003) tries to paint a picture of what it means to be a subject in a subordinated world. It is "to be from a minority, to live as the person who is always in the margins, to be the person who never qualifies as the norm, the person who is not authorized to speak" (2003, p. 1). This type of subordination of one group by another betrays the lop-sidedness inherent in the power relations of the modern world.

As Quijano (2000) observes, the imposition of the Western worldview on the rest of the cultures that lie outside the West is a reflection of what he calls the coloniality of the power matrix that enables the West to organise the world's population, borders and cultures into the hierarchical order of superior and inferior. In this type of arrangement, the epistemic formation of the West is promoted as acontextual, and the one and only standard. In this way, other cultures' scientific protocols and epistemic resources are repressed and marginalised, if not exterminated. Some decolonial scholars have exposed this ploy. Quijano (2007), as mentioned earlier, exposes the power matrix of constructed modernity spread across four spaces. Grosfoguel (2007) talks of a systematic transition from colonialism to coloniality, and recommends what he calls the "decolonial turn" to roll back the mechanisms of coloniality that imperialise the Western epistemic accumulation that translates to the residualisation of the knowledges in the normalised regions of the world. Leonhard Praeg (2019) warns that the epistemic hegemony of the West not only subalternises the rest of other cultures, but also drives a wedge into the wheel of global epistemic progress. This is because, as he puts it, creating such lopsided epistemic borders means that even the norm loses the capacity to view reality comprehensively as it can only advance and stand *at*, rather than *on* the border. There will always be prejudice against what comes out of the normalised cultures of the South.

Reflecting on the views described above, one might be puzzled as to how some actors in the West succeeded this long in deluding many people into seeing their scientific protocol as not only acontextual but incontestable, and the only possible option. Probing backwards into history, it is easy to see that the constructed modernity that spreads such an impression rests on three myths.

First is the myth of borderlessness: having universalised its cultural particularity, some scholars in the West create and sustain the myth that the universal is borderless. In other words, it is not the West's worldview but the objective universal category discovered by the West to whom the rest of the world must prostrate itself in gratitude. But "the central concern of the discourse on decoloniality consists of revealing the politics at work in this aspiration, particularly the history and logic of those processes that allowed the Western episteme to erase or conceal the contextual and temporal dimensions of its own origin so that it could present itself as acontextual, ahistorical and universal" (Praeg, 2019, p. 1). Bruce Janz has correctly described this as "the pretentions of Western philosophy" (2009).

Second is the myth of *thesis aeternalis*: some scholars in the West treat the protocol of modern science that they have constructed as the only possible epistemic formation — the only thesis. There is no anti-thesis because no epistemic formation from any culture can challenge it. And there cannot be a synthesis because it is self-sufficient and requires no further enrichment. Transmission of this false idea was the singular goal of coloniality of knowledge that Praeg (2019, 1) suggests "was premised on the notion that one episteme, the western, was superior to others, not because other systems of knowing and articulating the meaning of the world were somehow wrong, but because they could not aspire to universality […]". This is, however, false. What has happened is that the West has presented its own particularity as acontextual while at the same time defining what others have to offer from the prism of their local world-views alone. It is the West, through coloniality of power and knowledge, that defines both sides. Admittedly, as Achille Mbembe (1999, 2002) has argued, there are folds of thought in the decolonial movement, specifically those that he described as the nationalist and the Marxist, who want to isolate Africa from the rest of humanity by ghettoising the epistemic resources in Africa. For Mbembe, these groups propagate false philosophies. But the pontifications of what he calls false philosophers who try to de-universalise their postulates cannot be taken as representative of the ideas of the decolonial movement.

Third is the myth of the absolute: having successfully extended and imposed its epistemic hegemony on the rest of the world, some scholars in the West felt emboldened to treat the universal as absolute. Conceptually, an epistemic category is universal if applicable in many cultures, and it is absolute if applicable in all cultural areas and in all contexts and topics without exception. But there is no such thing as an absolute epistemic category. Every universalisable category has contextual or topical limitations that ensure its consistency. The recognition of this fact is the first lesson in intellectual humility that some scholars in the West ought to learn. Unfortunately, some are unwilling to do so and therefore feel the need to exercise the coloniality of power. What cannot be achieved through reasoned discourse, can be achieved through force. In this way, *a* scientific protocol developed in the West is promoted as *the* protocol, thereby marginalising other cultures and their contributions to knowledge.

Interestingly, for each of these three myths, Praeg has noticed that there are master tropes that are being used to confront coloniality. I am of the view that the advocates of these tropes, even though unconsciously, were addressing the three myths I identified above. For the myth of borderlessness, Miranda Fricker (2007) and others use the master trope of epistemic injustice to confront the myth. If the epistemic formation of the West is promoted as objective and acontextual, then this does an injustice to the accumulated knowledge by others, which automatically becomes subalternised. As regards the myth of thesis aeternalis, Boaventura Santos (2014, 2018) uses the master trope of epistemicide to confront it. If what the West has imposed on the world is eternal and unchallengeable, then it exterminates the knowledge visions of other cultures. Jonathan Chimakonam (2018a) and others also employ the master trope of epistemic marginalisation to challenge the myth of the absolute. This is because, if what comes from the West is absolute, then whatever actors are doing in the South is a waste of time. Epistemologies of the South are by default marginalised.

The power to determine scientific knowledge: This problem is not about the intellectual capacity of individuals but about cultural politics (see Rorty, 2007; Chimakonam & Nweke, 2018). This politics is controlled by the West which having gifted the world with modern civilisation seems to proclaim itself the model and not only arrogates but exercises the power to imperialise its epistemic accumulation while subordinating the knowledge of other cultures. Any system of knowledge that does not align with the established canons and approved methodologies of Western episteme is not just unorthodox but unscientific, supernatural, superstition and folk wisdom. The West rubber-stamps what counts as scientific knowledge, and no system, no matter its utility, is regarded as scientific if the West does not authenticate it.

To appreciate the existence and exercise of this power, one would have to look at the scientific community run by various western societies and institutions working harmoniously according to set criteria and rules to issue awards, prizes and recognition. Award and recognition panels are given criteria to work with that never differ on structure even if they differ on the list of items. On matters of structure, the scientific societies everywhere in the West are at agreement. Underlying this structure is logic, chiefly of the Aristotelian tradition with its principle of bivalence and orientation to determinism. This system of logic is rigid in its laws for defining reality. The laws of identity, contradiction and excluded middle are strict on the nature and characterisation of things.

It is perhaps more straightforward, however, to say that these laws make provision for one way of looking at reality, but the dominant Western culture insists it is the only credible way. This is where imperialisation begins. In this context, imperialisation refers to the imposition of a structure of epistemic interpretation belonging to or developed by a culture that self-styles itself as the model, on all the other cultures that it gazes upon as imitation or inferior to itself. This imperial culture is the West which uses subordinate categories such as non-standard, non-western and non-classical to segregate the nameless *restness* from itself. The Western epistemic formation is not only the standard that the rest struggle in vain to ape, but is also the classic, prior to which there could be none other to compare to it in glory and after which there can be none other to equal it in grandeur. It is a closed epistemic dictatorship. These manifestations of coloniality of knowledge, power, and being are now being confronted using decoloniality, a movement that confronts imperialism, colonialism, neocolonialism and othering. For Sabelo Ndlovu-Gatsheni, decoloniality is important in the struggle to emancipate the victims of othering "because the domains of culture, the psyche, mind, language, aesthetics, religion, and many others have remained colonized" (Ndlovu-Gatsheni, 2015, p. 485). But what is decoloniality?

According to Maldonado-Torres (2011, p. 117) "By decoloniality it is meant here the dismantling of relations of power and conceptions of knowledge that foment the reproduction of racial, gender, and geo-political hierarchies that came into being or found new and more powerful forms of expression in the modern/colonial world." Decoloniality then involves a systematic attempt to dislodge the barriers of racial chauvinism and cultural politics that the dominant Western culture by means of coloniality has systematically imposed on the rest of the world in order to subordinate them. Thus, underlying our idea of decoloniality is the quest to dislodge the bivalent logic of modernity and of coloniality that promotes binary opposition, and replace it with a trivalent logic that engenders binary complementarity. In the same way, the South African philosopher Praeg (2019, p. 1) conceives decoloniality as "nothing but the systematic dismantling of all the intellectual Disciplines or Subjects that combined to form the Western grid of intelligibility imposed on the world." The grid of intelligibility can refer to the underlying logic that shapes modernity.

As Praeg put it, contemporary discourse on decoloniality involves some:

> master tropes which play a central role: epistemicide, epistemological marginalisation or exclusion, epistemological disobedience and epistemic justice. What all the tropes have in common is not only a concern with episteme (from the Greek for 'knowledge'), but also with what it means to know, the knowledge of knowledge (epistemology), which in decolonial discourse manifests primarily as a concern with the politics that regulate and determine what it means to know. (Praeg, 2019, p. 1.)

Praeg (2019, p. 1) goes on to explain that the phrase "what it means to know" can no longer be understood only in a simple sense of "a neutral self-reflexivity (for example: how can I be sure that what I know is true? What is the source of my knowledge?)," but must also be understood as "the power at work in those processes through which one knowledge system—for example, Western philosophy—asserted and in many ways continues to assert itself over others, either by eliminating them (epistemicide), marginalising or excluding them from the 'canon' of legitimate knowledge, or assimilating them [...]". This exclusion is a demonstration of the dominant culture's power to determine who can be part of knowledge production processes, which translates to coloniality of being and then lastly of knowledge. These three structures that are the manifestations of the bordering problem have remained in effect in science, and it is the duty of the philosopher of science to dismantle them. Unfortunately, this does not seem to be getting serious attention from the philosophers of science. In the next section, I will argue that there is a need for philosophers of science to disborder scientific knowledge.

4 The need to disborder scientific knowledge

The CEK is a programme that simulates Hall's ideas of cultural identity recognition. Hall has identified two senses of cultural identity. The first is the recognition of what is shared (1990, pp. 222–224) and the second is the recognition of what is unique to each group (1990, p. 225). While for Hall it was the first that played a major role in post-colonial theorising, for me it is the second that plays an important role in decolonial thinking. By this I mean that the recognition of epistemic formations from different cultures is the first bold step towards a new epistemology that is balanced, and which for lack of a better expression I have described in this work as the Collective Ecology of Knowledge. I define bordering as the systematic discounting other epistemologies by direct appeal to intellectual geolocation, and here I am restricting this argument to the context of scientific knowledge.

So, to address the bordering problem in the philosophy of science, I propose a decolonial strategy of intellectual disbordering. By disbordering, I do not mean collapsing the borders of different knowledge formations; that would defeat the goal of CEK, which recognises such borders but aims at the conversation or creative struggle of different ecologies of knowledge. Creative struggle means the horizontal engagement of diverse ecologies of knowledge or a competition that leads to progress, the opening of new vistas for thought and the extension of the frontiers of human knowledge (Chimakonam, 2017b). The conversation is creative because it births new ideas, and it is a struggle because it is critical, objective and rigorous. So, we are promoting mutual respect and recognition without sacrificing epistemic rigour. By disbordering I mean dismantling the gatekeeping orientation of a dominant culture that empowers Western philosophers of science to stand *at* the border and gaze at the rest through distorting that residualises otherness and imperialises the self. Disbordering eliminates the divisive line between the model and the constructed otherness and paves the way for the horizontalisation of knowledge ecologies. With this new orientation, philosophers of science as critics and evaluators would be able to stand on the border and not at the border to compare ideas and methods from different scientific traditions.

Praeg (2019, p. 2) draws our attention to the implications of being *at* and *on* the border. Being at the border, for him, could mean standing at one side and gazing at what lies on the other. This offers a narrow vision that denies the observer the advantage of a comprehensive picture of reality. Meanwhile, in his view being on the border evokes the more balanced image of one "standing on the line of differentiation: neither on this side nor the other side. Formulated differently, whereas being 'at' the border suggests a difference between what lies on either side, being 'on' the border suggests less of a differentiation between this and that and more a de-differentiation of this and that, of being in difference." Being on the border, then, offers the advantage of multiple perspectives, but beyond that, it offers the opportunity for bringing those diverse views into a conversation. No epistemic edifice can approximate truth or certainty if it is not balanced.

However precise and productive its protocol, modern science still leaves another space, to the protocols of other epistemologies. It is a space that is perhaps humble and lowly, but on that they nonetheless can, and do occupy to the benefit of humanity. As unchallenged as the method of science has been since the assimilation of Einsteinian physics into the Western intellectual legacy and the development of three-valued logic by Jan Lukasiewicz (1920), experience still informs us that it is not the exhaustive measure of knowledge. It is easy for those who are too committed to Western intellectual hegemony to point to David Hilbert's statement that glorifies reason as having the capacity to solve all problems given sufficient time and resources (1901/2000), but that only exposes their arrogant ignorance of new knowledges. In the context of coloniality of knowledge, the concern is not with reason as one abstract reality, but with the various manifestations of reason and its journey from the particular to the collective ecology of knowledge. Modern science silences the voice of reason in its diverse cultural echoes, and this is imperialistic.

In the scheme of things, we are supposed to have the courage to say that some things are bad or incorrect, but this would not mean that some others are not good or correct. It is also true that we are supposed to have the courage to say that among good or correct things, A is better than B, but that would not mean that B is bad or incorrect. Similarly, as the West celebrates and glorifies the gains of science, it will serve us well if we strive to understand what goes on in other climes.

5 The prospect of a new logic

What does a new logic[4] entail for a decolonial project in philosophy of science? It entails a shift from a bivalent and deterministic logic that divides us into racial and cultural blocs to a trivalent and complementary logic that can bring all epistemic formations into a Collective Ecology of Knowledge (CEK). The logic of complementarity promotes horizontal cross-cultural interaction ideal for the realisation of CEK. Signs of change

---

[4] I see the two-valued logic system as the defining feature of all principal tropes of modernity. For example, it is the bivalent logic of modernity that shapes coloniality. To decolonise, then, necessarily should involve a change of this divisive logic.

began to emerge when Albert Einstein (1915) developed his theory of relativity, which could not be axiomatised straight-forwardly using two-valued logic. Jan Lukasiewicz (1920) then developed his three-valued logic, which provided a basis for the theory of quanta and attempted to explain the notorious future contingent propositions identified by Aristotle. These two developments were special moments in the history of thought, both indicating that the collective ecology of knowledge is not only possible but also plausible. What is required is a system of three-valued logic that can extend bivalence into trivalence and contradiction into complementarity by adjusting the laws of thought. Prototypes of such a logic have already been worked out in Chris Ijiomah's Harmonious Monism (2006, 2014), Innocent Asouzu's Complementarity Logic (2004, 2013), Chimakonam's Ezumezu Logic (2012, 2017c, 2018b, 2019). For lack of space, I will not go into a detailed discussion of these systems. Instead, I will simply highlight what is common to the three systems and relevant to the argument of this essay. To do this, I will employ two diagrams.

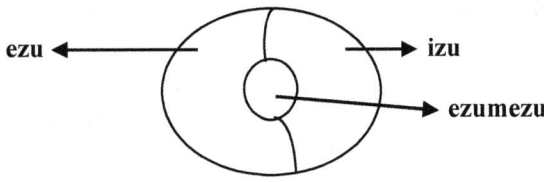

Figure 1. Diagram of Trivalence (source: Chimakonam, 2015)

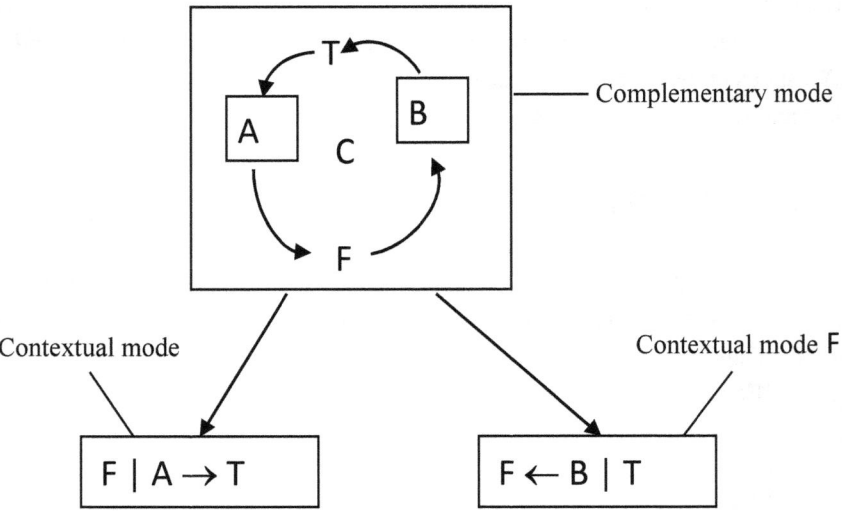

Figure 2. Diagram of contexts and modes (source: Chimakonam, 2015)

What will it cost a philosopher of science to conceive or imagine the possibility of different scientific protocols? The truth is that it will probably cost him a lot, perhaps more than we can ever imagine, but one thing that readily comes to mind is that—at the very least—it will cost him the convenience and the full complement of deterministic logic that has evolved and developed from Aristotle through Frege to Russell, Gödel, Quine and to this day. This tradition of logic, if not logics gives us the power to predict precisely, but a more liberal system would expand our horizon and grant us the power to imagine more subtly.

The protocol of modern science rests on the framework of Aristotelian logic tradition built on the traditional three laws of thought. The Polish logician Lukasiewicz challenged Aristotelian tradition on the basis of its commitment to determinism (Betti, 2001; Kachi, 1996; see also Lukasiewicz, 1918 and 1922 works [1970, pp. 110–128]). The theory of determinism can be stated as 'every statement is either necessary or impossible'. We can easily see the connection between this and the theory of bivalence, which states that 'every statement is either true or false'. But if every statement is to be judged as either true or false, then there is no middle position. A question might then arise: what about peculiar contexts that admit a middle position? Aristotle himself observed that future contingent propositions are neither true nor false. Realistically, they are both true and false.[5] They are like the butterfly clutched inside the palm. It is not neither dead nor alive, it is both dead and alive because the death and the life of the butterfly is in the hands of the one who clutches it. Figure 1 above enables the system of three-valued logic like Ezumezu to extend bivalence into trivalence. With this principle, in addition to being true and false, a proposition is complementary.

Furthermore, we can add the idea of complementary relationships and inferences of that kind where seemingly opposed variables can be brought into a complementary relationship rather than strict contradiction. All of these provide hope that a different system of logic that prioritises complementarism[6] over determinism can offer a different, if somewhat flexible, protocol for generating a richer and safer scientific knowledge. All that is required is a shift in the logical paradigm of science. Figure 2 above shows that propositions that are seemingly opposed (contradictions) are in contextual modes from where they could also move into an intermediate mode to complement each other. The other mode is the complementary mode. In other words, the system of three-valued logic

---

[5] Graham Priest (2018) in his paraconsistent logic discusses the idea he calls dialetheism or true-contradictions. This stipulates that there are statements that are both true and false.
[6] This states that seemingly opposed variables can be complementary besides being contradictory.

developed in the African tradition like Ezumezu is not against contradiction but identifies it as one of two possible contexts of relationships of variables.

As part of the on-going decoloniality project, some workers mentioned earlier are developing prototypes in the African tradition of logic. Their systems seek to undermine the deterministic property in contradiction, and extend bivalence into trivalence. Overall, the prospect of a new logic could inaugurate a new paradigm in science where different approaches, rather than a single approach, can be endorsed.

I will give one brief example of the scientific practice in Africa prior to colonial conquest. Before scientists in the west thought of brain surgery, African scientists had already developed an efficient model said to have a 96% success rate.[7] As a video documentary shot over 70 years ago explains, the protocol of the science of brain surgery that removes tumours and cleanses the brain in Africa is not only different from the one developed and practiced in the West much later but is more effective. Unfortunately, colonialism through epistemicide sought to destroy that knowledge formation, and through coloniality of power, being and knowledge, misinform the world that Africa had nothing prior to the arrival of the West.

6 Conclusion

There are many important parts in the corpus on scientific knowledge, but I would like to focus briefly on 'result' and the 'protocol' that yields it. I will therefore conclude this essay by saying that as regards the 'hypothesis' as driver in the scientific quest, the results that are its objective can be the same no matter where an experiment is carried out and if the conditions are the same, all things being equal. But as regards knowledge as a whole, the protocol, which is the procedure, should not be expected to be the same. As philosophers of science, we must then make it part of our concern to study different protocols, bring them into conversation and creative struggle and note what makes each effective or more effective than others. In this way, competition of ideas among different traditions and protocols of science can be encouraged and the Collective Ecology of Knowledge (CEK) can be realised.

To create a new orientation for confronting the colonial order entwined in the bordering problem, I propose a decolonial strategy of disbordering. Such a strategy for eliminating the divisive intellectual border between the norm and the normalised would enable

---

[7] See "A Kisii Traditional Healer Performing Brain Surgery", https://www.youtube.com/watch?v=6ireXE3hYdI.

philosophers of science to overcome the coloniality of knowledge that promotes the ideology of one standard protocol for science. I believe that the principles of Aristotelian logic that undergird modernity as it is constructed also provide the foundation for the ideology that the testable, demonstrable, empirical protocol of modern science is the best that the world can have. To fight this ideology, therefore, decolonial scholars must confront the logic that informs it.

Bibliography

Asouzu, I. (2013). *Ibuanyidanda (Complementary Reflection) And Some Basic Philosophical Problems in Africa Today: Sense Experience, "Ihe Mkpuchi Anya" and the Supermaxim*. Zurich: Lit Verlag GmbH and Co. Kg Wien.

Asouzu, I. (2004). *The Method and Principles of Complementary Reflection in and beyond African Philosophy*. Calabar: University of Calabar Press.

Bacon, F. (1629/1855). *The Novum Organon or a True Guide to the Interpretation of Nature*. Transl. G. W. Kitchin. Oxford: Oxford University Press.

Betti, A. (2011). The incomplete story of Łukasiewicz and bivalence. In T. Childers & O. Majer (Eds.), *The Logica Yearbook* 2001 (pp. 21–36). Prague: Filosofia.

Chimakonam, J. O. (2012). *Introducing African Science: Systematic and Philosophical Approach*. Bloomington, IN: Authorhouse.

Chimakonam, J. O. (2015). The criteria question in African philosophy: Escape from the horns of jingoism and Afrocentrism. In J. O. Chimakonam (Ed.), *Atuolu Omalu: Some Unanswered Questions in Contemporary African Philosophy* (pp. 101–123). Lanham: University Press of America.

Chimakonam, J. O. (2017a). Conversationalism as an emerging method of thinking in and beyond African philosophy. *Acta-Academica, 49*(2), 11–33.

Chimakonam, J. O. (2017b). What is conversational philosophy? A prescription of a new doctrine and method of philosophy, in and beyond African philosophy. *Phronimon, 18*, 114–130.

Chimakonam, J. O. (2017c). The question of African logic: Beyond apologia and polemics. In T. Falola & A. Afolayan (Eds.), *The Palgrave Handbook of African Philosophy* (pp. 106–128). New York: Palgrave Macmillan.

Chimakonam, J. O. (2018a). Addressing the epistemic marginalization of women in african philosophy and building a culture of conversations. J. O. Chimakonam & L. Du Toit (Eds.), *African Philosophy and The Epistemic Marginalisation of Women* (pp. 207–218). London: Routledge.

Chimakonam, J. O. (2018b). The philosophy of African logic: A consideration of Ezumezu paradigm. In Jeremy Horne (Ed.), *Philosophical Perceptions on Logic and Order* (pp. 96–121). Hershey, PA: IGI Global.

Chimakonam, J. O. (2019). *Ezumezu: A System of Logic for African Philosophy and Studies*. Cham: Springer.

Chimakonam, J. O., & Nweke, V. C. (2018c). Why the 'politics' against African philosophy should be discontinued. *Dialogue, 57*(2), 277–301.

Chimakonam, J. O., & Chimakonam, A. E. (2022). Two problems of comparative philosophy: Why conversational thinking is a veritable methodological option. In S. Burik, R. Smid, & R. Weber (Eds.), *Comparative Philosophy and Method Contemporary: Practices and Future Possibilities* (pp. 223–240). London: Bloomsbury.

Descartes, R. (1637/1968). *Discourse on method and the meditations*. Trans. F. E. Sutcliffe. London: Penguin.

Egbai, U. O., & Chimakonam, J. O. (2019). Why conversational thinking could be an alternative method for intercultural philosophy. *Journal of Intercultural Studies*, 40(2), 172–189.

Einstein, A. (1916). *Relativity: The Special and General Theory*. New York: H. Holt and Company.

Fricker, M. (2007). *Epistemic Injustice: Power and the Ethics of Knowing*. Oxford: Oxford University Press.

Fukuyama, F. (1992). *The End of History and the Last Man*, New York: The Free Press.

Grosfoguel, R. (2007) The epistemic decolonial turn: Beyond political-economy paradigms. *Cultural Studies, 21*(2–3), 203–246.

Gunderman, R. (2015). When science gets ugly – the story of Philipp Lenard and Albert Einstein. *The Conversation*. Pubblished June 16, 2015. Retrieved September 18, 2018. http://theconversation.com/when-science-gets-ugly-the-story-of-philipp-lenard-and-albert-einstein-43165.

Hall, S. (1990). Cultural identity and diaspora. In J. Rutherford (Ed.), *Identity, Community, Culture and Difference* (pp. 222–237). London: Lawrence and Wishart Limited.

Hebga, M. (1958). Logic in Africa. *Philosophy Today, 2*(4), 221–229.

Hilbert, D. (2000). Mathematical problems. *Bulletin (New Series) of the American Mathematical Society, 37*, 407–436. Reprinted from Bulletin of the American Mathematical Society, *8* (1901), 437–470.

Ijiomah, C. (2006). An excavation of a logic in African worldview. *African Journal of Religion, Culture and Society, 1*(1), 29–35.

Ijiomah, C. (2014). *Harmonious Monism: A Philosophical Logic of Explanation for Ontological Issues in Supernaturalism in African Thought*. Calabar: Jochrisam Publishers.

Janz, B. B. (2009). *Philosophy in an African Place*. Lanham: Lexington Books.

Kachi, D. (1996). Was Lukasiewicz wrong? Three-valued logic and determinism. Paper presented at *Lukasiewicz in Dublin – An international conference on the works of Jan Lukasiewicz*, July 7, 1996.
Kennedy, D. (2016). *Decolonization: A Very Short Introduction*. New York: Oxford University Press.
Kimmerle, H. (1994). *Die Dimensionen des Interkulturellen: Philosophie in Afrika – afrikanische Philosphie*. Amsterdam: Supplimente und Verallgemeinerungsschritte.
Łukasiewicz, J. (1920/1970). On three-valued logic. In L. Borkowski (Ed.), *Selected works by Jan Łukasiewicz* (pp. 87–88). Amsterdam: North-Holland.
Łukasiewicz, J. (1922/1970). On determinism. In L. Borkowski (Ed.), *Selected works by Jan Łukasiewicz* (pp. 110–128). Amsterdam: North-Holland.
Łukasiewicz, J. (1918/1970). Farewell lecture by Professor Jan Lukasiewcz delivered in the Warsaw University Lecture Hall on March 7, 1918. In L. Borkowski (Ed.), *Selected works by Jan Łukasiewicz* (pp. 84–86). Amsterdam: North-Holland.
Maldonado-Torres, N. (2011). Thinking through the Decolonial Turn: Post-continental Interventions in Theory, Philosophy, and Critique – An Introduction. *Transmodernity: Journal of Peripheral Cultural Production of Luso-Hispanic World*, 1(2), 1–15.
Maldonado-Torres, N. (2007). On coloniality of being: Contributions to the development of a concept. *Cultural Studies*, *21*(2–3), 240–270.
Mall, R. A. (2000). The concept of an intercultural philosophy. Translation from the German by Michael Kimmel. *Polylog: Forum for Intercultural Philosophy*, *1*.
Mbembe, A. (2002). African modes of self-writing. *Public Culture*, 14, no. 1, 239–273.
Mbembe, A. (1999). Getting out of the ghetto: The challenge of internationalization. *CODESRIA Bulletin*, *3*(4), 1–10.
Ndlovu-Gatsheni, S. (2015). Decoloniality as the future of Africa. *History Compass*, *13*(10), 485–496.
Praeg, L. (2019). Philosophy on the Border. In L. Praeg (Ed.), *Philosophy on the Border: Decoloniality and the Shudder of the Origin*. Pietermaritzburg: University of KwaZulu-Natal Press.
Priest, G. (2018). Dialetheism. In E. N. Zalta (Ed.), *The Stanford Encyclopedia of Philosophy* (*Spring 2018 Edition*).
Quijano, A. (2007). Coloniality and modernity/rationality. *Cultural Studies*, *21*(2–3), 168–178.
Quijano, A. (2000). Coloniality of power, eurocentrism and Latin America. *Nepantla: Views from the South*, *1*(3), 533–579.
Rorty, R. (2007). *Philosophy as Cultural Politics*. Cambridge: Cambridge University Press.

Santos, S. B. (2014). *Epistemologies of the South: Justice against Epistemicide*. New York: Routledge.

Santos, S. B. (2016). Epistemologies of the South and the future. *The European South, 1*, 17–29.

Santos, S. B. (2018). *The End of the Cognitive Empire: The Coming of Age of the Epistemologies of the South*. Durham: Duke University Press.

Serequeberhan, T. (1991). African philosophy: The point in question. In T. Serequeberhan (Ed.), *African Philosophy: The Essential Reading* (pp. 3–28). New York: Paragon House.

Seroto, J. (2018). Dynamics of decoloniality in South Africa: A critique of the history of swiss mission education for indigenous people. *Studia Historiae Ecclesiasticae, 44*(3), 1–14.

Wimmer, F. (2007). Cultural centrisms and intercultural polylogues. *Philosophy. International Review of Information Ethics, 7*, 1–8.

Wimmer, F. (1996). Is intercultural philosophy a new branch or a new orientation in Philosophy? In D. Gregory (Ed.), *Interculturality of Philosophy and Religion* (pp. 45–57). Bangalore: National Biblical Catechetical and Liturgical Centre.

Young, R. J. C. (2003). *Postcolonialism: A Very Short Introduction*. New York: Oxford University Press.

Author biography. Jonathan O. Chimakonam is a professor at the Department of Philosophy, University of Pretoria, South Africa. His teaching and research interests include African Philosophy, Logic, Environmental Ethics and Decolonial thinking. He aims to break new grounds in African philosophy by formulating a system that unveils new concepts and opens new vistas for thought (Conversational philosophy); a method that can drive theories in African philosophy and beyond (Conversational method); and a system of logic that grounds both (Ezumezu). His articles have appeared in several refereed and accredited international journals. He has authored, co-authored, edited and co-edited several books in African philosophy, including the latest *African Metaphysics, Epistemology and a New Logic: Decolonial Approach to Philosophy*. He is a second-generation member of the prestigious Calabar School of Philosophy.

Philosophy in Science Teacher Education

MICHAEL R. MATTHEWS[1]

Abstract. All science teachers, curriculum writers and school administrators need to engage with philosophical questions. Some of the philosophical questions are internal to teaching science and might be called 'philosophy for science teaching'. For example: Is there a singular scientific method? What is a scientific explanation? And so on. Other philosophical questions are external to the subject and might be called 'philosophy of science teaching'. For example: Can science be justified as a compulsory school subject? What characterises scientific 'habits of mind'? Is science the sole rational method for understanding nature and society? There are many reasons why study of history and philosophy of science should be part of preservice and in-service science teacher education programs. Increasingly school science courses address historical, philosophical, ethical, and cultural issues occasioned by science. Increasingly, Nature of Science (NOS) written into national curricular documents. For all of this, HPS is needed in science teacher education programmes.

Keywords: teacher education, curriculum, philosophy.

1 Introduction

Philosophical questions are ubiquitous in education: in deciding on education policies, in administrating schools, in classroom teaching and engagement with students, in choosing and appraising textbooks, in curriculum writing, in teachers' professional decision making, and many other areas. Teachers are confronted by many questions that require philosophical input: How can specific educational aims be elaborated and defended? How can education be distinguished from indoctrination and does the latter have any legitimate place in teaching? What are the mutual requirements of respect between teachers and students? What are the grounds for inclusion and exclusion of topics in curricula? What is the legitimate role of the state, church, business, school boards, parents, and other stakeholders in curriculum construction, textbook selection, and examinations? What are the legitimate versus illegitimate claims of culture and tradition on educational processes? How can schools resolve competing imperatives for the transmission of culture and the reform of culture? Which of the myriad cultural values present in most societies should

---

[1] School of Education, University of New South Wales, Sydney, Australia.

public schools promote? What are the ethical and political justifications for the countless funding, staffing and class-grouping decisions in schools—graded classes, single sex schools, selective schools, grammar schools, trade schools? How is the distribution of students between these types of schools to be justified? Should the state fully, partially or not at all, fund private education? And much else.

Such normative, non-empirical, philosophical questions arise for all teachers regardless of the subject they teach, whether it be mathematics, music, economics, history, literature, theology, or physical education. Answering them in an informed and thoughtful manner amounts to philosophising about education, and when sustained, this constitutes engagement in the discipline or practice of *general* philosophy of education. It is a discipline that has a long and distinguished history, to which many important philosophers and educators have contributed: Plato, Aristotle, Aquinas, Locke, Kant, Rousseau, Mill, Newman, Whitehead, Russell, and Dewey (to name just a Western First XI). All cultural traditions have comparable lists. Good education is central to the health of all cultures, and in all cultures first-rate philosophers have addressed educational principles, and often enough, specific policies.[2]

Teacher education programmes should introduce students to such philosophical, evaluative, and political questions, and provide practise in informed and serious engagement with them. Historically, this has been the function of general philosophy of education courses in teacher education.

2 Philosophy in science education

As well as *general* philosophy of education, teachers are also required to engage with *discipline-related* philosophy of education, or philosophical questions arising in the teaching of specific disciplines. For science teachers, such questions lead directly to the history, philosophy, and sociology of science. For mathematics teachers, they lead to the history, philosophy, and sociology of mathematics. And so on across school curriculum.

For science teachers, such omnipresent discipline-related questions include: What is the scientific method? Is there a single scientific method or a multitude? What is the scope of science? What is a scientific explanation? What are the criteria for good scientific theories? How might good science be distinguished from pseudoscience? Can there be observation without theory? Can observational statements be separated from theoretical statements? Do experimental results bear inductively, deductively or abductively upon

---

[2] See 45 contributions to Curren (2003).

hypotheses being tested? What are legitimate and illegitimate ways to rescue theories from contrary evidence? What are the characteristics of a good experiment? Are biological processes teleological? Can intentions be attributed to animals? Can Intelligent Design explanations in biology be replaced with adaptation explanations?

Philosophical questions have a special resonance when science courses include 'Socio-Scientific Issues' (Sadler, 2011). These include climate change, nuclear power, logging old growth forest, species preservation, genetically-modified (GM) foods, vaccination mandates, and the debate on the continuation of coal mining. Some such SSI are local, others universal. Philosophy, including values, is unavoidable in identifying, framing and discussing the issues. Although beginning as a *discipline-related* philosophical question, such issues soon enough lead into *general* philosophy of education. On such issues, should teachers indicate their own positions? Should teachers seek to convert students to their own opinions? What are the limits to the importation of political causes into classrooms?[3]

There are a wide range of borderline general/disciplinary philosophical questions that engage science teachers. Can science be justified as a compulsory school subject beyond primary school? What characterises scientific 'habits of mind' or 'scientific temper'? How might the competing claims of science and religion be reconciled? Should local or indigenous knowledge be taught in place of orthodox science, or taught alongside it, or not taught at all? These last questions have very serious contemporary educational, cultural and economic resonance in New Zealand, Canada and Australia. And, of course, in many other countries.

Michael Martin, a philosopher of science, had explored these issues in his *Concepts of Science Education* (Martin, 1972). Martin's book is infused with philosophy of science. The references include books by Agassi, Bridgman, Carnap, Feyerabend, Hanson, Hempel, Kuhn, Lakatos, Nagel, Popper, Quine, Reichenbach and Scheffler. He uses their illustrations, arguments, analyses to explicate, in his book's five chapters, the core topics of Scientific Inquiry, Explanation, Definition, Observation and Goals of Science Education. These topics are part of all science curricula and teaching and their pedagogical relevance would be obvious to trainee teachers.

---

[3] One UNSW student was removed from his Practice Teaching school because he had a 'Save the Whales' sticker on his briefcase. Some years earlier, an Australian teacher was dismissed for having a 'Stop the War' sticker on his car. In the USA the current (2021–22) raft of state laws prohibiting the teaching of a wide range of subject matters, puts these questions right in front of teachers and administrators.

Robert Ennis wrote a comprehensive review of the available literature on philosophy of science and science teaching (Ennis, 1979). His review listed six questions that science teachers constantly encounter in their classrooms and staffrooms and that could be illuminated by researchers in the philosophy and history of science. These questions were:

- What characterises the scientific method?
- What constitutes critical thinking about empirical statements?
- What is the structure of scientific disciplines?
- What is a scientific explanation?
- What role do value judgments play in the work of scientists?
- What constitute good tests of scientific understanding?

These questions are of perennial concern to science teachers and should be raised in science teacher education programmes. How can a science course, let alone a science education course, progress without attention to these core questions? Yet Ennis was moved to make the melancholy observation that, 'With some exceptions philosophers of science have not shown much explicit interest in the problems of science education' (Ennis, 1979, p. 138). The converse was also true, and science educators showed minimal, if any, interest in the history and philosophy of science. Hence the title of Richard Duschl's article of the period: 'Science education and philosophy of science. Twenty-five years of mutually exclusive development' (Duschl, 1985).

3 Philosophy in teacher education

It has long been argued that HPS should form part of the education of science teachers (Matthews, 2020). Forty years ago in the USA, Israel Scheffler, who had a joint appointment at Harvard in philosophy and education, argued for the inclusion of philosophy of science courses in the preparation of science teachers. It was part of his wider argument for the inclusion of courses in the philosophy of the discipline in programmes training people to teach that discipline. His suggestion was that 'philosophies-of constitute a desirable additional input in teacher preparation beyond subject-matter competence, practice in teaching, and educational methodology' (Scheffler, 1970, p. 40).[4] He summarised his argument:

> I have outlined four main efforts through which philosophies-of might
> contribute to education: (1) the analytic description of forms of thought

---

[4] For elaboration and appraisal of Scheffler's arguments about science teaching, see Matthews (1997); for wider appraisals see contributions to Siegel (1997).

represented by teaching subjects; (2) the evaluation and criticism of such forms of thought; (3) the analysis of specific materials so as to systematise and exhibit them as exemplifications of forms of thought; and (4) the interpretation of particular exemplifications in terms accessible to the novice. (Scheffler, 1970, p. 40.)

Each of these four contributions can only be made on the basis of historical and philosophical studies of the relevant teaching subject.[5]

Although the connection between HPS and ST may not have been explicit, it was frequently implicit in the writing of more thoughtful educators. Lee Shulman, whose National Teacher Assessment Project rejected behaviourist, managerial measures of teacher competence, asked about the 'missing paradigm', i.e., the command of subject matter, and the ability to make it intelligible to students. For Shulman:

Teachers must not only be capable of defining for students the accepted truths in a domain. They must also be able to explain why a particular proposition is deemed warranted, why it is worth knowing, and how it relates to other propositions, both within the discipline and without, both in theory and in practice. (Shulman, 1986, p. 9.)

But to explain why a particular proposition is deemed warranted—for instance a proposition about genetic inheritance, or the conservation of energy, or the valency of sodium, or the shape of the earth—assumes an epistemology of science. This epistemology will include standard matters of immediate evidence, both empirical and non-empirical, and considerations about testimony, as the bulk of what anyone knows in science comes by virtue of reliance on the testimony of others. Teachers are a responsible link in the chain of testimony. Teachers who have thought through some basic epistemological questions and know something about the history and sociology of science, will be much better able to explain why a proposition is deemed warranted than those who have no such training.

The American Association for the Advancement of Science (AAAS) affirmed the importance of HPS for science education in its landmark 1989 publication *Science for All Americans* (AAAS, 1989), its 1990 *The Liberal Art of Science* (AAAS, 1990), and its 1993 *Benchmarks for Science Literacy* (AAAS, 1993). The *Liberal Art* stated:

---

[5] For biology teachers, see contributions to Kampourakis (2013); for physics teachers, Cushing (1998); for chemistry teachers, Erduran (2013), Scerri & McIntyre (2015).

> The teaching of science must explore the interplay between science and the intellectual and cultural traditions in which it is firmly embedded. Science has a history that can demonstrate the relationship between science and the wider world of ideas and can illuminate contemporary issues. (AAAS, 1990, p. xiv.)

The AAAS conviction was that learning about science—its history and methodology—would have a positive impact on the thinking of individuals and consequently enrich society and culture. This had also been the conviction of the Enlightenment philosophers and educators (Matthews, 2015, Chap. 2). It was repeated in the *US National Science Education Standards*, which has a separate content strand devoted to 'History and Nature of Science Standards' (NRC, 1996).

In the UK, the *Thompson Report* of 1918 stated that 'some knowledge of the history and philosophy of science should form part of the intellectual equipment of every science teacher in a secondary school' (Thompson, 1918, p. 3). Between the wars, the educator, philosopher and historian Frederick Westaway was the outstanding embodiment of the *Report's* ideal. He was an 'HM Inspector of Schools' in the 1920s and also authored substantial books on history of science and philosophy of science.[6] In a widely used textbook for teacher education programmes, he wrote that a successful science teacher is one who:

> knows his own subject [...] is widely read in other branches of science [...] knows how to teach [...] is able to express himself lucidly [...] is skilful in manipulation [...] is resourceful both at the demonstration table and in the laboratory [...] is a logician to his finger-tips [...] is something of a philosopher [...] is so far an historian that he can sit down with a crowd of [students] and talk to them about the personal equations, the lives, and the work of such geniuses as Galileo, Newton, Faraday and Darwin. More than this he is an enthusiast, full of faith in his own particular work. (Westaway, 1929, p. 3.)

This set a high bar for the following decades, and it was a bar most British teachers went under rather than over. A 1981 review of the place of philosophy of science in British science teacher education stated:

---

[6] An extensive account of the life, writings and achievement of Westaway can be found in Brock & Jenkins (2014).

This more philosophical background which is being advocated for teachers would, it is believed, enable them to handle their science teaching in a more informed and versatile manner and to be in a more effective position to help their pupils build up the coherent picture of science appropriate to age and ability which is so often lacking. (Manuel, 1981, p. 771)

It further noted that 'The lack of reflection was most apparent in the neglect of the cultural, moral and philosophical aspects of science'.

One venture that cleared the Westaway bar was a new optional Upper-Level *Perspectives on Science* course that was introduced in 2007 (Swinbank & Taylor, 2007). The course has four parts:

> Pt.1 Researching the history of science
> Pt.2 Discussing ethical issues in science
> Pt.3 Thinking philosophically about science
> Pt.4 Carrying out a research project

The textbook for this course, on its opening page, states:

> *Perspectives on Science* is designed to help you address historical, ethical and philosophical questions relating to science. It won't provide easy answers, but it will help you to develop skills of research and argument, to analyse what other people say and write, to clarify your own thinking and to make a case for your own point of view. (Swinbank & Taylor, 2007, p. vii.)

The Philosophy section begins with about 16 pages outlining standard matters in philosophy of science—the nature of science, induction, falsifiability, paradigms, revolutions, truth, realism, relativism, etc. Importantly, the book then introduces the subject of 'growing your own philosophy of science' by saying:

> Having learned something about some of the central ideas and questions within the philosophy of science, you are now in a position to evaluate the viewpoints of some scientists who were asked to describe how they viewed science. The aim here is to use these ideas as a springboard to develop and support your own thinking. (Swinbank & Taylor, 2007, p. 149.)

4 Nature of science (NOS) research in science education

For more than twenty years, NOS has been a significant component of science education research.[7] The field would clearly benefit from more cooperation between science educators, historians and philosophers.

Susanne Lakin and Jerry Wellington in their study of UK science teachers' understanding of the nature of science, observed that:

> Teachers' lack of knowledge about the nature and history of science emerged strongly in the study [...]. As well as verbally recognising that their knowledge was patchy and their ideas not well formulated, non-verbal signals reflected an insecurity when the issues were probed in depth. (Lakin & Wellington, 1994, p. 186.)

Norm Lederman's NOS research group, based first at Oregon State University and then at the Illinois Institute of Technology, has been the most prolific and influential in science education over the past twenty years. Lederman received the NARST Distinguished Contribution to Research Award in 2011. His group, along with most science educators, has typically taken a broad and fairly relaxed view of the nature of science. This 'relaxed' position bears upon the validity of test instruments and of informed assessment of what counts as NOS learning (Lederman, Bartos & Lederman, 2014). The group's definition of NOS is characteristically catholic:

> Typically, NOS refers to the epistemology and sociology of science, science as a way of knowing, or the values and beliefs inherent to scientific knowledge and its development. (Lederman et al., 2002, p. 498.)

Significantly, in this definition both the epistemological *and* sociological aspects of science are subsumed under the NOS umbrella. This adds an extra dimension to specification of NOS; it is no longer be just a philosopher's domain. Such a widened socio-philosophical understanding of NOS is beneficially employed when answering, for instance, a central HPS question, such as the demarcation of science from non-science and pseudoscience (Pigliucci & Boudry, 2013).

---

[7] See contributions to Khine (2012), McComas (1998, 2020), Flick & Lederman (2004) and the eleven chapters in Part XIII of Matthews (2014, vol. 3).

The Lederman group maintains that 'no consensus presently exists among philosophers of science, historians of science, scientists, and science educators on a specific definition for NOS' (Lederman, 2004, p. 303). Although recognising no across-the-board consensus on NOS, the group does claim that there is sufficient consensus on central matters for the purposes of NOS instruction in K-12 classes. The group has elaborated and defended seven elements of NOS (the 'Lederman Seven' as they might be called) that they believe fulfil the criteria of: (i) accessibility to school students, (ii) wide enough agreement among historians and philosophers, and (iii) being useful for citizens to know (Lederman, 2004, Schwartz & Lederman, 2008).

The seven elements are:

1. The *empirical nature of science*, where they recognised that although science is empirical, scientists do not have direct access to most natural phenomena.
2. *Scientific theories and laws*, where they hold that 'laws are descriptive statements of relationships among observable phenomena'.
3. The *creative and imaginative nature of scientific knowledge*, where they hold that although science is empirical, its theorising is creative and imaginative.
4. The *theory-laden nature of scientific knowledge*, where it is held that scientists' theoretical and disciplinary commitments, beliefs, prior knowledge, training, experiences, and expectations actually influence their work.
5. The *social and cultural embeddedness of scientific knowledge*, where it is held that science as a human enterprise is practiced in the context of a larger culture and its practitioners are the product of that culture.
6. The *myth of scientific method*, where it is held that there is no single scientific method that would guarantee the development of infallible knowledge.
7. The *tentative nature of scientific knowledge*, where it is maintained that scientific knowledge, although reliable and durable, is never absolute or certain.

This list was widely adopted and reproduced in articles, books and science teacher education programmes around the world where, against the explicit intentions of the Lederman group, it was often reduced to an unthinking mantra, a catechism Understandably, the list has also attracted criticism on various grounds and to different degrees.[8]

Clearly these seven features of science, or NOS elements, need to be much more philosophically and historically refined and developed in order to be useful to teachers

---

[8] See Matthews (2011, 2015, pp. 390–400), Erduran & Dagher (2014), McComas (2020b), Irzik & Nola (2014).

and students. To say this is not simply to make the obvious point that when seven matters of considerable philosophical subtlety, and with long traditions of debate behind them, are dealt with in a few pages, then they will need to be further elaborated. Rather it is to make the more serious claim that at crucial points there is an ambiguity that compromises the list's usefulness as curricular objectives, assessment criteria, and as goals of science teacher education courses.

It is easy for an NOS list of this kind to become all things to all people. Each of the seven points can be pressed into the service of militantly anti-scientific programmes, for alternative or complementary sciences, and for a range of science-sceptical ideologies. A good grounding in HPS is needed for teachers to be able to recognise and avoid these 'list-informed' avenues. Unfortunately, such HPS grounding is the exception rather than the norm.

5 The philosophical poor health of science teacher education

The paucity of serious HPS input into science teacher education is depressingly well documented in Peter Fensham's book *Defining an Identity: The Evolution of Science Education as a Field of Research* (Fensham, 2004). The book opens an authoritative window onto international science teacher education and the ethos of science education graduate schools. The book is built around Fensham's interviews with 79 prominent, extensively published science educators from 16 countries. The interviewees were asked to respond to two questions:

> *# Tell me about two of your publications in the field that you regard as significant.*
> *# Tell me about up to three publications by others that have had a major influence on your research work in the field.*

It is clear from the answers that the science education community has little engagement with serious history and philosophy of science. Apart from Thomas Kuhn, no historian or philosopher is mentioned.[9] Jay Lemke, one of the interviewees, writes:

> Science education researchers are not often enough formally trained in the disciplines from which socio-cultural perspectives and research methods derive. Most of us are self-taught or have learned these matters

---

[9] The deleterious impact of Kuhn on science education is documented in Matthews (2004).

second-hand from others who are also not fully trained. (Lemke, 2001, p. 303.)

Fensham remarks on many occasions that the pioneer researchers came into the field either from a research position in the sciences or from senior positions in school teaching. For both paths, training in psychology, sociology, history or philosophy—the foundation disciplines essential for most serious research in education—was exceptional. He mentions Joseph Schwab, 'a biologist with philosophical background' as an exception (Fensham, 2004, p. 20).[10] James Robinson is another 'pioneer' who could have been mentioned (Robinson, 1968, 1969). Or Walter Jung, the German physics educator (Jung, 1983/1994).

6 Conclusion

The inclusion of some science-teaching focussed HPS course in the preparation of science teachers is a necessity. Ideally it means the creation of specific courses presenting tangible theoretical, curricular and pedagogical topics in science teaching that teachers can identify and recognise as genuine problems, and then showing HPS considerations can contribute to the better understanding and resolution of the issues.

The following diagram—which is an elaboration of an informative comparable diagram in Roland Schulz (2014)—shows the constellation of subjects that support the formation of well-prepared science teachers.[11]

---

[10] For the life and achievements of Schwab, see DeBoer (2014).
[11] This is elaborated in Matthews (2020).

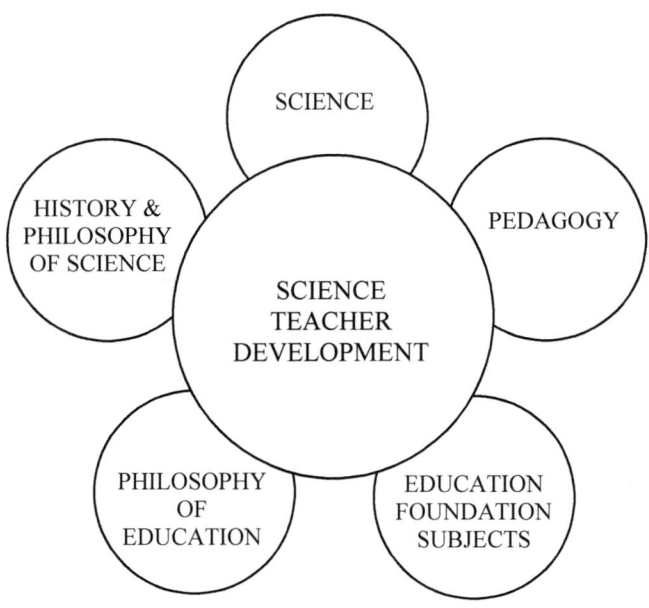

Figure 1. Science teacher development

SCIENCE: undergraduate and/or postgraduate science degree, etc.

HISTORY AND PHILOSOPHY OF SCIENCE: internal curriculum-based HPS and external education-related HPS studies, etc.

PEDAGOGY: practice teaching, educational technology, instructional theory, local curricular, assessment theory and practice, administrative matters, special-needs education, etc.

PHILOSOPHY OF EDUCATION: aims of education, personal and social goals of education, ethical standards for classroom teaching and teacher-student interactions, and for school systems, conceptual analysis of teaching and learning, etc.

EDUCATION FOUNDATION SUBJECTS: sociology of education, history of science education, psychology and cognitive science, developmental psychology, curriculum theory, etc.

The inclusion of both philosophy of education and history and philosophy of science in teacher training programmes can do something to broaden the vision of teachers, and so ensure that their students not only arrive at destinations (scientific competence), but arrive with broader horizons. In the long run this contributes to the health of science and society.

Tragically, and the word is chosen carefully, during the past 30 or more years, the status and presence of philosophy in teacher education programmes, and in graduate studies of education, has progressively diminished.[12] In many institutions, including my own, it has simply disappeared. The philosophical questions and issues have not gone away. What has gone away is the adequate preparation of teachers, researchers, administrators, and policy makers for sensibly and informatively dealing with philosophy-related questions.

Bibliography

(AAAS) American Association for the Advancement of Science: 1989, *Project 2061: Science for All Americans*, Washington, DC: AAAS. Also published by Oxford University Press, 1990.

(AAAS) American Association for the Advancement of Science: 1990, *The Liberal Art of Science: Agenda for Action*. Washington, DC: AAAS.

(AAAS) American Association for the Advancement of Science: 1993, *Benchmarks for Science Literacy*. New York: Oxford University Press.

Brock, W. H., & Jenkins, E. W. (2014). Frederick W. Westaway and science education: an endless quest. In M. R. Matthews (Ed.), *International Handbook of Research in History, Philosophy and Science Teaching* (pp. 2359–2382). Dordrecht: Springer.

Colgan, A. D., & Maxwell, B. (Eds.). (2020). *The Importance of Philosophy in Teacher Education*. New York: Routledge.

Curren, R. (Ed.). (2003). *A Companion to the Philosophy of Education*. Oxford: Blackwell.

DeBoer, G. E. (2014). Joseph Schwab: his work and his legacy. In M. R. Matthews (Ed.), *International Handbook of Research in History, Philosophy and Science Teaching* (pp. 2433–2458). Dordrecht: Springer.

Duschl, R. A. (1985). Science education and philosophy of science twenty-five, years of mutually exclusive development. *School Science and Mathematics, 87*(7), 541–555.

Erduran, S., & Dagher, Z. R. (2014). *Reconceptualizing the Nature of Science for Science Education*. Dordrecht: Springer.

Erduran, S. (2013). Philosophy, chemistry and education: an introduction. *Science & Education, 22*(7), 1559–1562.

---

[12] See contributions to Colgan & Maxwell (2020).

Fensham, P. J. (2004). *Defining an Identity: The Evolution of Science Education as a Field of Research*. Dordrecht: Kluwer.
Flick, L. B., & Lederman, N. G. (Eds.). (2004). *Scientific Inquiry and Nature of Science: Implications for Teaching, Learning and Teacher Education*. Dordrecht: Kluwer.
Irzik, G., & Nola, R. (2014). New directions in nature of science research. In M. R. Matthews (Ed.), *International Handbook of Research in History, Philosophy and Science Teaching* (pp. 999–1021). Dordrecht: Springer.
Jung, W. (1983). Toward preparing students for change: a critical discussion of the contribution of the history of physics to physics teaching. In F. Bevilacqua & P. J. Kennedy (Eds.), *Using History of Physics in Innovatory Physics Education* (pp. 6–57). Pavia: Pavia University. Reprinted in *Science & Education*, 1994, *3*(2), 99–130.
Kampourakis, K. (Ed.). (2013). *The Philosophy of Biology: A Companion for Educators*. Dordrecht: Springer.
Lakin, S., & Wellington, J. (1994). *Teaching the Nature of Science: Project Report*. Sheffield: Education Research Centre, University of Sheffield.
Lederman, N., Abd-el-Khalick, F., Bell, R. L., & Schwartz, R. S. (2002). Views of nature of science questionnaire: towards valid and meaningful assessment of learners. Conceptions of the nature of science. *Journal of Research in Science Teaching*, *39*, 497–521.
Lederman, N. G, Bartos, S. A., & Lederman, J. (2014). The development, use, and interpretation of nature of science assessments. In M. R. Matthews (Ed.), *International Handbook of Research in History, Philosophy and Science Teaching* (pp. 971–997). Dordrecht: Springer.
Lederman, N. G. (2004). Syntax of nature of science within inquiry and science instruction. In L. B. Flick & N. G. Lederman (Eds.), *Scientific Inquiry and Nature of Science* (pp. 301–317). Dordrecht: Kluwer.
Lemke, J. L. (2001). Articulating communities: sociocultural perspectives on science education. *Journal of Research in Science Teaching*, *38*(3), 296–316.
Manuel, D. E. (1981). Reflections on the role of history and philosophy of science in school science education. *School Science Review*, *62*(221), 769–771.
Martin, M. (1972). *Concepts of Science Education: A Philosophical Analysis*. New York: Scott, Foresman & Co.
Matthews, M. R. (1997). Scheffler revisited on the role of history and philosophy of science in science teacher education. *Studies in Philosophy and Education*, *16*(1–2), 159–173. Republished in Siegel, H. (Ed.). *Reason and Education: Essays in Honor of Israel Scheffler* (pp. 159–173). Dordrecht: Kluwer.
Matthews, M. R. (2004). Thomas Kuhn and science education: what lessons can be learnt? *Science Education*, *88*(1), 90–118.

Matthews, M. R. (2011). From nature of science (NOS) to features of science (FOS). In M. S. Khine (Ed.), *Advances in the Nature of Science Research: Concepts and Methodologies* (pp. 1–26). Dordrecht: Springer.

Matthews, M. R. (2012). Philosophical and pedagogical problems with constructivism in science education. *Tréma, 38*, 41–56.

Matthews, M. R. (Ed.). (2014). *International Handbook of Research in History, Philosophy and Science Teaching*, 3 volumes. Dordrecht: Springer.

Matthews, M. R. (2015). *Science Teaching: The Contribution of History and Philosophy of Science: 20th Anniversary Revised and Enlarged Edition*. New York: Routledge.

Matthews, M. R. (2020). The contribution of philosophy to science teacher education. In A. D. Colgan & B. Maxwell (Eds.), *The Importance of Philosophical Thinking in Teacher Education* (pp. 121–142). New York: Routledge.

McComas, W. F. (Ed.). (1998). *The Nature of Science in Science Education: Rationales and Strategies*. Dordrecht: Kluwer.

McComas, W. F. (Ed.). (2020). *Nature of Science in Science Instruction: Rationales and Strategies*. Dordrecht: Springer.

McComas, W. F. (2020b). Considering a consensus view of nature of science content for school science purposes. In W. F. McComas (Ed.), *Nature of Science in Science Instruction: Rationales and Strategies* (pp. 23–34). Dordrecht: Springer.

Nadeau, R., & Destautels, J. (1984). *Epistemology and the Teaching of Science*. Ottawa: Science Council of Canada.

National Research Council (NRC) (1996). *National Science Education Standards*. Washington, DC: National Academies Press.

Olson, J. K. (2018). The inclusion of the nature of science in nine recent international science education standards documents. *Science & Education, 27*(7–8), 637–660.

Pigliucci, M., & Boudry, M. (Eds.). (2013). *Philosophy of Pseudoscience: Reconsidering the Demarcation Problem*. Chicago: University of Chicago Press.

Robinson, J. T. (1968). *The Nature of Science and Science Teaching*. Belmont, CA: Wadsworth.

Robinson, J. T. (1969). Philosophical and historical bases of science teaching. *Review of Educational* Research, *39*, 459–471.

Sadler, T. D. (Ed.). (2011). *Socio-Scientific Issues in the Classroom: Teaching, Learning and Research*. Dordrecht: Springer.

Scerri, E. R., & McIntyre, L. (Eds.). (2015). *Philosophy of Chemistry: Growth of a New Discipline*. Dordrecht: Springer.

Scheffler, I. (1970). Philosophy and the curriculum. In *I. Scheffer: Reason and Teaching*, 1973 (pp. 31–44). London: Routledge. Reprinted in *Science & Education*, 1992, *1*(4), 385–394.

Schulz, R. M. (2014). Philosophy of education and science education: a vital but underdeveloped relationship. In M. R. Matthews (Ed.), *International Handbook of Research in History, Philosophy and Science Teaching* (pp. 1259–1315). Dordrecht: Springer.

Schwartz, R., & Lederman, N. (2008). What scientists say: scientists's views of nature of science and relation to science context. *International Journal of Science Education, 30*(6), 727–771.

Shulman, L. S. (1986). Those who understand: knowledge growth in teaching. *Educational Researcher, 15*(2), 4–14.

Siegel, H. (Ed.). (1997b). *Reason and Education: Essays in Honor of Israel Scheffler*. Dordrecht: Kluwer.

Swinbank, E., & Taylor, J. (Eds.). (2007). *Perspectives on Science: The History, Philosophy and Ethics of Science*. Harlow, Essex: Heinemann.

Thompson, J. J. (Ed.). (1918). *Natural Science in Education*. London: HMSO. (Known as the *Thompson Report*.)

Westaway, F. W. (1929). *Science Teaching*. London: Blackie and Son.

Author biography. Michael R. Matthews is an honorary associate professor in the School of Education at the University of New South Wales. He has published extensively in the fields of philosophy of education, history and philosophy of science, and science education. He was Foundation Editor (1990–2015) of the Springer journal *Science & Education: Contributions from the History and Philosophy of Science*. His publications include: seven books, 50 articles, 40 book chapters, and 10 edited anthologies and handbooks including the 3-volume, 76-chapter *Handbook of Research in History, Philosophy and Science Teaching* (Springer, 2014). Two recent books are: *Feng Shui: Teaching About Science and Pseudoscience* (Springer, 2019), *History, Philosophy and Science Teaching: A Personal Story* (Springer, 2021).

Mathematical Understanding by Thought Experiments

GERHARD HEINZMANN[1]

Abstract. The goal of this paper is to answer the following question: Does it make sense to speak of thought experiments not only in physics, but also in mathematics, to refer to an authentic type of activity? One may hesitate because mathematics as such is the exercise of reasoning par excellence, an activity where experience does not seem to play an important role. After reviewing some results of thought experiments in physics, we look at what lessons they can teach us. Based on this, we turn our attention to thought experiments in mathematics, especially in fundamental mathematics, and investigate what thought experiments can teach us there. If we accept the principle that mathematical practice can sometimes be best described by a pragmatist approach where the concept of experience in mathematics is considered from a new perspective, thought experiments can in some cases be a useful instrument different from both 'mathematical experiments' and 'formal' proofs: they are mathematical experiments using deviant methods.

Keywords: thought experiments, mathematics, mathematical experiments, epistemic intuition, pragmatic approach.

1 Introduction

Let us imagine a world populated only by beings with no depth and let us suppose that these 'infinitely flat' animals are all in the same plane from which they cannot escape. Let us also suppose that this world is far enough from the other ones to be free of their influence. While we are in the process of making these hypotheses, it is no more trouble for us to endow these beings with the ability to reason and to imagine them to be capable of doing geometry. They will certainly attribute only two dimensions to space in this situation.
H. Poincaré, *Science and Hypothesis* (1902, p. 35)

---

[1] Université de Lorraine/Université de Strasbourg/CNRS – Archives Henri-Poincaré (UMR 7117), Nancy.

1.1 General characterisation of thought experiments

A physicist, unlike a mathematician, has no trouble giving an intuitive example of a thought experiment. Poincaré's example quoted above is emblematic of this. But how can thought experiments in physics be justified and what, if any, is their cognitive contribution? Can the justifications be transposed from physics to procedures in mathematics in a way that would justify our calling them 'mathematical thought experiments'?

According to Ernst Mach, the most prominent author of thought experiments, such experiments are to be thought of in analogy with real experiments. They are imagined experiments (or inferences of a special sort) in contexts that explore hypothetical or counterfactual states ('scenarios') in order to understand something, for instance the possibility of a two-dimensional world. They generally introduce into the hypothesis of a real experiment some variation, a 'deviant' element, which differs in an aspect (here a world peopled with beings of no thickness) 'that does not have an independent effect on the outcome of interest' (Reiss, 2016, p. 123). This is what distinguishes a hypothesis as a simple counterfactual from a thought experiment as a particular counterfactual. Thought experiments are generally performed by constructing particular cases that incorporate concrete elements which are in principle reproducible in specific spatiotemporally individuated situations (e.g., 'infinitely flat' animals). One accepts or rejects a thought experiment when the consequences derived from it do, or do not, coincide with our 'data', that is, with our understanding and expectations of such a scenario, which typically include tacit knowledge and background assumptions ('infinitely flat' animals cannot have the third dimension because we assume that they think exactly the way we do) (Buzzoni, 2008, p. 65, 67, 93).

Nevertheless, when a thought experiment fails, it does not fail because of some undesirable empirical outcome but because it does not work as a thought experiment, that is, as a thought that helps us access and explore certain opaque intuitions or nonintuitive formalisms (Lenhard, 2018, p. 485). For instance, Poincaré's example makes the formal possibility of a 'real' two-dimensional space accessible. On the other hand, the fact that a thought experiment 'does violate in imagination not only some but all of the most fundamental natural laws, so that it is physically or biologically unthinkable, looks like a good reason to forget about […] it' (Marconi, 2017, p. 121). Thought experiments are not simply fictions: they are very special fictions, namely counterfactuals whose degree of counterfactuality is not maximal.

Below, I put the expressions 'data' and 'real things' in quotes. This is because I consider Sellars's argument (1956/1997) to the effect that there is no theory-independent, mind-independent, or conceptually independent given convincing. The argument that experiments are about 'real things' can only mean that these things are 'real' relative to a language-dependent articulation of the world. This claim has an important consequence: the interface between an experiment and a physical thought experiment is vague, and so the interface between mathematical experiments and mathematical thought experiments may be vague as well.

Now, it is quite right to claim that Sellars's attack on the 'Myth of the Given' was a decisive move in turning analytic philosophy away from the foundationalist motivation of logical empiricism. But neither Sellars (nor Quine) were the first in the twentieth century to abandon the foundationalist approach: Ferdinand Gonseth's *open philosophy*, adopted in the 1950 by David Hilbert's collaborator Paul Bernays, came significantly earlier. But we shall return to this later.

1.2 Two celebrated examples in physics

Most likely the best-known thought experiments in physics are Galileo's physical thought experiment concerning the 'natural falling speed' and Simon Stevin's (1548–1620) thought experiment with a chain, whose aim was to elucidate the balance of forces on an inclined plane. Galilei's thought experiment targeted the claim, originally put forth by Aristotle, that a falling object in the vacuum has a particular 'natural falling speed' proportional to its weight. The thought experiment then ran as follows: Let us imagine a heavy ball (H) attached to a light ball (L). What would happen if they were released together? The lighter ball would slow up the heavy ball, so that the speed of the combined balls would be slower than the speed achieved by the heavy ball alone (i.e., $H + L < H$). But because the combined balls are heavier than the heavy ball alone, the combined object should in fact fall faster than the heavy one (i.e., $H + L > H$). But this is a straightforward contradiction! Its solution consists in a thesis, proposed without any new empirical evidence, that all objects fall at the same speed. It can be seen that the thought experiment differs from a real experiment by assuming a completely abstract surrounding environment.

Stevin's thought experiment proceeded as follows: at the beginning, we have a context of rules of thumb such as were often used in construction. They feature levers and the return of forces by pulleys. One can also place here the findings regarding weights placed on inclined planes. Then we have an established body of knowledge, mainly concerning geometry including the notion of symmetry. To this, we add the notion of weight or mass (the two, and especially their additivity, had not been differentiated yet in Stevin's time). This theoretical corpus has the content of the aforementioned rules as the experimental field of application.

Stevin imagines a triangular prism whose edges are placed horizontally and whose cross section is represented by *ABC*. *AC* is likewise assumed to be horizontal. On the inclined faces *AB* and *BC* of the prism, Stevin places a closed ball chain. Is the chain in balance or not? He assumes that the chain is weightless and the motion is frictionless (see Figure 1).

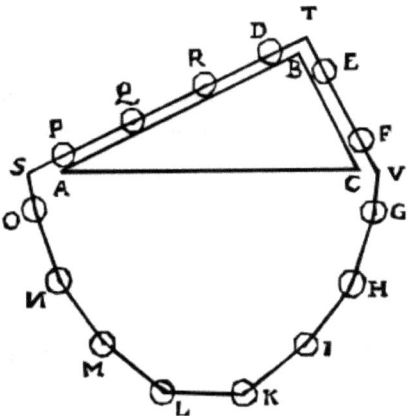

Figure 1.

Stevin remarks that supposing that the chain can slide freely on the inclined planes, it must nevertheless remain at rest, for if it began to slide of its own accord in any one direction, it would have to continue sliding indefinitely because the same cause of motion would remain. In other words, the chain would, due to the uniformity of its parts, be placed always in the same way on the triangle, which would lead to perpetual motion, which is absurd. He can then conclude that on inclined planes of equal height, equal weights act in inverse proportion to the lengths of the planes.

This solution rests on transferring to the imagined motion a property according to which all unassisted motion tends to cease. This principle was proposed already by Aristotle, who claimed it applied to all terrestrial motions (but not to the motions of celestial bodies). To transfer this principle to the abovementioned case, one has to invoke a piece of purely instinctive knowledge that rules out perpetual motion. In fact, our understanding of the impossibility of perpetual motion and correct statement of the principle, as well as the development of Stevin's argument, ultimately rely on the notion of mechanical energy, which was described much later.

Both Galilei's and Stevin's thought experiments offer a real cognitive gain without recourse to a real experiment whose execution may even be impossible due to physical, technological, or financial reasons. Of course, one could think of experimenting with the equilibrium of a chain placed on two inclined planes but even in Stevin's time, one would have suspected that any balance thus achieved would be at best unstable. With some solid friction, the chain can be moved a little without disrupting the equilibrium. And a perfectly oiled chain will inexorably fall to one side and then to the other. The outcome would be indecisive. In Galileo's experiment, meanwhile, a real vacuum is difficult to produce and 'the rate of fall, even *in vacuo*, could logically depend on other parameters' (Atkinson & Peijnenburg, 2004, p. 122).

Both thought experiments moreover use a distortion (or variation) if the established experimental rules to achieve their conclusion. If the reasoning remained within the limitations of experimental rules throughout, one would not speak of a thought experiment. It is also important that the distortion be more than merely superficial.

The main point is that we seem to be able to grasp 'nature' or a reality simply by thinking—and therein lies thought experiments' great potential for physics. How can we learn apparently new things about nature without new empirical data, while just sitting in the armchair?

1.3 The classical analysis of thought experiments

Literature on thought experiments is vast but there is no agreement on either their epistemological role or their analysis. In fact, it seems that the trouble with thought experiments is not captured by debating what they are. They core of the problem is better captured by an epistemological debate about the source of epistemic authority of thought experiments.

Most frequently debated within the context of epistemology of thought experiments are (1) Brown's realistic position according to which thought experiments have a purely non-propositional (objectual) character, (2) Norton's empiricist position according to which they are purely propositional (conceptual), and (3) Buzzoni's Kantian position according to which they may have a transcendental character for real experiments (cf. Brown, 1986; Norton, 1996, 2004; Buzzoni, 2011).

According to Brown, the cognitive gain of a thought experiment is provided by the hypostatic assumption that we can, via intuition, 'perceive' abstract objects. But what does this mean? According to a solution in the spirit of Norton, the cognitive contribution of thought experiments concerns our conceptual apparatus (Kuhn, 1977, Chapter 10) and/or our discovery of empirical elements in the premises of inductive or deductive arguments (Norton, 1966, 2004). Thought experiments are then nothing but arguments dressed up in a narrative rhetoric with mostly presupposed hidden empirical premises. This rhetoric can be eliminated by logical transformations.

Buzzoni (2011) proposes a 'Kantian solution' as the third option: A thought experiment is a precondition of possibility of real experiments. A real laboratory experiment must be conceivable in thought because we must assess its feasibility even before attempting to implement it. In this approach, the epistemic authority of thought experiments is explained as follows: one should conceive of scientific thought experiments as experiments that could be really performed if the assumed conditions could be brought to obtain. Personally, I am not sure if this last condition actually applies to various important thought experiments in contemporary physics.

In mathematics, the interpretation of thought experiments is even less clear: in this field, it is far from evident it makes any sense to talk about thought experiments at all.[2] Marco Buzzoni is one of but a handful of experts in the field of thought experiments who published on *mathematical* thought experiments. In a recent article 'Mathematical vs Empirical Thought Experiments', he observes that 'mathematical thought experiments might be regarded as a subset of visual proofs based on diagrams. However [...] the use of this sense of "thought experiment" is harmless only under certain conditions' (Buzzoni, 2021, pp. 1–2). He lists three:

> 1) The distinction between logical-formal thinking and experimental operational thinking must not be underestimated.
> 2) The separation between the context of discovery and that of justification—a separation advocated by logical empiricists but difficult to support after the pragmatic turn—must be rejected.
> 3) The distinction between mathematical thought experiments and formal proofs must be regarded as one of degree, and not as a distinction

---

[2] Nevertheless, according to Imre Lakatos (1976, p. 11) 'Thought experiment was the most ancient pattern of mathematical proof.' Lakatos, as Ernst Mach before him, contrasted informal thought experiments in number theory and geometry—which start from observations—with formal proofs. Here, thought experiments are also linked to informal proofs or to the starting point of mathematics from experience.

in principle, however important this distinction may be de facto for particular or local purposes. (Buzzoni, 2021, p. 9.)

In this paper, motivated by Buzzoni's result, we agree with his second condition and elaborate it. Conditions (1) and (3), on the other hand, are reformulated as consequences of our pragmatist position as follows:

1') The distinction between formal logical thinking and experimental operational thinking may frequently be expressed in mathematics as a change of perspective from logical stringency to mathematical understanding.
3') The distinction between mathematical thought experiments and 'formal' proofs must be ultimately viewed as a difference of degree but for the moment being, let us view it a difference in logical and intuitive rigor.

This reformulation then also makes Buzzoni's restriction of mathematical thought experiments to 'a subset of visual proofs, based on diagrams' too narrow.

2 Mathematical experiments

The main difficulty with the treatment of thought experiments in mathematics is that in mathematics, distinctions between thought experiments and real experiments disappear, fade, or seem at least obscured or transformed. Even for a pragmatist like Peirce, mathematics is not about facts but about the necessary consequences of hypotheses (cf. e.g., Peirce, 1898, 3.558sq).

In our examples, we will see that sometimes also in mathematics one speaks of thought experiments. But it could be merely a usurpation of the term. It is for this reason that we should ask again Buzzoni's question concerning the conditions for thought experiments in mathematics and not simply 'Is there anything like thought experiments in mathematics?' (which would best be answered only after the conditions are known). If there is to be a definite meaning to the concept of 'thought experiments in mathematics', we ought to say something about justification of claims to validity.

With respect to the former question, it should be noted that if we are to define thought experiments in mathematics by analogy with the best-known kind of thought experiments, i.e., those in the natural sciences, it makes no sense to view all proofs in mathematics as thought experiments or to refrain from drawing a distinction between mathematical *thought experiments* and mathematical *experiments*. It should further be taken into account that in the natural sciences, one must distinguish between mere imagination of the conditions and procedure of an experiment that precede any real experiment and fruitful thought experiments such as those proposed by Galileo.

In short, one must draw a distinction between

- mathematical experiments (which can be themselves be imagined without becoming thought experiments) and
- mathematical thought experiments.

Experimentation implies the conception of a possible situation against which we test our hypotheses and from which we reason about a 'real' case, which we oppose to a possible case. A 'real' experiment thus seems to be based on confirmation or falsification by 'real states of affairs'. One might think that experiments would feature in the context of argumentation but—in contrast to thought experiments—they are not arguments, as evidenced by the fact that we conduct experiments without arguing. In fact, scientific experimentation demonstrates both the investigators' ability to produce correct data that contain relevant information about the phenomenon in question and it aid further development of the experimental protocol and the use of appropriate instruments.

In fact, the first way to make sense of the expression 'mathematical experiment' is to have recourse to domains outside of mathematics: to physics, diagrams, or calculations (often automated). It is especially the last possibility that stood at the beginning of an entire domain of mathematics called 'experimental mathematics'. According to Borwein and Bailey (2004), mathematics is not characterised by 'formal' proofs in a broad sense (one should not view what can be formalised in a logical theory as a reference to 'formal' discourse) but by certain mathematical knowledge. In other terms, the correctness of a judgement is not reduced to the truth of its propositional content and validity of an inference is not reduced to its logical consequences.[3]

In the following, by 'formalism' in a broad sense we mean a mathematical discourse that fits into the canon of methods accepted at the relevant time as rigorous. This being the

---

[3] With respect to the mode of comprehension, this point was actually already put forth in arguments against the logicisation of mathematical proofs: according to Poincaré, a proof in its logical form is conceptually insufficient for understanding the extensive character of the result with respect to the premises. For experimental mathematicians, 'formal' proofs are sometimes incomprehensible due to their excessively abstract character. Poincaré wishes to supplement the standard 'formalism' of proof by an architecture (cf. Detlefsen, 1992), while the experimental mathematicians would supplement it by a calculation.

case, we take 'formalism' to denote a language with its special grammar, a combination of symbols and propositions, whereby a mathematician is a person who is familiar with that language. In other words, mathematician is someone who has experience with its syntax and semantics, that is, with the rules of that symbolic mathematical language and, naturally, also a person who has some mathematical 'ideas' and 'perception'. Experimental mathematics uses *computational* evidence as 'an exploratory tool to discover mathematical truths' and concrete paths for formalisms (Borwein & Bailey, 2004, p. 7). On the other hand, we should also bear in mind that experimental mathematics does not cover the entire field of mathematics.

However, mathematics can limit itself to just its own 'cultural' sphere. A mathematician does not necessarily draw inspiration from sciences outside of mathematics: when that is the case, mathematical practice and thought experiments in mathematics (if it makes sense to speak of them) might be said to belong to the same sphere *a priori*. Moreover, mathematical objects are not necessarily given independently of a mathematical theory. So then, in mathematics, where should the 'concrete data' normally used to test our hypotheses come from? Perhaps they are *instances* of proof ideas, that is, mathematical proofs? This solution would be either too narrow or too broad. Too narrow because mathematical proofs in their standard form do not look at all like experiments (save for the well-known special case of diagrammatic proofs in geometry, category theory, knot theory, graph theory, etc.) and too broad because if we were to consider any instance of a regulative idea as an experiment, then every rational activity would count as experimental.

Now, at least since the practical turn in the philosophy of mathematics and logic, efforts have been made to study mathematical processes as opposed to just mathematical products, and to gain a better understanding of the modes of inference used by actual active mathematicians. This suggests a return to the 'pragmatic' critique of formal inference that had been put forth by Poincaré and Peirce. One of the main challenges for linking pragmaticism and mathematical practice is to explain the dynamics of mathematical reasoning. In this process, effective (but not necessarily logically rigorous) argumentation is supposed to appeal to intuition or other cognitive faculties. It is thus epistemic in its nature. In other words, we can say—using the terminology of Göran Sundhom—that one dispenses with 'blind (ungrounded) correct judgments—be [they] mediate or not'. This signifies the abandonment of two Bolzanian reductions, in particular '(i) that of the correctness of the judgement to that of the truth of the propositional content and (ii) that of the validity of an inference between judgements [demonstration or non-formalized proof] to a corresponding logical consequence [proof] among suitable propositions' (Sundholm, 2012, p. 945).

This is why I endorse a philosophical background indebted not only to Poincaré and Peirce but above all to Ferdinand Gonseth's 'open philosophy'. This constitutes what I call a 'pragmatic approach'.[4]

Let us now consider the rational learning process of mathematical idealisation that gives us an experience *sui generis*, in other words that which Gonseth calls 'intellectual experience'. The claims put forth by Gonseth's 'open philosophy' in the forties, which paradigmatically exemplified in all details the genesis of geometry (Gonseth, 1945–55), are quite different from a theory whose justification and confirmation relies on syntactic (theoretic) or operative evidence based on an isomorphism between a construction or pre-existent reality (realism/empiricism). In this approach, the construction is itself a constitutive element of mathematical reality. In fact, 'construction' is, in this context, used in three distinct senses (as later recorded by Sundholm). It can refer to '(i) a construction-act (or process); (ii) the construction object constructed in the construction act; (iii) the process i) considered as an object' (Sundholm, 2008, p. 187).

In the first step, a 'doctrine préalable', which is similar to Wittgenstein's notion of common sense, acts of all as a substitute for the *a priori* or 'free disposition' adopted with respect to the means used by the construction. The antifoundational tenet of the open philosophy is itself justified only by future developments, namely by its usefulness within the general scientific network. Theoretical terms are then partially defined in terms of observational vocabulary ('horizon of experience'): a vague understanding of observational terms delivers the construction of partially defined theoretical terms ('horizon of theory'). This principle states what is today called an 'unstable epistemological equilibrium' (Goffi & Roux, 2018, p. 442).

The construction process between an uncompleted 'real' and the theory in the making *en devenir* leads to a schematic synthesis of different levels n, where levels under n are

---

[4] According to a tradition spanning from Peirce to the Erlangen School of Lorenzen/Lorenz, the pragmatic approach can be characterised by a set of at least five theses:
(1) It is a method, not a system.
(2) Abandoning every absolute empirical criterion or absolute rationale, it takes as its basis observations and statements of common sense.
(3) It admits a convergence between ontology and the theory of knowledge and considers fruitfulness as an important criterion of the acceptability of propositions.
(4) It rejects a separation in principle between the context of justification and the context of invention as well as between the context of justification and the context of understanding.
(5) It admits a correlation between knowledge and value.
By forming this picture, my aim is neither to articulate a particular historical position held by various thinkers, nor do I intend to present the features shared by such positions.

engaged in level *n* as its external aspects but each can be viewed as having its own reality, i.e., its own internal structure. In an extension of this Gonsethian construction, I introduce a mathematical hierarchy using the following terminological definitions from the philosophy of language (see Lorenz, 2010, p. 154): A statement of the type 'I see a cat' is called an *expression by external perception*. It appears as a claim to *knowledge by acquaintance*. The statement of 'I see something as a cat' thus transforms the expressive statement about an external perception into a *cognition*. The language in this statement then has not an *expressive* but *representative* function: it appears as a claim of *knowledge by description*.

Now, at the first level 1 of a mathematical hierarchy, we begin with the construction of an informal mathematical language (language in a *representative* function) with respect to our natural language of common sense, which is of level 0. This natural language is an intuitive language whose function is expressive: its 'perception–object (verification–object)', that is, the object of the 'fourfold iteration construction of strings' makes the first level proposition that for instance 'four is a number' true. The construction object serves as a truth-maker (cf. Sundholm, 2008, p. 209). In this way, 'language' in a perceptive function (*knowledge by acquaintance*) is transformed into a language with a representative function that enables *knowledge by description*.

On higher levels, criteria governing the hierarchy are *generalisation by symbolisation* and *success*. Success does not consist in arriving at a true translation of one language into the other but in reaching a particular goal. This success with respect to a goal is expected to be reached by a *mathematical simplification* when comprehension of a language of a lower level fails.

On the nth level, the obtained symbolic language, called formalism, provides opportunities for further generalisations, which are only in special cases expressed in a strictly formalised language by a working mathematician. To use a formalism means two things: on the one hand, the proofs are anchored in the grammar and standard logical and semantic norms of the time, which makes them rigorous, and, on the other hand, they are based on mathematical results of the lower level. The mathematical content and the logical and semantic norms of any non-zero level n constitute an 'experimental' domain of level $n + 1$.[5] The thus constructed hierarchy is open to further developments. This allows us the formulation of the following theses concerning mathematical experiments:

---

[5] Let us take the construction of numbers as an example of such a hierarchy: The beginning consists in an intuitive iteration of a unit. Level 1 is then a recursive definition of natural numbers with addition, multiplication, and order relation. Level 2 is the formulation of an axiom system in a formal language (Peano). Level 3 is an algebraic construction where one notices that subtraction

Thesis 1: The way in which instances of proofs and constructions at level $n-1$ downwards can be viewed as realisations of mathematical experiments with respect to the 'rigorous' proofs and constructions of level n is a variant of the way in which physical experiments can be viewed as realisations of a theory in the natural sciences. What we have in mind is that the realisation relation between a language of a level under n and a language of level n is an evidence-based judgment or inference. In other words, believing it has an epistemic, not alethic character.

Thesis 2: The pivotal element of experimentation is the attempt to confirm a 'theoretical' language use by 'observational' language use, that is, to ensure the validity of level n by an evidence-based inference from both the construction object and constructions as objects of level $\leq n-1$. The model-theoretical and ontological view of the theory and an independent universe is abandoned.

According to these theses, *understanding* of the truth of a mathematical sentence is a re-conceptualisation of its 'logical' proof, not a meta-action that would provide 'secondary standards' with respect to logic. Rather, it offers an answer to another why-question than 'why is the proof logical rigorous?'. This new question concerns the explanatory content but, of course, the answer will not necessarily go deeper with respect to the question than logical reasoning did with respect to the question regarding logical rigor.

Proofs of irrationality of $\sqrt{2}$ exemplify mathematical experimentation in the abovementioned sense. The traditional logical proof involves the concept of parity and proceeds by indirect reasoning:

Assume integers p and $q \neq 0$ do exist and (a) $\sqrt{2} = p/q$, where (b) p/q are reduced to the lowest terms. Then you get a contradiction from (c) $2q^2 = p^2$.

A short proof of irrationality of $\sqrt{2}$ can be obtained from the rational root theorem, that is, if $p(x)$ is a monic polynomial with integer coefficients $x^n + a_{n-1}x^{n-1} + \cdots + a_0$, then any rational root of $p(x)$ is necessarily an integer.

---

of natural numbers cannot be executed without restrictions: by the introduction of integers as equivalence classes over natural numbers, one arrives with respect to addition to a commutative group, and with respect to addition and multiplication to a ring. Division as inversion of multiplication is, however, not unrestrictedly executable in the domain of integers. On level 4, one finally arrives at the field of rational numbers.

From $\sqrt{2} = x$, and $2 = x^2$, we obtain the polynomial $p(x) = x^2 - 2$.

It follows from the root theorem that $\sqrt{2}$ is either an integer or irrational. Because $\sqrt{2}$ is not an integer (2 is not a perfect square), $\sqrt{2}$ must be irrational.

The polynomial $x^2 - 2$ exemplifies the unit general polynomial with integer coefficients $x^n + a_{n-1}x^{n-1} + \cdots + a_0$.

This proof seems to give a better explanatory view than the first proof: it proceeds by interdependence between the global polynomic of level n and a local polynomic of level $\leq n - 1$. The polynomial equation is instantiated by a *monic* polynomial equation which *exemplifies* it and is finally subjected to a special valuation. It does not 'explain' by a logical relation but by the fact that the particular is *interdependent* with the general, that it is its *exemplification*. As its instance, it *presents* the general.

3 Mathematical thought experiments

If this definition of mathematical experiments as an *exemplificatory* recourse to a lower level is plausible, the epistemic questions are closely similar but not the same for mathematical experiments and mathematical thought experiments, that is, experiments with a deviant element. As Pascal Engel remarks, thought experiments:

> …imply an extension of our conception or imagination to new possibilities, and the question is whether these possibilities are genuine ones. However, in order to answer the question whether a given situation is possible or not, we first have to answer another one: how can we have access to genuine possibilities? The latter question is epistemological; it asks what sort of faculty or cognitive capacity gives us access to the possible situation: Imagination? Conceptual understanding? Intuition? A priori reasoning? Empirical reasoning? These are not necessarily the same. So the modal question and the epistemological question are closely associated. For if we answer that the envisaged possibility is just a fiction of our imagination, we reject thought experiments as "mere" exercises of imagination, and not as genuine possibilities. […] So the main issue about any thought experiment is: does conceivability imply possibility? (Engel, 2011, p. 147.)

Concerning mathematical inferences, the standard gradient between conceivability and genuine possibility is expressed as the difference between a proof idea and the possibility of a logically stringent proof. While a mathematical experiment breaks the logical stringency by an exemplificatory reference to previously constructed objects, proofs, or methods, mathematical thought experiment goes one step further:

Thesis 3: In mathematical thought experiments, we take recourse—based on mathematical experiments—to a modal deviation using further semiotic means. *Epistemic intuition* provides us with access to these deviations as 'genuine possibilities' of mathematical inferences as opposed to 'pure fictions'. This is the justification of the validity claim concerning mathematical thought experiments.

Nevertheless, if the relevance of mathematical experiments and mathematical thought experiments takes precedence over logical rigor, one must know what kind of error and what degree of failure one can tolerate. Some errors may be useful because their correction would advance knowledge, but that is not the point. With respect to logical rigour, it would be better to speak of *imperfections* rather than errors. In thought experiments, one always encounters something that challenges the reliable, accepted usage. One moves on thin ice, on a ground one can call apocryphal or deviant, which is why one can imagine some tolerance, leeway, with respect to logical rigour.

To clarify the nature of the deviation, and to see how the framework of epistemic intuition introduced in Heinzmann (2013) works, let us describe the functioning of examples within mathematics:

Example 1: Euler's solution of the Basel Problem (example exposed by Philippe Lombard). One is interested in the limit of the series

$$\sum_{n=1}^{\infty} \frac{1}{n^2} = \frac{1}{1} + \frac{1}{4} + \frac{1}{9} + \frac{1}{16} + \cdots \frac{1}{n^2} + \cdots$$

which converges very slowly. In the 1730s, Leonhard Euler published different papers in which he calculated an approximation of the sum with the 6 exact decimals (1731) and in 1734, he recognised in this result the approximate value of $\pi^2/6$ (cf. Ayoub, 1974).

Euler's idea was the following: It is known (since 1715) that the 'Taylor development' of a function as a series. In particular, the function sin x is written in the form:

$$\sin x = \sum_{n=1}^{\infty} \frac{(-1)^n}{(2n+1)!} = x - \frac{x^3}{3!} + \frac{x^5}{5!} - \frac{x^7}{7!} + \cdots$$

$$\frac{\sin x}{x} = 1 - \frac{x^2}{3!} + \frac{x^4}{5!} - \frac{x^6}{7!} + \frac{x^8}{9!} - \frac{x^{10}}{11!} + \cdots$$

But we know the roots of this function: they are the non-zero values of x that cancel $\sin(x)$:

$x = \pm \pi; x = \pm 2\pi; x = \pm 3\pi; x = \pm 4\pi$, etc.

Let us 'boldly' assume that we can express this infinite series as a product of linear factors given by its roots:

$$\frac{\sin x}{x} = \left(1 - \frac{x}{\pi}\right)\left(1 + \frac{x}{\pi}\right)\left(1 - \frac{x}{2\pi}\right)\left(1 + \frac{x}{2\pi}\right)\left(1 - \frac{x}{3\pi}\right)\left(1 + \frac{x}{3\pi}\right)\cdots$$

$$= \left(1 - \frac{x^2}{\pi^2}\right)\left(1 - \frac{x^2}{4\pi^2}\right)\left(1 - \frac{x^2}{9\pi^2}\right)\cdots$$

In the product expansion, the term in x^2 is precisely:

$$-\frac{1}{\pi^2} - \frac{1}{4\pi^2} - \frac{1}{9\pi^2} - \frac{1}{16\pi^2} - \cdots = -\frac{1}{\pi^2}\sum_{n=1}^{\infty}\frac{1}{n^2}$$

But from the original infinite series development of $\sin(x)/x$, the coefficient of x^2 is:

$$-\frac{1}{3!} = -\frac{1}{6}$$

The two coefficients must be equal, therefore:

$$-\frac{1}{6} = -\frac{1}{\pi^2}\sum_{n=1}^{\infty}\frac{1}{n^2}$$

Multiplying both sides of this equation by $-\pi^2$, we get the sum of inverses of the squares of positive integers:

$$\frac{\pi^2}{6} = \sum_{n=1}^{\infty}\frac{1}{n^2}$$

In 1741, Euler gave a completely different solution to the Basel problem, one that did not depend on the mysteries of infinite products.

Here is the analysis of this example:

1. The preliminary calculations constitute the first level of mathematical experience. This helps Euler guess, within the approximate values obtained, the approximation of the exact value that he advances.

2. The general theory involved includes the definition of a polynomial and decomposition into factors involving the zeros. Its field of application includes all polynomials.

3. The transgression by transfer in Euler's 'proof' is the following (see Figure 2):

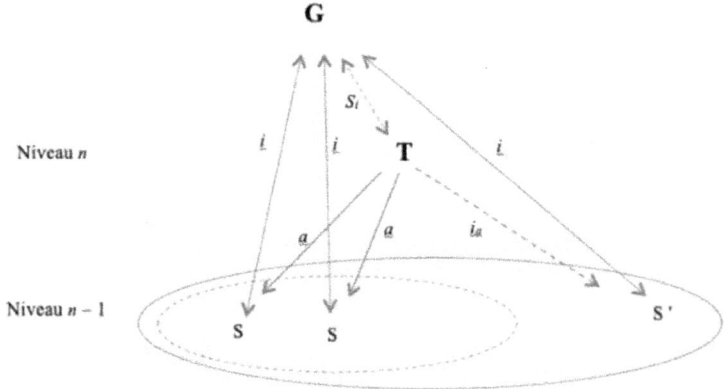

Figure 2.

Each polynomial S of level $n-1$, considered as singular element, is interdependent (i) with the general aspect G, which is that of a 'combination of powers'. We say that they are linked by epistemic intuition, or intuitively linked. In this case, there is an element T of level n, for which all the polynomial S of level $n-1$ to which it applies (a) are intuitively linked to the general element G. We can then say that T, as an object construction, is in symbolic interdependence s_i with the general aspect G. But the series development of the function $S' = \sin x/x$ of level $n-1$ that Euler considers is also interdependent (i) with the aspect G. We will then say that the series is intuitively a polynomial to which the decomposition T into factors according to the zeros can metaphorically be applied (i_a), but to which T does not apply *in principle*. The deviation is a transfer that makes T apply to S'.

4. The method used by Euler 'sees' in an apocryphal way the relevance of a result without being able to prove it by logical means: it is a thought experiment.

5. A modern mathematician will obviously be able to draw a rigorous proof at the 'higher' level of analysis of what Euler did within the framework of Eulerian products.

Example 2: Poincaré's reconstruction of geometric space. Poincaré's reconstruction of geometric space proceeds from the imagination of certain muscular sensations (level 0). Similar to Carnap's *Aufbau*, the starting point (guided by experience) is for Poincaré the definition of two two-place relations in the range of muscular sensations. If a muscular sensation can be compensated by an 'inverse' muscular sensation, it is called a 'change of position'. Poincaré defines the equivalence class of changes of position and calls it a *displacement* (level 1).

But Poincaré is not an empiricist. He is convinced that experience can only teach us 'that the compensation has approximately been effected', that it gives the mind merely the 'occasion to perform this operation', but 'classification is not a crude datum of experience' (Poincaré, 1898, p. 9). In fact, Poincaré is very precise about the vagueness he has in mind at this level of reconstruction:

> When experience teaches us that a certain phenomenon does not correspond *at all* to these laws, we strike it from the list of displacements. When it teaches us that a certain change obeys them only approximately, we consider the change, *by an artificial convention*, as the resultant of two other component changes. The first is regarded as a displacement *rigorously* satisfying the laws [...], while the second component, which is small, is regarded as a qualitative alteration. (Poincaré, 1898, p. 11.)

The composition of sensations is in a *vague* sense transitive and displacements are the result of a 'conventional' classification. They form a transformation group in a mathematical sense, and it depends on the choice of its subgroups whether the group corresponds to Euclidean or non-Euclidean geometry (level 2).

Where does the concept of a group come from? According to Poincaré, the ability to create the general concept G of a transformation group is the expression of a form of our understanding that 'exists in our mind'. The set of relations S that satisfy group axioms T, i.e., the set-theoretic model (transformation group), is an expression of a universal (structure) *exemplified* (in a Goodmanian sense) by the sensation system, which is a *vague part S'* of extension of the structure G. In other words, the form in the mind is a special kind of an epistemologically accessible universal that is available to use although we do not have the possibility of obtaining the particular form of the universal by purely logical means. The universal is exemplified, and exemplification uses epistemic intuition (Heinzmann, 2013).

Exemplification differs from denotation by its direction: it goes from the object to the predicate. That is, *the object exemplifies a predicate that applies to it.* While a strictly denotative sign does not necessarily have what it designates, an exemplification sign must possess what it exemplifies. But that does not mean that an object exemplifies all the traits it possesses. A 'red' inscription written in blue denotes all red objects but does not meet the conditions of the predicate 'to be red'. On the other hand, it can exemplify 'to be an inscription of "red"' (Goodman, 1968).

We can easily see that this construction has the form of a mathematical thought experiment:

A sensation system S' is in the experimental realm and interdependent (i) with the group structure G (universal). A *set* of relations S satisfies (a) the group axioms T, symbolically interdependent (s_i) with G, which can be applied (a) to S and metaphorically applied (i_a) to S'. Poincaré 'sees' in an apocryphal way the relevance of sensation compensation for the formation of a transformation group structure without being able to prove it by logical means: *it is a thought experiment.*

Example 3: The construction of elementary arithmetic following Poincaré and Hilbert/Bernays. In §2 of volume I of Hilbert and Bernays' *Grundlagen der Mathematik* (1934), we read:

> In the field of arithmetic, we are not concerned with problematic issues connected with the question of the specific character of geometrical knowledge; and it is indeed also here, in the disciplines of elementary number theory and algebra, that we find the purest manifestation of the standpoint of direct contentual thought that has evolved without axiomatic assumptions.

> This methodological standpoint is characterized by thought experiments with things that are assumed to be concretely present, such as numbers in number theory, or expressions of letters with given numerical coefficients in algebra. Let's take a closer look at the procedure and methodically tighten up the basics [*Anfangsgründe*].[6]

At the end of his analysis, in which the reader is left alone to discover where Bernays brings a mathematical thought experiment into play, he notes:[7]

> Our treatment of the basics of number theory and algebra was meant to demonstrate how to apply and implement direct contentual inference that takes place in thought experiments [*Gedanken-Experimente*] on intuitively conceived objects and is free of axiomatic assumptions.
>
> Let us call this kind of inference 'finitist' inference for short, and likewise the methodological attitude underlying this kind of inference as the 'finitist' attitude or the 'finitist' standpoint. [...] With each use of the word 'finitist', we convey the idea that the relevant consideration, assertion, or definition is confined to objects that are conceivable in principle, and processes that can be effectively executed in principle, and thus it remains within the scope of a concrete treatment.[8]

Before analysing Bernays's thought experiment about arithmetic bases, let us briefly recall his philosophical position on which our definition of mathematical hierarchy presented above is based.

According to Bernays (1953, p. 32), we need 'a revision in relation to the delimitation of rational and empirical elements': in a systematic corpus, the rational element must be, of course, present on the level of expressions of a scientific theory, but it is there in a particularly developed stage. Nevertheless, although less developed, the rational element is present already in the myth such that the myth can be considered, *modulo* certain intentions, a true doctrine of reality. If there is a distinction between myth and scientific theory, it is gradual and not principled. In other words, there is no clearcut divide. In this way, we are led to a pragmatic transformation of the correlation between

---

[6] Hilbert & Bernays (1934, p. 20). Quotation proposed and translated by Andrew Arana.
[7] In all probability and confirmed by oral information from Bernays' last doctoral student, Gert-Heinz Müller, these passages were written by Paul Bernays.
[8] Hilbert & Bernays (1934, p. 32). Quotation proposed and translated by Andrew Arana.

the terms in question: on the one hand, the empirical (in its revised meaning) is not opposed to the rational in general. Instead, it is contrasted with systematisation, that is, 'reason no longer asserts itself in *a priori* principles, but through the progress of the formation of concepts and methods of explanation' (Bernays, 1937, p. 289). In parallel, the rational (in Bernays's revised conception) is not opposed to the empirical, but to the capacity to conceive of impressions with content (*inhaltliche Eindrücke*); in other words, to the ability to conceive of the domain of data. This is how Bernays (1952, p. 134) can speak of experience being 'geistig' (spiritual, in the mind) in relation to the domain of mathematics. The Kantian contrast between a concept and intuition is replaced by a distinction between the form and content, which is along neo-Kantian lines viewed as a functional relationship (cf. Cassirer, 1910, p. 343). The invariance on which the analogy between the concepts of experience and thought experience rests consists in the fact that both are thought of as contents in a functional relationship with their respective form.

In his 'Quelques points de vue concernant le problème de l'évidence' (1946), Bernays gives examples of evidence on three different levels: the purest form of evidence is evidence of the combinatorial or formal kind. Such kinds of evidence are used 'in applying the usual rules of elementary algebra'. The intuition of a general concept of number in conjunction with the principle of complete induction and reasoning by recursive definitions form a second level. An even weaker degree on evidence is then found in infinitesimal analysis which, although it does not have a purely arithmetical character, still produces 'evidence *sui generis*' (Bernays, 1946, p. 324). In fact, Bernays sets no limits on the process of weakening of evidence in the foundations of sciences. In doing so, he shows that Hilbert's original project to limit intuition to primitive evidence was insufficient.

Michael Detlefsen remarks in his 'Hilbert's program' that

> Hilbert and Bernays described finitary reasoning as a form of mental experimentation with concretely conceived objects, where the experiments conducted consist in envisioning what happens to a concrete object when one applies certain constructive operations on it. Viewed on a certain level, it would seem that every variety of constructive reasoning could be described as consisting in thought experiments wherein one performs various sorts of operations on objects of a given type. (Detlefsen, 1982, p. 49.)

It is not clear whether Detlefsen and Hilbert and Bernays make a distinction between mathematical experiments and mathematical thought experiments. Let us therefore return to § 2 of Hilbert and Bernays's work to find out what in particular they might concretely consider a thought experiment in the construction of elementary arithmetic. Their construction is the best-known expression of a tradition which, revived by Wittgenstein, is a version of an idea that goes back at least to Kant and Poincaré. While using the Kantian concept of consciousness of a successive repetition (*Critique of Pure Reason*, A 103–104, B 300), Poincaré transforms the Kantian 'sensible intuition' into an 'intellectual intuition'. This is the starting point of an interpretation by Hilbert and Bernays which retransformed Poincaré's intellectual intuition into a perceptual intuition regarding the particular configuration of a process G of threading 'numerals': I, II, III, IIII, ...

Now, 'to know oneself capable' of successive repetition is equivalent to being able to pass from a concrete successive repetition to an indefinite iteration, a fact that can be expressed by the following diagram S:

a) $\Rightarrow$ I [we can construct I]
b) $n \Rightarrow n$I [if we have n, we can construct nI]

which is itself suggested by the experience of being able to add I to I: II; I to II: III; I to III: IIII and so on, as long as experience confirms it (without use of the schematic parameter n). The pattern S is thus confirmed by the experiment without being experimental itself. As in the previous examples, we have here a difference between an empirical repetition (level 0) and a schematic repetition S, which is suggested by mathematical *experiments* (they are naturally experiments in the mind, not thought experiments in the genuine sense) which are instantiations of S. According to Hilbert and Bernays, 'the method of proof of complete induction [...] is obtained from the concrete construction of numbers'. They do not say any more. How should we thus imagine this further step?

In fact, what is still missing is the final clause:
c) We can obtain all natural integers by application of the scheme S.

This clause c) does not follow analytically from clauses (a) and (b) of S. According to Poincaré, it is the result of a nonlogical reflection upon the construction S. We have thus returned to what is called a recursive definition of natural integers:

S': a) $N(\mathrm{I})$ [I is a number]
 b) $N(n) \Rightarrow N(n\mathrm{I})$ [if n is a number, its successor also is a number]
 c) We can obtain all numbers by application of (a) and (b).

Yet, as Poincaré correctly observes:

> the principle of induction does not mean that every integer number can be obtained by successive additions; it means that for all numbers that one can obtain by successive additions, one can prove any property by way of recurrence'. (Poincaré, 1905, p. 835.)

Indeed, complete induction means for any property E:

$$T: [E(I) \land \forall n(N(n) \land E(n) \rightarrow E(nI))] \rightarrow \forall n(N(n) \rightarrow E(n))$$

that the justification of T is correlative to S',[9] which means that the final clause (c) cannot be deduced from the clauses (a) and (b): an application of T is needed. In other words, T implies S', and S' implies T. On the other hand, it is impossible to give a schematic proof for T. For an arbitrary number n, constructed according to S', we can certainly indicate an operation that leads to a proof of $E(n)$, but it is impossible to indicate a uniform form of proof. This is because the length of the proof, that is, the number of applications of the modus ponens, depends on n in such a way that a singular proof for a particular argument cannot be viewed, given its internal structure, as an instance of the scheme of proof:

> If instead of showing that our theorem is true for all numbers, we only wish to show that it is true, for example, for the number 6, it will be sufficient to establish the first five syllogisms of our cascade. [...] however far we go in this manner, we will never reach the general theorem, applicable to all numbers, which alone can be the object of science. To do so, we would need an infinite number of syllogisms, we would need to cross an abyss that could never be bridged by the patience of the analyst who is restricted solely to the resources of formal logic. (Poincaré, 1902, p. 13.)

According to Poincaré, to cross this abyss we need pure intuition. In fact, the closure (c) of the scheme S' is not created but only *presented* by an indefinite repetition relating to different levels: the scheme consists, first of all of an overview of potential reiteration on the level of objects (clauses (a) and (b)) and, secondly, in an overview in the sense of a current totality of potential reiteration. This sort of a highly deviant capacity of pure intuition allows Poincaré to say that reasoning by recurrence is the expression of an

---

[9] See for example Kleene (1952, pp. 21–22).

infinite construction, 'condensed to be thus said in a single formula' (Poincaré, 1902, pp. 38–39).

This apocryphal faculty to exclude any non-standard model of arithmetic cannot be ascribed to or made evident by sensible intuition but could be in principle applied if we are granted a powerful intellectual intuition.

Expressed in the terminology introduced in Euler's example, numerals constructed using the rule S of iteration constitute the experimental realm (i) belonging to the general Kantian scheme G (universal) of string repetition, to the rules S and S' of iteration and to the principle of compete induction T. S is interdependent (i) with G and the iteration rule S' (= S + the final clause) is symbolically interdependent (s_i) with G; limiting the field of numbers, S' is metaphorically (by means of a very far-reaching intuition) related (i_a) to S and interdependent (i) with T. Poincaré 'sees' in an apocryphal (intuitive) way the relation between the iteration principle S' and iteration rule S without being able to prove it by logical means: *it is a genuine thought experiment.*

4 Conclusion

In summary, the general flow of a mathematical thought experiment independent of the physical world, figures, diagrams, or calculi can be described like this:

- We have a realm which we view as an experimental field of mathematical elements on a lower level.
- A theory—in a very broad sense: it can be a proof, a definition, or a statement—is applied to it according to established rules in order to validate a mathematical result. We then speak of a mathematical experiment.
- If in the experiment one uses deviant tools, it is called a mathematical thought experiment, which perhaps later becomes a simple experiment with respect to the standard formalism of the time. The mathematical content of a mathematical experiment and mathematical thought experiment can be identical: both refute validity based on 'blind' (ungrounded) inferences. Their cognitive functioning is identical but mathematical thought experiments make use of epistemically non-standard norms.

In a mathematical thought experiment, an idea that does not fit well into the established language seems sufficiently promising to deserve inclusion in the family although we do not know exactly one should integrate it into the accepted formalism or how to clearly formalise it to facilitate an acceptable implementation of this formalism. A mathematician who uses a thought experiment is aware of deviating from the norm.

We saw that what distinguishes a mathematical experiment and mathematical thought experiment is not a difference of principle but a difference of perspective: If the method used by Euler in 1734 can be viewed as a thought experiment in comparison with the mathematical experiment of the calculus of 1731, from a retrospect, this thought experiment looks like a simple mathematical experiment compared to the later formalism which respects the norm.

Acknowledgements. I would like to thank my Nancy colleagues Jean-Pierre Ferrier, Philippe Lombard, and Andrew Arana for their advice and numerous discussions of this article first presented in the FFIUM seminar (French–German ANR/DFG project), where the mathematical example of the Basel problem was presented by Phillipe Lombard, the example of construction of an elementary arithmetic proposed by Andrew Arana, and the example of Stevin analysed by Jean-Pierre Ferrier, to whom I am indebted for several other analyses. I am also grateful to Marco Buzzoni (Macerata), who sparked my interest in this topic, directed me to relevant literature, and shared with me many ideas during our discussions in Macerata and Nancy. Finally, I would like to thank Anna Pilátová for her linguistic corrections.

Bibliography

Atkinson, D., & Peijnenburg, J. (2004). Galileo and prior philosophy. *Studies in History and Philosophy of Science*, *35*, 115–136.
Ayoub, R. (1974). Euler and the zeta function. *The American Mathematical Monthly*, 81, 1067–1086.
Bernays, P. (1937). Grundsätzliche Betrachtungen zur Erkenntnistheorie. *Abhandlungen der Friesschen Schule*, *6*(3–4), 279–290.
Bernays, P. (1946). Quelques points de vue concernant le problème de l'évidence. *Synthese*, *5*, 321–329.
Bernays, P. (1952). Dritte Gespräche von Zürich. *Dialectica*, *6*, 130–136.
Bernays, P. (1953). Diskussionsbemerkungen zum Referat von Hernn Husson. *Dialectica*, *7*, 32–34.
Borwein, J., & Bailey, D. (2004). *Mathematics by Experiment: Plausible Reasoning in the 21st Century.* Natick, MA: A. K. Peters.
Brown, James R. (1986). Thought experiments since the scientific revolution. *International Studies in the Philosophy of Science*, *1*, 1–15.

Buzzoni, M. (2008). *Thought Experiment in the Natural Sciences. An Operational and Reflective-Transcendental Conception*. Würzburg: Königshausen & Neumann.
Buzzoni, M. (2011). Kant und das Gedankenexperiment. Über eine kantische Theorie der Gedankenexperimente in den Naturwissenschaften und in der Philosophie. *Deutsche Zeitschrift für Philosophie, 59*, 93–107.
Buzzoni, M. (to appear). Mathematical vs empirical thought experiments.
Cassirer, E. (1990). *Substanzbegriff und Funktionsbegriff*. Darmstadt: Wissenschaftliche Buchgesellschaft. (Original work published 1910)
Detlefsen, M. (1986). *Hilbert's Program. An Essay on Mathematical Instrumentalism*. Dordrecht-Boston: Reidel.
Detlefsen, M. (1992). Poincaré against the logicicians. *Synthese, 90*, 349–378.
Engel, P. (2011). Philosophical thought experiments: in or out of the armchair? In K. Ierodiakonou, & S. Roux (Eds.), *Thought Experiments in Methodological and Historical Contexts* (pp. 145–163). Leiden-Boston: Brill.
Goffi, J.-Y., & Roux, S. (2018). A dialectical account of thought experiments. In J. R. Brown, Y. Fehige, & M. Stuart (Eds.), *A Routledge Companion to Thought Experiment* (pp. 439–453). London-Boston: Routledge.
Gonseth, F. (1945–55). *La géométrie et le problème de l'espace*, vols. I–VI, Neuchâtel: Griffon.
Goodman, N. (1969). *Languages of Art. An Approach to a Theory of Symbols*. London: Oxford University Press.
Heinzmann, G. (Ed.). (1986). *Poincaré, Russell, Zermelo et Peano. Textes de la discussion (1906–1912) sur les fondements des mathématiques: des antinomies à la prédicativité*. Paris: Blanchard.
Heinzmann, G. (2013). *L'intuition épistémique. Une approche pragmatique du contexte de justification en mathématiques et en philosophie*. Paris: Vrin, collection Mathesis.
Hilbert, D., & Bernays, P. (1934). *Grundlagen der Mathematik I*, zweite Auflage (1968). Berlin-Heidelberg-New York: Springer.
Kleene, S. C. (1980). *Introduction to Metamathematics*. Groningen-Amsterdam-New York: North-Holland. (Original work published 1952)
Kuhn, T. S. (1977). *The Essential Tension. Selected Studies in Scientific Tradition and Change*. Chicago-London: University of Chicago Press.
Lakatos, I. (1976). *Proofs and Refutation – The Logic of Mathematical Discovery*. London: Cambridge University Press.
Lenhard, J. (2018). Thought experiments and simulation experiments. Exploring hypothetical worlds. In J. R. Brown, Y. Fehige, & M. Stuart (Eds.), *A Routledge Companion to Thought Experiment* (pp. 484–497). London-Boston: Routledge.

Lorenz, K. (2010). *Logic, Language and Method. On Polarities in Human Experience*. Berlin: De Gruyter.

Marconi, D. (2017). Philosophical thought experiments: the case for engel. *Philosophia Scientiae*, *21*(3), 111–124.

Norton, J. D. (1996). Are thought experiments just what you thought? *Canadian Journal of Philosophy*, *26*, 333–366.

Norton, J. D. (2004). Why thought experiments do not transcend empiricism. On C. Hitchcock (Ed.), *Contemporary Debates in the Philosophy of Science* (pp. 44–66). Oxford: Blackwell.

Peirce, C. S. (1933–1958). *Collected Papers*. Edited by C. Hartshorne, & P. Weiss, Volume I–VI. Cambridge, MA: Belknap Press.

Poincaré, H. (1898). On the foundations of geometry. *The Monist*, *9*, 1–43.

Poincaré, H. (1902). *La Science et l'hypothèse*. Paris: Flammarion. English translation *Science and Hypothesis: The Complete Text*. New York-London: Bloomsbury, 2017.

Poincaré, H. (1905). Les mathématiques et la logique. *Revue de métaphysique et de morale*, 13, 815–835. Reprinted in Heinzmann, 1986, 11–34.

Reiss, J. (2016). Thought experiments in economics and the role of coherent explanations. *Studia Metodologiczne*, *36*, 113–130.

Sellars, W. (1997). *Empiricism and the Philosophy of Mind* (2nd ed.). Cambridge, MA: Harvard University Press. (Original work published 1956)

Sundholm, G. (2008). Proofs as acts and proofs as objects: some questions for Dag Prawitz. *Theoria*, *64*(2–3), 187–216.

Sundholm, G. (2012). "Inference versus consequence" revisited: inference, consequence, conditional, implication. *Synthese*, *187*, 943–956.

Author biography. Gerhard Heinzmann is Emeritus Professor of Philosophy of Mathematics at the Department of Philosophy, University of Lorraine at Nancy. Founder of the Research Center *Archives Henri-Poincaré*, he is the editor of the *Publications of the Henri-Poincaré Archives* (Springer/Birkhäuser) and founder of the journal *Philosophia Scientiae* (Kimé). Author of numerous articles and books on the philosophy of Henri Poincaré and on philosophy of mathematics and logic, among them *Zwischen Objektkonstruktion und Strukturanalyse. Zur Philosophie der Mathematik bei Henri Poincaré* (1995); *L'Intuition épistémique. Une approche pragmatique du contexte de justification en mathématiques et en philosophie* (2013), he was the President of the *International Academy for Philosophy of Science,* Brussels (2014–21); he is member of the *Academia Europaea,* the *European Academy of Sciences* and the *International Institut of Philosophy.*

Objective Description in Physics

HANS HALVORSON[1]

Abstract. I identify two ideals for objective description in physics, one represented by Albert Einstein and the other represented by Niels Bohr. I show that the Einsteinian ideal, which aims for the god's eye view of reality, is explicitly endorsed by philosophers such as Bernard Williams and Ted Sider, and is tacitly adopted by many other analytic metaphysicians. I argue, however, that this ideal is incoherent and empty of practical content. I then explicate the Bohrian ideal for objective description in terms of covariance relative to context; and I argue that covariance provides a standard of correctness without the incoherent assumption of a god's eye view.

Keywords: objectivity, quantum mechanics, Einstein, god's eye view.

1 Introduction

Scientists and philosophers are not afraid to talk about "good theories" or "bad theories". A theory might be good because it is simple, empirically adequate, or consistent with other well-established theories. Conversely, there are many ways in which a theory can be bad, for instance if it is vague, inconsistent, needlessly complex, or has a bloated ontology. Regardless of what one thinks about these particular characteristics, there is no doubt that judgments of this kind do play an important role in practical decisions about how to do science. For example, if one judges a theory to be bad (which is what, e.g., Einstein and Bell thought about quantum mechanics), then one has a prima facie reason to look for another theory.

Philosophers have been less prone to judging scientists or scientific practices as good or bad. There are some exceptions to this rule, especially in the aftermath of the practice turn in the philosophy of science. Nonetheless, large swaths of the literature are devoted to evaluating the abstract products of scientific theorizing.

One interesting borderline case is the virtue of objectivity or of describing a situation objectively.[2] Is it people who can be objective? Or is objectivity attached primarily to abstract things such as theories or descriptions? For example, a description such as "Brussels sprouts taste bad" might be viewed as lacking objectivity, because we tend to

---

[1] Department of Philosophy, Princeton University, Princeton, NJ 08544.
[2] See Douglas (2004) for a discussion of the multifaceted nature of the virtue of objectivity.

think that there are no objective facts about taste. But does that statement lack objectivity in itself or is it just that a person who asserts that statement is not being objective? These questions are not easy to settle, and I will not try to tackle them head on.

What I will be concerned with is how the ideal of "objective description" can and should function to steer scientific practice. In particular, I identify two contrasting ideals of what physics should deliver in terms of objective descriptions of the world:

- (Einstein) Physics aims to provide a description of the world as it is in itself.
- (Bohr) Physics aims to enable humans to make correct descriptions and to communicate these descriptions to each other.

Einstein's ideal is shared by many philosophers, both past and current, including Baruch Spinoza, G. W. F. Hegel, Bernard Williams, and Ted Sider. Einstein's ideal is also tacitly assumed by many metaphysicians in their search of "artifact-free representations", and by many philosophers of physics in their search of "coordinate-free" or "intrinsic" formulations of theories. I will argue, however, that the ideal of describing things as they are in themselves is incoherent. I will instead advocate an ideal of objectivity that is more like Bohr's. What's more, I argue that this ideal does *not* involve a retreat from the belief in a shared objective reality, which is expressed in terms of objective standards for a correct translation between descriptions.

2 Einstein's ideal

In a first approximation, Einstein believed that the aim of physics is to "know God's thoughts", or, to use a similar metaphor, to describe the world from God's eye view. Einstein thought that quantum mechanics does not supply such a God's eye view, and on that basis, he judged it to be a bad theory. Bohr, in contrast, believed that the aim of physics is to harmonize the experiences of different finite observers, and he thought that quantum mechanics is a good theory precisely because it does that. And thus what might seem to be an abstruse philosophical question—What is an objective description?—had a decisive influence on the choices that Bohr and Einstein made in their scientific careers.

To the best of my knowledge, Einstein never gives an explicit account of his understanding of objectivity. For example, he never explicitly says "the goal of physics is a God's eye view description of reality". It would be therefore unfair of me to turn Einstein's vague ideal into something precise, and then criticize the details of it. Instead, I will point out a common thread in the thoughts of Einstein and some contemporary philosophers who are clearer about their ideal of objective description. I will then direct my criticism at the views of these philosophers.

What was Einstein's beef with quantum mechanics? While he sometimes expresses negative sentiment about indeterminism or about non-locality, his summative judgment is that quantum mechanics fails to describe a reality that is independent of the perceiving subject.

> Fragt man, was unabhängig von der Quanten-Theorie für die physikalische Ideenwelt charakteristisch ist, so fällt zunächst folgendes auf: die Begriffe der Physik beziehen sich auf eine reale Außenwelt, d.h. es sind Ideen von Dingen gesetzt, die eine von den wahrnehmenden Subjekten unabhängige (reale Existenz) beanspruchen (Körper, Felder, etc.), welche Ideen andererseits zu Sinneseindrucken in möglichst sichere Beziehung gebracht sind. (Einstein, 1948, p. 321.)

> If one asks what, irrespective of quantum mechanics, is characteristic of the world of ideas of physics, one if first of all struck by the following: the concepts of physics relate to a real outside world, that is, ideas are established relating to the things such as bodies, fields, etc., which claim a 'real existence' that is independent of the perceiving subject—ideas which, on the other hand, have been brought into as secure a relationship as possible with the sense-data. (English translation by Born, 2004, p. 170.)

Einstein repeats the criticism in his autobiographical account.

> Physics is an attempt conceptually to grasp reality as it is thought independently of its being observed. In this sense one speaks of 'physical reality'. In pre-quantum physics there was no doubt as to how this was to be understood. In Newton's theory reality was determined by a material point in space and time; in Maxwell's theory, by the field in space and time. (Einstein, 1949, p. 81.)

Here Einstein is explaining what he thinks is bad about quantum mechanics by pointing out what he considers to be good about other theories of physics: they describe a reality whose existence is *independent* of any perceiving subject.[3]

Einstein's claim makes a lot of intuitive sense, but it contains a hidden ambiguity. In particular, if D is a description and X is the state of affairs described, then is it D or X that is supposed to be independent of the describer? Einstein cannot have intended to say that X must be independent of the describer, because that would not impose any requirement on the description D, nor on the describer herself. What's more, the describer does not have any say about whether reality is independent of her, so she could hardly be to blame if it is not. So, what Einstein must have intended is that an objective description is independent of the describer. But what kind of thing could a description be such that it is independent of the describer?

Einstein believed that quantum mechanics fails to underwrite any categorical claims about reality but only conditional claims of the form: "if a subject makes a measurement, then such and such outcomes are possible, with such and such probabilities". But this picture misconstrues the role of the subject in producing quantum-mechanical descriptions. The subject does not play a causal role in bringing reality into existence but a semantic role in determining the context of her description.[4] When Bohr spoke of the "epistemological lesson" of quantum mechanics, he sometimes reverted to vague formulations, such as "in the drama of existence we are ourselves both actors and spectators". However, his point is not that the drama of existence is created, in a causal sense, by the perceiving subject, but rather that the describer is entangled in the drama, and that imposes limits on her ability to describe it objectively.[5] In a certain sense, Bohr's claim is completely obvious: if a person is entangled in something, then she might have to take special measures to say anything objective about it.

---

[3] Einstein's view of objective description presupposes the physical separability of the describer and the described. On this point, Bohr agrees with Einstein—and then Bohr wrestles with the fact that, according to our best physics, describers may be entangled with the physical systems they are attempting to describe. For more details, see (Howard, 1979, 1989; Clifton & Halvorson, 2001).

[4] Throughout this article, I use "context" in the sense of Kaplan (1989). Where I differ from standard accounts of indexicals, is in extending the notion of context to include things like frames of reference, or in the case of quantum mechanics, the classical experimental context (see Halvorson & Clifton, 2002; Landsman, 2017). I read Bohr as saying that the context of utterance might include the setup of an experiment, described in terms of "ordinary language supplemented with the terminology of classical physics". For example, in Bohr's example of the walking stick (Klein, 1967, p. 93), there are two distinct descriptive contexts: one where the stick is part of the subject, and one where the stick is part of the object.

[5] I owe this point to Howard (1979).

Thus, while Einstein sought a formalism that gives a picture of reality from the God's eye view, Bohr sought a formalism that gives correct descriptions of reality relative to contexts within that very reality. For example, in the Einstein–Podolsky–Rosen thought experiment, EPR ask (in my paraphrase): What is the real condition (i.e., from the God's eye view) of the second system? Bohr's reply is (in my paraphrase): Ask not how God would describe the second system; ask how a finite subject would describe the second system. What's more, finite subjects have specific contexts which determine the concepts that can meaningfully be employed. For example, the context could include a fixed frame of reference, allowing the describer to employ the concept of position; or the context could assume that the system under study is closed, allowing the describer to employ the concept of momentum. For Bohr, it is not known a priori whether our familiar concepts will continue to be applicable in contexts beyond those for which they originally were adapted.

3 Bernard Williams and the absolute conception

I suspect that many contemporary philosophers have a view of objective description that is similar to Einstein's. But even so, few of them have articulated or defended the view. One exception is the moral philosopher Bernard Williams, for whom the notion of "the absolute conception of reality" plays a central role. Williams articulates this idea in his creative recounting of Descartes' philosophical project:

> What God has given us, according to Descartes, is an insight into the nature of the world as it seems to God, and the world as it seems to God must be the world as it really is. (Williams, 1978, p. 196.)

In other words, Reason allows humans to transcend their finite, limited perspectives, and to see things as God himself does. What's more, this God's eye view is given concretely by mathematical physics, i.e., by the science of matter in motion. While humans experience objects in terms of secondary qualities, such as colors and temperatures, physics sees only geometrical configurations.

For both Descartes and John Locke after him, secondary qualities do not "inhere" in objects themselves but arise from how those objects relate to peculiarly human modes of perception. For example, an apple is not red in itself, but is only red for a subject in a certain context, i.e., from a particular point of view. In contrast, primary qualities are absolute—they inhere in the objects themselves and are independent of the point of view of the describer. Thus, the absolute conception is supposed to be "a conception of reality as it is independent of our thought, and to which all representations of reality can be related" (Williams, 1978, p. 196).

In later philosophical work, Williams uses the idea of the absolute conception to develop a sophisticated moral relativism (see Williams, 1985). In particular, he claims that apparently conflicting systems of moral claims both amount to knowledge only if they are different perspectives on reality in itself.

> If what they both have is knowledge, then it seems to follow that there must be some coherent way of understanding why these representations differ, and how they are related to each other. (Williams, 1978, p. 49.)

Schematically: if one person knows that T_a while another person knows that T_b, then there must be a third conception T such that T_a is a correct account of T in context a, and T_b is a correct account of T in context b. In this way, T provides a consistency check for T_a and T_b: there is a way a world could be such that both T_a and T_b are correct.

One reason that Williams' picture is alluring—and difficult to refute—is because it is just that: a picture. We are supposed to imagine what it would be like to know the world as God would, and we are supposed to think of this blessed state as the telos of "pure inquiry". We are not, however, encouraged to ask questions such as: What exactly is a conception, and what concretely can we do to purify our conceptions of subjectivity and perspective? So, while Williams' idea of the absolute conception might serve as an inspiration, it cannot, without significant supplementation, serve as a concrete guide for scientific practice.[6]

But now I am going to be even more harsh and state that it is not just that the absolute conception is an unclear idea: it is incoherent. To see this, let us look at some of the kinds of examples that are supposed to motivate the notion of the absolute conception.

1. Suppose that Alice believes that the water in a certain bucket is warm while Bob believes that the same water is cold. In that case, the absolute conception might be a description of the (objective) temperature of the water, or even better, a description of the position and velocity of the atoms that make up the water—as well as a description of Alice and Bob, the states of their brains, their histories etc., that predicts that they would feel the way they do.

---

[6] Putnam (1992, Chap. 5) contains a sustained critique of the absolute conception. The debate is then continued in (Williams, 2000; Putnam, 2001). See Moore (1997) for an elaboration and defense of the absolute conception.

2. Alice stands directly above a coin on the ground, and she sees it as a disc. Bob is standing at some distance from the coin, and he sees it as an oval. The absolute conception describes Alice, Bob, and the coin as occupying regions in three-dimensional space. The projection of the coin on Alice's retina is a disc and the projection of the coin on Bob's retina is an oval.
3. Alice is sitting on a boat traveling at a constant velocity, while Bob is sitting on the shore. According to Alice's theory T_a, the boat is stationary. According to Bob's theory T_b, the boat is traveling at four knots east. The absolute conception T describes Alice, Bob, and the boat as spacetime worms.
4. Alice is holding a meter stick. Bob is flying past Alice in a spaceship, and he measures the stick as one-half meter. The absolute conception—provided by the special theory of relativity—describes the stick as a four-dimensional spacetime worm with projections of differing lengths onto Alice and Bob's simultaneity hyperplanes.

These are standard examples which are taken to support the metaphor of absolute and relative conceptions, but they actually uncover an ambiguity in Williams' notion of a 'conception'.

On the one hand, a conception could be a sort of picture, without any specification of how to apply that picture to concrete reality. Let us call this a *conception in the non-descriptive sense*. For example, van Gogh's *Starry Night* is a conception in the non-descriptive sense, as is Tolkien's *Lord of the Rings*, a map of Middle Earth, the number 42, or Minkowski spacetime. On the other hand, a conception could be a specific attempt to describe physical reality. Let's call this a *conception in the descriptive sense*. For example, I have a conception in the descriptive sense of The Netherlands as a flat country where lots of people ride bicycles. Similarly, I have a conception in the descriptive sense of my coffee cup as topologically homeomorphic to a doughnut.

Williams equivocates between descriptive and non-descriptive notions of "conception", and this makes the notion of an absolute conception seem initially plausible. But every conception in the descriptive sense is put forward by a person with a particular point of view, i.e., a person in a context, while the absolute conception is supposed to describe reality in perspective-free or context-insensitive way. In other words, the absolute conception is supposed to have the miraculous property of being a conception in the descriptive sense while lacking the features that every conception in the descriptive sense has.

To see the problem more clearly, imagine you asked me what my conception of physical reality is, and I answered "42". Obviously, thinking of a number is not yet having a conception in the descriptive sense. Moreover, it would not much improve the situation if I said that my conception is that reality is represented by 42. The obvious next question would be "how is reality represented by 42?" because a mathematical object gives rise to a conception in the descriptive sense only when a person specifies how that mathematical object is intended to latch onto concrete reality. This moral holds not just for numbers but also for the mathematical objects that play a starring role in contemporary physics. For example, "reality is represented by the manifold M" is not a conception in the descriptive sense, nor is "reality is represented by the wavefunction ψ". It takes more work to produce a conception in the descriptive sense than to consider a mathematical model and think "reality is like this".

It is only by equivocating between descriptive and non-descriptive notions of "conception" that the classical examples seem to support the idea of an absolute conception. In the second example above, the absolute conception is supposed to be given by a mathematical object such as $(\mathbb{R}^3, a, b, C)$, where $a, b \in \mathbb{R}^3$ represent Alice's and Bob's locations and $C \subseteq \mathbb{R}^3$ represents the coin. However, $(\mathbb{R}^3, a, b, C)$ is a mathematical object, and thus a conception only in a non-descriptive sense. Similarly, in the third example, the absolute conception is supposed to be given by Galilean spacetime M and a couple of points $a, b \in M$ with velocity vectors v_a, v_b in their respective tangent spaces. But once again, the mathematical object (M, v_a, v_b) is a conception only in a non-descriptive sense. Nor could we get a conception in the descriptive sense by plugging the relevant mathematical objects into the sentence "there are things in physical reality to which X corresponds". This latter statement still lacks the determinate content that proper descriptive claims must have.

I conclude that the above examples do not support the idea that there could be an absolute conception of reality, much less the idea that the aim of physics is to achieve such an absolute conception. In none of these examples is there anything that could qualify both as a conception in the second sense (a descriptive claim) and as absolute (perspective-free). Based on such examples, I am more inclined to think that descriptive claims are, by their very nature, contextual; and that the abstract objects (propositions, geometric shapes, manifolds, wavefunctions, etc.) we employ to relate our relative descriptions to each other are not themselves "conceptions" in any epistemically relevant sense.

4 Metaphysics and the third theory

According to Bernard Williams, finding the absolute conception of reality is the objective of physics and philosophers cannot be expected to contribute much to the achievement of this objective. But not all philosophers share Williams' modest view of their enterprise. For example, Hegel thought that he was in a better position than physicists—with their narrow focus on inert matter—to see reality as God sees it. Similarly, many analytic metaphysicians take themselves to be on the hunt for a description of the fundamental structure of reality.

Let us set aside the question of whose job it would be to find the absolute conception. What I want to understand is how adopting a certain ideal of objective description might influence the decisions that people make vis-a-vis research programs. It is my strong sense that Einstein's ideal of objective description played a central role in his rejection of quantum mechanics and his search for a grand unified theory. But I will leave it to better historians than myself to evaluate whether my sense about that is correct.

I also have a sense that analytic metaphysics is often driven by an ideal of objective description that is similar to Einstein's. Indeed, Ted Sider (2020) states explicitly that if a description is true in a fundamental sense, then it must be free from every arbitrary contribution of the describer. This requirement leads to a sort of imperative:

> (Imperative of the third theory) If there are distinct theories T_a and T_b that correctly describe the same domain, then one ought to search for a third theory T that is free from the conventional features of T_a and T_b, and that explains why T_a and T_b are correct.

Sider himself applies this imperative to the following example:

Example (Mass scale). There is a book on the table. One person, Kilo, says that the book weights one kilogram. Call her description T_a. Another person, Pound, says that the book weighs (approximately) 2.2 pounds. Call his description T_b. Which description of the situation should be adopted, T_a or T_b? How can we rationally decide between them?

Sider's answer to this question is that we cannot rationally decide between these two descriptions, and indeed, that both of them are defective for having conventional elements that are not part of their representational content.

> The fact that the number 1 is used isn't part of the representational content of the model; it's an artefact of the choice to use one scale rather than another for measuring mass. The objects aren't objectively 1 in mass, assuming there is no distinguished unit. (Sider, 2020, p. 192.)

T_a and T_b are thus apparently not objective descriptions, because a book is not objectively related weight-wise to either the number 1 or to the number 2.2. An objective description, says Sider, would need to be 'unit free'. He suggests, in particular, that a more objective description is provided by a theory of mass comparisons in terms of a binary relation $\geq$.

Whenever a philosopher uses a specific example to make a general point, we should ask whether the example is paradigmatic of the phenomenon in question. Unfortunately, Sider's example of different mass scales is not paradigmatic of cases of "different perspectives on the same fundamental facts". Indeed, the mass scale theories T_a and T_b result from taking theory T, adding constant symbols corresponding to non-negative real numbers, and adding axioms for the semifield of nonnegative real numbers. Thus, T_a and T_b are simply rigidifications of T in the sense that all elements of a model of T_a (or T_b) are labelled with constant symbols, and such a model has no non-trivial symmetries.[7]

In a more typical example of "different perspectives on the same fundamental facts", a theory T of the fundamental facts contains more information than any one of the perspectival theories, and, in fact, the perspectival theories can be deduced from T and information about context. Consider, for example, the case of a theory T describing a three-dimensional shape, where T_a is the theory of projection of this shape onto the xy plane, and T_b is the theory of projection of this shape onto the yz plane. In that case, T_a and T_b are derivable from T and information about context; and T is not in any sense the "common content" of both T_a and T_b.

Although the example of mass scales does not generalize, I do think that Sider has put his finger on a general pattern of reasoning among metaphysicians—and, indeed, on a sort of common understanding of what it takes for a description to be objective. As Sider himself says:

> I think that many metaphysicians tend to assume (perhaps implicitly) something like the following: It's fine to construct models with artefacts. But there must always be some way of describing the

---

[7] Another possible regimentation of T_a and T_b would have them as two-sorted theories with one sort for physical objects and another sort for positive real numbers.

> phenomenon in question that (in some sense) lacks artefacts. There must be some way of saying what is really going on. For example, although we can model mass with real numbers, there must be some underlying artefact-free description, such as the $\geq$ and C description, from which one can recover a specification of which numerical models are acceptable, and a specification of which features of the models are artefacts. (Sider, 2020, p. 192.)

If Sider is correct about this, it explains a lot about the projects that metaphysicians choose to work on. One particular example that Sider could have discussed is the case of nonsymmetric relations.

Example (Nonsymmetric relations). Peter Geach (1957): There is a teacup a on top of a table b. This state of affairs can be described by the sentence Rab, where R is the relation "is above". The very same state of affairs can be described by the sentence R^*ba, where R^* is the relation "is below". Which theory should be adopted: the theory T_a stated in terms of the relation R, or the theory T_b stated in terms of the relation R^*?

Geach raised this issue over sixty years ago—and it continues to vex the best metaphysicians (see Williamson, 1985; Fine, 2000). Their responses range from rejecting the very notion that there can be asymmetric relations at the fundamental level (see Dorr, 2004) to attempts to construct a new formal logic that collapses the distinction between a relation R and its converse R^*. (For more on this issue, see MacBride, 2020.) For the present discussion, the interesting question is why metaphysicians believe that it is imperative to do anything.

We see here a striking similarity between the visions of Einstein, Williams, and these convention-averse metaphysicians. For all of them, the aim is to find a description without any contribution from the describer. As Sider says, the aim is to find a representation that is free from "representational artifacts", i.e., any feature of the representation that is accidental to its role qua representation.

As with Williams, Sider has many uses for the absolute conception. This conception is not just the most perspicuous representation of reality, it is also needed to establish the equivalence of perspectival theories.

> To support a claim of equivalence between a pair of theories [...] we brought in a third language, a language in which mass is described in a unit-free way, using the concepts $\geq$ and C. This third, more

fundamental, language gave us a perspective on the fundamental facts. (Sider, 2020, p. 187, notation adjusted.)

It is ironic that Sider says that the third theory T gave us a "perspective" on the fundamental facts, because of course he intends T to be non-perspectival. Thus, the third theory plays essentially the same role for Sider as the absolute conception plays for Williams—the only difference between them is that Sider, like Hegel before him, thinks that philosophy has something to contribute to the search for the absolute conception.

5 Spacetime is not the absolute conception

According to a common way of thinking, spacetime theories—such as Einstein's special and general theories of relativity—reconcile the various frame-relative descriptions of states of affairs by embedding them in a God's eye view picture of the contents of spacetime. For example, Alice describes a boat (on whose deck she is sitting) as stationary, while Bob describes the same boat as traveling to the east at four knots. Alice's description is correct relative to her context and Bob's description is correct relative to his context, but neither is correct in an absolute sense. For a description that is correct in an absolute sense, we should think of the boat as a four-dimensional extended object in spacetime.

This story is so commonplace that I am tempted to call it the orthodoxy. One finds this point of view assumed by almost every philosopher who discusses special relativity—except for those who reject the special theory of relativity in favor of a Lorentzian theory (see Craig, 2001) and those who reject the idea that there is a single objective reality (see Fine, 2005). For example, Balashov (2010) argues that the three-dimensional appearances are projections of four-dimensional objects onto our respective hypersurfaces of simultaneity. The central idea is that a mathematical spacetime M with some contents Γ is supposed to yield a conception of reality *sub specie aeternitatis*; and our local conceptions, that is, our respective worlds of appearances, can be obtained by deducing perspectival information from (M, Γ).

When discussing Bernard Williams, I argued that a mathematical model M is only a conception in a non-descriptive sense. To use M to form a conception in the descriptive sense requires that one relate the parts of M to parts of physical reality, and that presupposes a specific context, namely a location (in a general sense) in physical reality. For example, if M is a rectangle, then I can use M to form a conception in the descriptive sense of a piece of paper on my desk. I can do this, for example, by imagining a context c that is located directly above my desk and looking down at it. From that context c, it is correct to say that the piece of paper is rectangular—and that can be cashed out roughly as saying that the projection of M onto the visual field of c is a rectangle.

Or consider the example where M is a map of Paris. Then I could correctly say that M describes Paris, if, for example, I imagine a context that is 5,000 feet directly above Paris and looking down. Moreover, this context must include an orientation for the map, because if I change context by turning the map upside down, then it is no longer true that M describes Paris.

The situation is slightly more complicated for three-dimensional objects. Suppose now that M is a mathematical cube. If I say that M describes the box that is sitting on the floor of my office, then what context am I implicitly assuming? Or is it the case that "M represents the box" is intended to be true independent of context?

Such a statement cannot be true independent of context. Just as a person might misalign a two-dimensional mathematical model with a two-dimensional slice of physical reality (e.g., if the map is turned upside down then "the map represents Paris" changes truth-value), so a person might misalign a three-dimensional mathematical model with a three-dimensional physical object. For example, if "M represents the box" is true in one context, but then the context is changed by rotating M, then "M describes the box" may no longer be true. It follows that "M represents the box" is implicitly contextual even when M is a three-dimensional mathematical object.

There is no reason to think that the context-dependence of mathematical modelling suddenly ceases when we come to four-dimensional objects, or to the entirety of spacetime. Supposing that there is a mathematical object M that can be used correctly to represent spacetime, this same object M could be used incorrectly to represent spacetime. Whether M does or does not represent spacetime depends on the context, in a broad sense, of the person using it to describe. If "M represents spacetime" is true in one context a, then it is false in another context b. Therefore, M does not provide an absolute conception of reality.

Let's think about what it means to say that spacetime itself is always described from a particular point of view. All the points of view that we human beings know have the feature that they are located at a particular place and a particular time. What's more, the person with that point of view has a particular state of motion. In short, that person has a frame of reference in the sense that is familiar from physics. Thus, while a person may think of space and time as a whole, her description of space and time as a whole still presupposes a frame of reference.

What does it mean, then, to say that "spacetime is described by M" is true relative to a frame of reference? In the first instance, we might think that the analysis of such statements should follow the same model as the analysis of statements such as "the office is described by C", where C is a cube in $\mathbb{R}^3$. Roughly speaking, the statement "the office is described by C" is true in context p (a location in space, represented by $\mathbb{R}^3$) just in case the distances between that point p and the various bits of the office are the same as the distances between that point p and the various elements of C.

The case of representing spacetime is a bit more subtle because there is little consensus about how we should understand statements about future times. But the point I would insist on is that "M represents spacetime" is to be analyzed into statements about three-dimensional spatial and one-dimensional temporal distances from a context c. In other words, context-relative statements form the explanatory basis for the apparently context-insensitive statement that M represents spacetime.

To reinforce this point, imagine that Γ describes some distribution of matter in spacetime. I claim, then, that the relationship between (M, Γ) and frame-dependent descriptions is not asymmetric in the way that it would need to be for (M, Γ) to be the absolute conception. Recall that the absolute conception is supposed to be more fundamental than the various relative conceptions and, moreover, this asymmetry is what gives the absolute conception its unique epistemic authority. However, in the case of the special theory of relativity, all features of a mathematical model (M, Γ) are deducible from any one of the frame-relative descriptions. For example, if Γ is a timelike line in M describing a massive particle on an inertial trajectory, then Γ determines a unique position x_a, energy e_a, and velocity v_a relative to any frame of reference a. Conversely, any reference frame a and triple (x_a, e_a, v_a) determines a unique timelike line Γ. In short, a frame-relative description of the content of spacetime is logically complete in the sense that it entails every fact about the content of spacetime.

One might object that the frame-relative facts entail all facts only if all objects are assumed to follow inertial trajectories. However, the argument can be strengthened by taking into account all frame-relative facts, and not just the facts relative to a single frame of reference. Obviously, any curve Γ in M is uniquely determined by the projection of its tangent vectors onto all simultaneity hypersurfaces; and hence, all facts are deducible from the logical sum of all frame-relative facts. In short, there is no reason to think that the facts represented by the four-dimensional spacetime model are more fundamental than the three-dimensional, frame-relative facts.

6 Objective description and coordinates

Another popular myth is the idea that we can increase the objectivity of our descriptions by passing from coordinate descriptions to intrinsic geometric descriptions. The metaphor that often gets brought out here is directly analogous to the one that motivates Williams and Sider: there are context-bound individuals $a, b, \ldots$ with their coordinate descriptions $T_a, T_b, \ldots$ Then there is a coordinate-free, geometric description T from which all the coordinate descriptions can be derived. This coordinate-free description T is supposed to represent reality as it is in itself, while the coordinate descriptions $T_a, T_b, \ldots$ involve arbitrary conventions, e.g., choosing to set the coordinate origin in one place rather than in another.

This picture exercises a strong grip, and yet I will argue that it is based on a confusion. Coordinate-free mathematical objects—such as affine spaces and manifolds—do not provide more perspicuous or more intrinsic descriptions of physical reality than coordinate descriptions do. A more accurate thing to say is that these abstract mathematical objects facilitate the harmonization of individual coordinate descriptions, or less metaphorically, these objects provide translation schemes between coordinate descriptions.[8]

Consider, for example, two distinct coordinate descriptions of space. For example, T_a might be a description of Princeton, NJ, where the origin of the coordinates is set at Nassau Hall, while T_b is a description where the origin of the coordinates is set at 1879 Hall. According to the Einstein-Williams-Sider picture, these two coordinate descriptions both have the flaw of involving an arbitrary choice of origin. Williams and Sider would then say that there is an epistemic imperative to find a third, coordinate-free description T of Princeton. In this case, the obvious candidate for T is simply a two-dimensional affine space A, which has no preferred origin. Then saying "Princeton is represented by A" involves no arbitrary choice, and so it can be taken as the sought-after, more objective description.

There is, however, a problem with this suggestion. An affine space A is a set consisting of infinitely many distinct points. For a person to represent physical space by A presumably requires that person to coordinate points of A to points of physical space. But then this person is once again faced with a problem of arbitrary convention: should a particular point $a \in A$ be assigned to Nassau Hall, or should a be assigned to 1879 Hall? The theorist has to choose one or the other coordinatization, but neither of them is

---

[8] For a different argument for the same conclusion, see Wallace (2019).

preferred by the physical situation. Thus, the problem of the arbitrariness of coordinate descriptions remains even if we replace numbers by other mathematical objects.

There is, of course, a precise sense in which an affine space A has less structure than a vector space V. In particular, for any vector space V, there is an affine space A such that V is isomorphic to $(A, 0)$. That is, a vector space is precisely an affine space A plus a designated origin $0 \in A$. What's more, the symmetries of $(A, 0)$ are the subset of the symmetries of A that fix 0. It might seem then that representing Princeton with $(A, 0)$ involves the postulation of more structure than representing Princeton with A. But it all depends on the intentions of the describer. If a describer is well aware that she assigned 0 arbitrarily, and could have just as well assigned 0 to any other location, then her representation via $(A, 0)$ does not postulate any more structure than a representation via A. The important point is what the describer herself intends to be the degree of arbitrariness in her description. A person who represents Princeton by A might be taken to be saying that whatever point $a \in A$ she assigned to Nassau Hall was arbitrary, that is, she herself is not committed to this choice being better than another. In contrast, a person who represents Princeton by $(A, 0)$ might be taken to be signaling that Nassau Hall has some theoretical significance, e.g., has some property that is going to play a role in explanations. The mathematical model does not itself determine how the describer intends to use the mathematical model to represent things.

7 Objectivity as covariance

I take it for granted that physics frequently succeeds in producing objective descriptions of physical states of affairs. However, *pace* Williams, physics has never gotten close to an absolute conception—and I'm not sure that the idea is even coherent. In that case, the burden is on me to explain what could be meant by an "objective description".

Recall that according to Einstein's ideal, an objective description is independent of the describer. We saw, however, that this ideal is caught on the horns of a dilemma: either it requires that the relevant state of affairs is independent of the describer (which places no requirement on the description), or it requires that the description is independent of the describer (which makes little immediate sense). Given that the first horn of the dilemma is a nonstarter as an account of objective description, I propose that we work on the second horn of the dilemma, i.e., to make sense of "description D is independent of the describer s".

There are two paths we could follow at this point: on the one hand, we could attempt to decouple the description D from the describer—to consider it as an abstract object, such as a proposition, that exists independently of any human subject. In that case, there would be a trivially simple answer to the question "what is an objective description?", namely: an objective description is a set of true propositions. Or, if there is a worry about the possibility of a description that is about describer herself (e.g., her preferences), then we could nuance the requirement as follows:

> (PROP) D is an objective description for subject s if D is a set of true propositions that make no reference to s.

But PROP has many problems. First, what does it mean for a proposition to make reference to a person? For example, does a proposition describing carbon atoms make reference to you or not? Second, PROP places a restriction on the description D, but not on the describer s, which conflicts with the intuition that a subject is essentially involved in cases where the notion of "objection description" is relevant.

An even worse problem for PROP is that falls prey again to the problem with the first explication of Einstein's objectivity requirement, in particular, it does not provide guidance about how to attain objective descriptions. It is generally supposed that when people say or write declarative sentences, then they assert propositions without further effort. So, what is it that PROP requires of a describer? The only guidance that PROP gives to a person is that he should speak truly and not about himself. That hardly seems like helpful guidance for scientific practice.

The second path we could follow is the path of practical implementability. To this end, consider again the example of two people, Alice who finds the room temperature to be cold, and Bob who finds the room temperature to be warm. If Alice says

> (S_1) The room is cold.

then we do not normally think of her as asserting an objective fact. But why not? For one, Bob would not directly affirm S_1. Nonetheless, many philosophers of language would say that S_1, asserted by Alice, might pick out a true proposition, which is more transparently represented by the sentence: (S_2) The room feels cold relative to context a. Does S_2 count as an objective fact? The answer to that question depends on what we understand by the context a. On the one hand, if "being in context a" simply means "being Alice", then S_2 might not be an objective fact. The problem in that case, I think, is that Bob might lack a rule for interpreting statements in context a into statements in his own context. On the other hand, the contexts a and b might be specified by parameters and there might be well-defined transformation rules for a description D_a in context a to a description D_b in context b. In that case, I would consider S$_2$ to be objective, despite the fact that it makes explicit reference to a specific context.

To spell this idea out more fully, I propose the following sketch of an ideal for objective description:

> (Descriptive Covariance) For any two contexts a and b, and for any physical transformation $f: a \to b$, there is a translation $D_f : D_a \to D_b$ from the contextual description D_a to the contextual description D_b. Moreover, these various translations "commute" with each other in a rule-governed way.[9]

If Descriptive Covariance holds, then descriptions "co-vary" with the contexts—in the sort of way that might be expected for perspectival descriptions of a single, coherent reality. In this case, objectivity is captured not by the existence of a context-free description but by the rule that connects the contextual descriptions.

Descriptive Covariance is just a sketch of an idea, and it raises many further questions. For example, what counts as a "rule" relating contextual descriptions of reality? Isn't the notion of "rule" so flexible that it is trivial to say that there is a rule relating different perspectival descriptions of reality? To this question, I answer that we do have some intuitions about what might count as a reasonable translation between two descriptions, and furthermore, about the notion of uniformity. For example, in special relativity, the Lorentz transformations are a uniform rule in the sense that the contextual parameters play the same role in determining the translation from one frame of reference to another. Similarly, in the logical theory of models, there is a precise definition of when a concept is definable uniformly across all models of a theory.

The ideal of objectivity as "covariance relative to context" finds inspiration in Niels Bohr's account of the aims of physics. First of all, Bohr explicitly rejects the idea of a god's eye view description of reality (see Favrholdt, 1994, 2015).[10] Nonetheless, Bohr

---

[9] For those familiar with category theory: Descriptive Covariance is basically the requirement that there is a functor from contexts to descriptions.

[10] In this regard, Bohr follows his teacher Harald Høffding, who adapted the view of his own teacher Rasmus Nielsen. Høffding (1909, Chap. 16) relays Nielsen's claim that an objective description presupposes an "objectifying subjectivity". Høffding then argues that Nielsen should have concluded from this fact that a God's eye view of reality is an incoherent notion.

maintains that the goal of physics is to provide objective descriptions of states of affairs, where objectivity is equated with a lack of ambiguity, i.e., with *Eindeutigkeit* (German) or *entydighed* (Danish).

> Every scientist is constantly confronted with the problem of objective description of experience by which we mean unambiguous communication. (Bohr, 1958, p. 67.)

> We must strive continually to extend the scope of our description, but in such a way that our messages do not thereby lose their objective or unambiguous character. (Petersen, 1963, p. 10.)

> [...] our task must be to account for such experience in a manner independent of individual subjective judgment and therefore objective in the sense that it can be unambiguously communicated in ordinary human language. (Bohr, 1963, p. 10.)

The idea here is that there should be a rule such that for any correct description D_a relative to context a, and for any other context b, there is a unique translation of D_a into D_b. In this sense, the descriptions are uniquely interpretable, *eindeutig*, which Bohr takes to be a necessary condition for objectivity.

8 Conclusion

"Being objective" is an important virtue for scientists, and producing objective descriptions is among the more important goals of science. Nonetheless, it seems to be quite challenging to give a precise characterization of this virtue of objectivity. Indeed, some of history's greatest scientists have had radically different views about the nature of objective description.

With due respect for Einstein's scientific genius, his view of objective description—as a description of reality as it is in itself—is ambiguous, and on some disambiguations, it is simply nonsensical. What's more, Einstein's view works against the cause of objective description, insofar as it would encourage scientists to produce descriptions that are completely detached from any context. But there are no such descriptions, and if there were, then contextual beings like you and me would not be able to understand them.

It is understandable that Einstein, Williams, and Sider, among others, would take the notion of objective description to presuppose the existence of an absolute conception. In particular, the absolute conception is supposed to provide an objective standard for measuring the correctness of other conceptions. If there were no such standard, then what would we even mean by saying that a conception is correct?

I have a simple, if deflationary, answer to that question: correctness of a description is an irreducibly relational notion. That is, "D is a correct description of X" is a claim about a relation between a description D and concrete reality X which cannot be reduced to a conjunction of claims about D and X separately. In particular, to say that D is a correct description of X is not a matter of D being "isomorphic" to X (which would be defined in terms of D and X having certain monadic properties in common).

Now, if correctness of a conception is an irreducibly relational notion, then there is no longer any place for the absolute conception as a standard by which correctness of conceptions is measured. What's more, there is no reason to think that two relative conceptions are equivalent only if they are related in the right way to the absolute conception. In fact, I see no evidence to suggest that reality admits, ultimately, of just one correct description. All the evidence points in the opposite direction: every true description is essentially contextual.

Acknowledgements. For feedback, I thank audiences at the University of Illinois and at ICLMPST 2019. Special thanks to Frederik Pedersen (Princeton '20) for many illuminating discussions about these issues.

Bibliography

Balashov, Yuri. (2010). *Persistence and Spacetime*. Oxford: Oxford University Press.
Bohr, Niels. (1958). *Atomic Physics and Human Knowledge*. New York: Wiley.
Bohr, Niels. (1963). The unity of human knowledge. In *Essays 1958–1962 on Atomic Physics and Human Knowledge* (pp. 8–16). New York: John Wiley and Sons.
Born, Max. (2004). *The Born-Einstein letters: 1916–1955*. New York: Palgrave Macmillan.
Clifton, Rob, & Halvorson, Hans. (2001). Entanglement and open systems in algebraic quantum field theory. *Studies in History and Philosophy of Modern Physics, 32*(1), 1–31.
Craig, William Lane. (2001). *Time and the Metaphysics of Relativity*. Dordrecht: Springer.
Dorr, Cian. (2004). Non-symmetric relations. *Oxford Studies in Metaphysics*, 1, 155–192.
Douglas, Heather. (2004). The irreducible complexity of objectivity. *Synthese, 138*(3), 453–473.

Einstein, Albert. (1948). Quanten-Mechanik und Wirklichkeit. *Dialectica*, 2, 320–324.
Einstein, Albert. (1949). Autobiographical notes. In P. A. Schilpp (Ed.), *Albert Einstein, Philosopher-Scientist* (pp. 1–94). Chicago: Open Court.
Favrholdt, David. (1994). Niels Bohr and realism. In Jan Faye & Henry Folse (Eds.), *Niels Bohr and Contemporary Philosophy* (pp. 77–96). Dordrecht: Kluwer.
Favrholdt, David. (2015). *Filosoffen Niels Bohr*. Copenhagen: Informations Forlag.
Fine, Kit. (2000). Neutral relations. *The Philosophical Review*, *109*(1), 1–33.
Fine, Kit. (2005). Tense and reality. In *Kit Fine: Modality and Tense: Philosophical Papers* (pp. 261–320). Oxford: Oxford University Press.
Geach, Peter. (1957). *Mental Acts: Their Content and Their Objects*. London: Humanities Press.
Halvorson, Hans, & Clifton, Rob. (2002). Reconsidering Bohr's reply to EPR. In Tomasz Placek and Jeremy Butterfield (Eds.), *Non-Locality and Modality* (pp. 3–18). Dordrecht: Kluwer.
Høffding, Harald. (1909). *Danske Filosoffer*. Copenhagen: Gyldendal.
Howard, Don. (1979). *Complementarity and Ontology: Niels Bohr and the Problem of Scientific Realism in Quantum Physics* (Doctoral dissertation, Boston University Graduate School, Boston).
Howard, Don. (1989). Holism, separability, and the metaphysical implications of the Bell experiments. In Jim Cushing & Ernan McMullin (Eds.), *Philosophical Consequences of Quantum Theory: Reflections on Bell's Theorem* (pp. 224–253). Notre Dame, IN: University of Notre Dame Press.
Kaplan, David. (1989). Demonstratives: An essay on the semantics, logic, metaphysics, and epistemology of demonstratives and other indexicals. In Joseph Almog, John Perry & Howard Wettstein (Eds.), *Themes from Kaplan* (pp. 481–563). Oxford: Oxford University Press.
Klein, Oskar (1967). Glimpses of Niels Bohr as scientist and thinker. In S. Rozental (Ed.), *Niels Bohr: His Life and Work as Seen by His Friends and Colleagues*. New York: John Wiley and Sons.
Landsman, Klaas. (2017). *Foundations of Quantum Theory: From Classical Concepts to Operator Algebras*. Cham: Springer.
MacBride, Fraser. (2020). Relations. In E. N. Zalta (Ed.), *The Stanford Encyclopedia of Philosophy* (*Winter 2020 Edition*).
Moore, Adrian W. (1997). *Points of View*. Oxford: Clarendon Press.
Petersen, Aage. (1963). The philosophy of Niels Bohr. *Bulletin of the Atomic Scientists*, *19*(7), 8–14.
Putnam, Hilary. (1992). *Renewing Philosophy*. Cambridge, MA: Harvard University Press.

Putnam, Hilary. (2001). Reply to Bernard Williams' "Philosophy as a humanistic discipline." *Philosophy*, 76(4), 605–614.

Sider, Theodore. (2020). *The Tools of Metaphysics and the Metaphysics of Science*. Oxford: Oxford University Press.

Wallace, David. (2019). Who's afraid of coordinate systems? An essay on representation of spacetime structure. *Studies in History and Philosophy of Modern Physics*, 67, 125–136.

Williams, Bernard. (1978). *Descartes: The Project of Pure Enquiry*. New York: Routledge.

Williams, Bernard. (1985). *Ethics and the Limits of Philosophy*. London: Fontana.

Williams, Bernard. (2000). Philosophy as a humanistic discipline. *Philosophy*, 75(4), 477–496.

Williamson, Timothy. (1985). Converse relations. *The Philosophical Review*, 94(2), 249–262.

Author biography. Hans Halvorson is Stuart Professor of Philosophy at Princeton University, and Research Professor at the Niels Bohr Archive, Copenhagen. He has published extensively in physics, logic, and philosophy, and is the author of (among other things) *The Logic in Philosophy of Science*.

Representation and Abstraction in Theories of Operations and Classes

RAYMOND TURNER[1]

Abstract. Representation and abstraction are two fundamental concepts that support the activities of specification and programming. While representation is the goal of programming, abstraction facilitates and enables problem solving at a level closer to the problem specification (Floridi, 2008; Dale & Walker, 1996; Bjørner & Jones, 1978; Abrial 1988; Jones, 1980; Spivey, 1992). However, aside from Turner (2021), Turner (2018), Colburn & Shute (2007) and Floridi (2008), there is little sustained logical or philosophical analysis of computational abstraction. In this paper we provide one inspired by contemporary abstractionism (Ebert & Rossberg, 2016; Wright, 1983; Hale & Wright, 2001; Linnebo, 2018; Mancosu, 2016; Heck, Jr., 1993). This will be enriched and linked to the formal work on representation found in the computer science literature (Jones, 1980; Dale & Walker, 1996; Liskov & Zilles, 1974; Thomas, Robinson & Emms, 1988). Formally, our account will be situated within Feferman's theories of operations and classes (Feferman, 1979; Feferman, 1975).

Keywords: abstraction, representation, operations, classes, types, specification.

1 Dual notions

The practice of computer science is underpinned by various processes or devices of abstraction (Colburn & Shute, 2007). Indeed, the term "abstraction" covers a multitude of notions. However, we shall concentrate on one of the central concepts namely "data abstraction" (Dale & Walker, 1996; Jones, 1980; Liskov & Zilles, 1974). It is the flip side of the better documented notion of computational representation (Jones, 1980; Dale & Walker, 1996). The latter begins with an abstract type (A) and seeks a more concrete one (C) that can stand proxy for it in computational contexts. As a witness to its representational status one must also locate a surjective representation function:

$$rep : C \Rightarrow A$$

that preserves the operations of the concrete type; in algebraic terms a surjective homomorphism. On the face of it, this is the reverse of data abstraction. Here one begins

---
[1] University of Essex, UK.

with a concrete type C and somehow locates a more abstract one (A). But how exactly is this to be achieved, i.e., given the concrete structure how we do "abstract" the more abstract one?

The approach we adopt has its roots in Frege (2003), and has its contemporary elaboration in Ebert & Rossberg (2016), Wright (1983), Hale & Wright (2001) and Fine (2008). We speak of the shape of a building, the direction of a line, the number of entries in a list. Many singular terms formed by means of functional expressions denote ordinary concrete objects: 'the father of Plato', 'the capital of France'. In contrast, the functional terms that pick out abstract entities are distinctive. In these cases, where $f(a)$ is such an expression, there is an equation of the following form:

$$f(a) = f(b) \text{ if and only if } R(a,b)$$

Here R is an equivalence relation on some domain. The following are stock examples:

The direction of a = the direction of b if and only if a is parallel to b
The number of F's = the number of G's if and only if there are just as many F's as G's

In the first, the equivalence relation is on the domain of lines; in the second it involves concepts. More formally, abstraction principles take the following form:

$$\forall x \in C. \forall y \in C. f(x) =_A f(y) \leftrightarrow R(x,y)$$

Here C is the more concrete domain and A is the abstracted one. Such abstraction principles function as implicit definitions that introduce new kinds of thing. They do so by providing explicit equality conditions for the new domain. For example, given the kind of things that are "lines", abstraction introduces the kind of thing that are "directions":

$$\forall x \in Lines. \forall y \in Lines. direction(x) =_{Directions} direction(y) \leftrightarrow x \parallel y$$

It is important to note that the new domain is not the set of equivalence classes of the old one. For instance, directions are not to be understood as equivalence classes of lines. Such an approach delivers set-theoretic representations not abstractions. In contrast, Frege style abstraction introduces new primitive objects, and a new primitive domain of those objects.

There are significant philosophical differences between representation and abstraction. It has been argued that "abstraction principles" have a special semantic status (Hale

& Wright, 2001). To understand the term 'direction' involves knowing that the direction of a and the direction of b are the same entity if and only if the lines a and b are parallel. Seemingly, the equivalence relation that appears on the right hand side of the biconditional is semantically prior to the functional expressions on the left. For instance, mastery of the concept of direction presupposes mastery of the concept of parallelism, but not vice versa. Consequently, abstraction principles are said to have semantic significance: the semantic interpretation of the functional expressions on the left hand side would appear to be given by the equivalence relation on the right hand side. This is the perspective of semantic abstractionism (Ebert & Rossberg, 2016; Hale & Wright, 2001): our ability to have singular thoughts about objects of a certain type is fixed by the truth-conditions of identity judgments about objects of that type.

There is no such semantic role for representation where the relationship between the abstract and concrete structures is an implementation one, i.e., the concrete one provides an implementation of the abstract one. Here the concrete structure does not provide a semantic interpretation of the abstract one: semantic interpretations supply "correctness conditions" whereas implementations have to be correct relative to some independent semantic account. This is the reverse of what happens with abstraction.

2 A theory of operations and classes

To formalize matters we employ a Feferman-style theory (Feferman, 1979; Beeson, 1985; Feferman, 1975) of operations and classes. These theories are concerned with a universe of "computational objects" and provide a suitable framework for formulating theories of abstract data types (Feferman, 2009; Feferman, 1991). We employ a (slight) reformulation of the theory T_0 (Feferman, 1979; Feferman, 1975; Beeson, 1985).

2.1. The language. It is a two-sorted language consisting of individual and class terms. The former are built from individual variables $(u, v, w, x, y, z, u_1, v_1, \ldots)$ and individual constants $(s, k, c_n, i, j, p, p_1, p_2, d, 0, succ, rec)$ by application (app). The constants include the combinators (s, k) of combinatorial logic (Hindley & Seldin, 1986), constants for comprehension terms (c_n), inductive definitions (i), the join operation (j), pairing and projections (p, p_1, p_2), definition by cases (d), and the operators for the natural numbers $(0, succ)$. We write $app(t,t)$ as tt. Class terms are constructed from class variables $(A, B, C, \ldots, X, Y, Z, \ldots)$ and the natural number class $\mathbb{N}$. We employ $e, e_1, e_2, \ldots$ as metavariables for terms of either sort. We use lower case Greek letters for formulas. These are generated from the atomic assertions of membership $(e_1 \in e_2)$, equality $(e_1 = e_2)$, definedness $(e \downarrow)$ and absurdity $(\bot)$. The complex formula are generated from these by the logical connectives $(\varphi \wedge \varphi, \varphi \vee \varphi, \varphi \rightarrow \varphi)$, individual $(\forall x. \varphi, \exists x. \varphi)$ and class quantification ($\forall X.\varphi, \exists X.\varphi$). Negation is defined in terms of implication and absurdity ($\neg \varphi \doteq \varphi \to \bot$). Restricted quantification is defined as: $\forall x \in s.\varphi \doteq \forall x. x \in s \to \varphi$. Partial equality is defined as follows:

$$e_1 \simeq e_2 \doteq (e_1 \downarrow \lor e_2 \downarrow) \to e_1 = e_2$$

For each (application) term t, we can associate a term t^* whose free variables are those of t minus x, and such that: $t^* \downarrow$ and $\forall x. t^*x \simeq t$. We write this in familiar lambda notation as follows:

$$\lambda x.t \doteq t^*$$

Containment is defined in the normal way: $a \subseteq b \doteq \forall x. x \in a \to x \in b$.

2.2. The logic. The individual logic is the logic of partial terms (Beeson, 1985; Feferman, 1995). There are classical and intuitionistic versions depending upon whether we adopt the law of excluded middle. Nothing in this paper depends upon this choice. In either case, the universal and existential quantifier axioms satisfy the following:

$$\frac{(\forall x. \varphi(x) \land t \downarrow) \to \varphi(t/x)}{\varphi \to \forall x. \psi}$$

$$\frac{(\varphi(x) \land t \downarrow) \to \exists x. \varphi(x)}{\exists x. \varphi \to \psi}$$

In the last two rules, x is not free in ψ. These are justified as the individual variables are taken to range over defined objects, i.e., $x \downarrow$.

Classes are taken to be defined so class quantification obeys the normal quantification rules.

2.3. Axioms of equality and definedness. All constants are defined. Membership and equality are strict:

$$e_1 \in e_2 \to e_i \downarrow$$
$$(e_1 = e_2) \to e_i \downarrow$$

Equality satisfies the usual axioms and rules of replacement:

$$\forall x. x = x$$

$$\forall x, y.\ x = y \to (\varphi(x) \to \varphi(y))$$

Partial equality satisfies:

$$(xy \simeq u \land xy \simeq v) \to u = v$$

The combinators, pairs and cases obey the following axioms:

$$kxy = x$$
$$sfgz \simeq fz(gz)$$
$$k \neq s$$
$$sxy \downarrow$$
$$p_1(pxy) = x \land p_2(pxy) = y$$
$$pxy \downarrow \land p_1 z \downarrow \land p_2 z \downarrow$$
$$(x = y \to dxyuv = u) \land (x \neq y \to dxyuv = v)$$

Finally, there is a basic ontological axiom that guarantees that every class is a defined individual:

$$\forall X.\ \exists x.\ x = X$$

2.4. Natural numbers. The natural numbers are taken to satisfy the standard axioms for zero and successor:

$$0 \in \mathbb{N}$$
$$succ \in \mathbb{N} \Rightarrow \mathbb{N}$$
$$\forall u \in \mathbb{N}.\ \forall v \in \mathbb{N}.\ succ(u) = succ(v) \to u = v$$
$$\forall u \in \mathbb{N}.\ succ(u) \neq 0$$

To complete matters we require an axiom of induction:

$$\forall X.\ (0 \in X \land \forall x \in X.\ succ(x) \in X) \to \forall x \in \mathbb{N}.\ x \in X$$

The recursion theorem provides a general form of primitive recursion with a recursion operator that has the following form:

$$rec \in ((X \Rightarrow Y) \otimes (X \otimes Y \Rightarrow Y)) \Rightarrow ((\mathbb{N} \otimes X) \Rightarrow Y)$$

Let $g \in X \Rightarrow Y$ and $f \in (X \otimes Y) \Rightarrow Y$ and $u \in X$. Then, given definition by cases, this is taken to satisfy the following recursion equations:

$$rec_{f,g}(0, a) = g(a)$$
$$rec_{f,g}(succ(n), a) = f(a, rec_{f,g}(n, a))$$

Here we write the operational arguments as suffixes.

2.5. Comprehension. Not all forms of comprehension are sanctioned. *Stratified formula* are a special case where the only atomic formula have one of the following forms: $t \in X$, $t \in \mathbb{N}$, $t = t$, $t \downarrow$, $\bot$. *Elementary formula* are a special case that permits no class quantification. To state the scheme of comprehension we introduce some notation. Let n be the Gödel number $\ulcorner \varphi(x, x_1, \ldots, x_n, X_1, \ldots, X_m) \urcorner$ with all its free variables specified. We put:

$$\{x. \varphi(x, x_1, \ldots, x_n, X_1, \ldots, X_m)\} \doteq c_n(x, x_1, \ldots, x_n, X_1, \ldots, X_m)$$

This shows that the process of class formation by comprehension as a uniform function of its parameters. The scheme of elementary comprehension can then be stated as follows. *Comp*: For each elementary formula δ:s

$$\exists X. X = \{x. \delta(x, x_1, \ldots, x_n, X_1, \ldots, X_m)\} \wedge \forall y. y \in X \leftrightarrow \delta(y, x_1, \ldots, x_n, X_1, \ldots, X_m)$$

The stratified version follows suite but permits stratified formula in the defining formula. T_0 has only the elementary version.

As an instance of comprehension we define a universal class of defined objects by $V = \{x. x = x\}$. We may introduce the Boolean class by:

$$\mathbb{B} \doteq \{z. z = tt \vee z = ff\}$$

where $tt \doteq \lambda x. \lambda y. x$; $ff \doteq \lambda x. \lambda y. y$; $cond \doteq \lambda z. \lambda x. \lambda y. zxy$.

$\mathbb{N}^+ = \{x \in N. x > 0\}$ represents the non-zero numbers. A representation of the integers provides another simple example:

$$\mathbb{Z} \doteq \{(n, b) : \mathbb{N} \otimes \mathbb{B}. n = 0 \rightarrow b = tt\}$$

The following instances of comprehension provide the definition of products, the class of operations from one class to a second, class union and intersection. Note that these are operations on classes:

$$X \otimes Y \doteq \{z. \exists u \in X. \exists v \in Y. z = pair(u,v)\}$$
$$X \Rightarrow Y \doteq \{f. \forall x \in X. fx \in Y\}$$
$$X \cap Y \doteq \{x. x \in X \wedge x \in Y\}$$
$$X \cup Y \doteq \{x. x \in X \vee x \in Y\}$$

With elementary comprehension we can only form these where class variables and constants appear in the class position in comprehension terms.

2.6. Inductive generation. Inductive generation plays a crucial role in representation and abstraction. In particular, it enables the representation of the natural numbers, lists, queues and, more generally, inductive data types such as trees. The two axioms provide the closure and induction principles for the class:

(Clo) $\quad \exists I. i(A,R) = I \wedge (\forall y. (y,x) \in R \rightarrow y \in I) \rightarrow x \in I$
(Ind) $\quad \forall x \in A. ((\forall y. (y,x) \in R \rightarrow y \in X) \rightarrow x \in X) \rightarrow \forall x \in i(A,R). x \in X$

The recursion theorem justifies the existence of the following form of general recursion:

$$rec_I(x,f) \simeq f(x, rec_I)$$

In lambda terms this can be expressed as:

$$rec_I = Y(\lambda h. \lambda f. \lambda x. f(x,h))$$

where Y is a fixed point operator. rec_I is defined in the following circumstance. Let $Pd_R(x) = \{y. (y,x) \in R\}$. Then:

$$\forall f. (\forall x \in I. \forall g \in Pd_R(x) \Rightarrow V. f(x,g) \downarrow) \rightarrow \forall x \in I. r_I(x,f) \downarrow$$

This will justify the special cases of recursion we shall employ for various abstract data types such as natural numbers, lists, queues, finite sets, etc.

2.7. Join. The join axiom enables the formation of families of classes and will also play a role in representation and abstraction.

Join $\quad (\forall x \in A. \exists Y. fx \simeq Y)$
$\quad \rightarrow \exists Z. Z = j(X,f) \wedge \forall z. z \in Z \leftrightarrow \exists u. \exists v. z = (u,v) \wedge u \in X \wedge v \in fu$

We shall frequently write $j(A,f)$ as $\Sigma x \in A. f$.

This completes the description of T_0. This only admits elementary comprehension. T_1 is obtained by allowing stratified comprehension, i.e., extending the comprehension axiom to all stratified formula.

2.8. Types. The Boolean class facilitates the expression of the following notion that identifies those relations that are expressible or "implementable" in the theory.

Definition 1. A class C is said to be a *type* if there is a Boolean operation $eq_C \in C \otimes C \Rightarrow \mathbb{B}$ that satisfies:

$$\forall x, y \in C. \, eq_C(x, y) = true \leftrightarrow x = y$$

$\mathbb{B}$ itself is a type. This is guaranteed by $e_\mathbb{B} \in \mathbb{B} \otimes \mathbb{B} \Rightarrow \mathbb{B}$:

$$e_\mathbb{B}(x, y) \doteq cond(x, y, cond(y, false, true))$$

We also have that

Proposition 2. $\mathbb{N}$ *is a type.*

Proof. This is witnessed by the following:

$$eq_\mathbb{N}(0,0) \doteq tt$$
$$eq_\mathbb{N}(0, succ(y)) \doteq false$$
$$eq_\mathbb{N}(succ(x),0) \doteq false$$
$$eq_\mathbb{N}(succ(x), succ(y)) \doteq eq_\mathbb{N}(x, y)$$

$\square$

We shall see many other instances of classes that are types. And these will arise by representation and abstraction.

3 Structures

Representation and abstraction operate on *structures* of the following form:

$$\mathsf{C} = \langle C, o_1, \ldots, o_n \rangle$$

Here C is a class and $o_1, \ldots, o_n$ operations on C. We shall often write these more succinctly as $\mathsf{C} = \langle C, O \rangle$ where $O = \{o_1, \ldots, o_n\}$. Structures are ordered tuples of objects where the first is a class, and the others are operations of the form $o \in X^n \Rightarrow X$, i.e.,

$$Structure(c, o_1, \ldots, o_n) \doteq \exists X . c = X \wedge \bigwedge_{1 \le i \le n} (o_i \in X^{n_i} \Rightarrow X)$$

A simple example is afforded by the Booleans:

$$\mathsf{B} = \langle \mathbb{B}, tt, ff, cond \rangle$$

We admit a slight generalization of the notion of structure that permits the operations to take parameters from other classes, e.g., $o \in E_1 \otimes \ldots \otimes E_n \Rightarrow E_{n+1}$ where possibly $E_i \ne C$. We could do this in a many sorted way, but to my taste this machinery is too clumsy for the meager applications we have in mind.

Structures may include a notion of relative equality, for example:

$$\mathsf{N} = \langle \mathbb{N}, =_{\mathbb{N}}, 0, succ \rangle$$

where $x =_{\mathbb{N}} y \doteq x = y$. But such relative equality will not always be that of the underlying system T_0. In particular, abstraction introduces equality via equivalence relations.

3.1. Homomorphism. Two associated algebraic notions for structures are those of *congruence* and *homomorphism*. The relationship between these two echoes the relationship between *abstraction* and *representation*. The following is pretty standard, but it will be central to the account of representation.

Definition 3. Let $\mathsf{C} = \langle C, O \rangle$ and $\mathsf{C}' = \langle C', O' \rangle$ be two structures with the same signature. *rep*, a total function from C to C', is a *homomorphism* iff for each function $o \in C^n \Rightarrow C$ in O we have:

$$\forall x_1 \in C \ldots \forall x_n \in C . (rep(o(x_1, \ldots, x_n)) =_{C'} o'(rep(x_1), \ldots, rep(x_n))$$

where the operations take parameters from other classes, *rep* acts as the identity on these classes, i.e., if $o \in E_1 \otimes \ldots \otimes E_n \Rightarrow E_{n+1}$ then for those $E_i \ne C$ then $\forall x \in E_i . rep(x) = x$. This covers relations treated as Boolean valued functions. We employ the usual notions of *surjection, isomorphism* and *automorphism*.

3.2. Congruence. The second notion plays a role in the account of abstraction; it is the standard algebraic notion of *congruence*.

Definition 4. Let $C = \langle C, O \rangle$. A relation $R \subseteq C \otimes C$ is an *equivalence* relation on C if it is reflexive, symmetric and transitive relative to C. It is a *congruence* relation iff for each $o \in C^n \Rightarrow C$ in O,

$$\forall x_1, y_1 \in C \ldots \forall x_n, y_n \in C. (R(x_1, \ldots, y_1) \wedge \ldots \wedge R(x_n, \ldots, y_n)) \\ \to o(x_1, \ldots, x_n) =_C o(y_1, \ldots, y_n)$$

where operations take parameters from other classes, we substitute the identity relation for R on these classes, i.e., if $o \in E_1 \otimes \ldots \otimes E_n \Rightarrow E_{n+1}$ then for those $E_i \neq C$ we have: $\forall x, y \in E_i. R(x, y) \leftrightarrow x = y$.

4 Definition and representation

In set theory the term "representation" is sometimes used to refer to the encoding of mathematical notions as sets. The classic instance is the encoding of numbers as von Neumann or Zermelo numerals, e.g., 0 is encoded or represented as the empty set ∅, 1 is represented by {0, {0}}, 2 by {1, {1}}, etc. But what is the ontological status of such encoding? Within ZF itself they function as stipulative definitions (Gupta, 2021). That is, within ZF we treat the *definiens* as the *definiendum*. But in absolute terms they may not be intended to be definitions of the actual entities. And there are compelling reasons why not. Do we select the von Neumann or Zermelo numerals as the actual numbers? Each attributes a different collection of non-arithmetic, "foreign" properties to numerals, and, while there maybe good mathematical reasons for a choice, these are internal mathematical reasons. There appears to be no good ontological reason why one account is superior to another. On the other hand, both accounts cannot both be "correct" since they contradict each other, e.g., over membership. This predicament is called *Benacerraf's identification problem* (Benacerraf, 1965). Moreover, in standard expositions of set theory, the non-arithmetical properties are never used in the development of mathematics inside set theory.

However, the important distinction here is the internal/external one. From an external perspective, internal encodings are not required to be the numbers: they are intended to stand proxy for them.

Moreover, within theoretical computer science the notion of "representation" has an additional component (Dale & Walker, 1996; Jones, 1980).

4.1. Representation in computer science. Within theoretical computer science representation involves two structures: the one that is taken to represent and the represented itself. Generally, this is an internal matter to the formalism employed to articulate the structures involved. What makes one a representation of the other is the existence of a representation function (a surjective homomorphism). The requirement for the function to be a homomorphism provides the "correctness" conditions for one structure to represent the other. And part of this, to avoid triviality, we need to ensure that every element of the target domain is represented (surjective). This is the notion of representation that we shall employ.

Definition 5. Let $C = \langle C, O \rangle$ and $C' = \langle C', O' \rangle$ be two structures with the same signature. C is a *representation* of C', iff there exists a surjective homomorphism from C to C'.

We shall still employ the former notion of internal definition. In this regard, notice there is also a sense in which being a representation may not be enough to function as an internal definition. In particular, the representing structure may have a much more fine-grained notion of equality. There is no guarantee in the notion of homomorphism that the equality of the two structures agrees. Linked to this is the observation that the representing structure is likely to be more "concrete" than the represented. For instance, the von Neumann representation of the numerals contains non-numerical information such as $1 \in 2$. In any axiomatic account of the natural numbers such notions would be absent, and they play no role in arithmetic practice.

The same story emerges in T_0 but now with respect to lambda representations. Actually, the way in which numbers are treated in the two papers (Feferman, 1975; Feferman, 1979) is slightly different. In Feferman (1975), they are derived from a general scheme of inductive definition; in Feferman (1979) they are taken as a primitive notion. We have adopted the approach of Feferman (1979).

4.2. Defining versus representing the rational numbers. An example of the distinction is provided by the rationals. In computational contexts a "representation" is a mapping that links the concrete structure with the abstract one. To set things up we employ the rational numbers and their representation in T_0. To enable this we first add the rationals numbers ($\mathbb{Q}$) as new primitive class in T_0. The following are the operators of the class/type:

$$/ \in \mathbb{Z} \otimes \mathbb{Z}^+ \Rightarrow \mathbb{Q}$$
$$0_\mathbb{Q} \in \mathbb{Q}$$
$$1_\mathbb{Q} \in \mathbb{Q}$$
$$+_\mathbb{Q} \in \mathbb{Q} \otimes \mathbb{Q} \Rightarrow \mathbb{Q}$$
$$\times_\mathbb{Q} \in \mathbb{Q} \otimes \mathbb{Q} \Rightarrow \mathbb{Q}$$

Here $\mathbb{Z}^+ = \{x \in \mathbb{Z}. z > 0\}$. The equality conditions for these operators are then given as follows:

$$\forall u, w \in \mathbb{Z}. \forall v, z \in \mathbb{Z}^+. u/v =_\mathbb{Q} w/z \leftrightarrow uz =_\mathbb{Z} vw$$

The next axioms fix the identities, addition and multiplication for these operators:

$$0_\mathbb{Q} = 0/1$$
$$1_\mathbb{Q} = 1/1$$
$$\forall u, v \in \mathbb{Z}. \forall z, w \in \mathbb{Z}^+. u/z +_\mathbb{Q} v/w = (uw + vz)/wz$$
$$\forall u, v \in \mathbb{Z}. \forall z, w \in \mathbb{Z}^+. u/z \times_\mathbb{Q} v/w = uv/zw$$

This is the internal target of any representation. One representation is in terms of pairs of integers of the form $\mathbb{Q} = \mathbb{Z} \otimes \mathbb{Z}^+$. Addition and multiplication are defined as usual: $(u, z) + (v, w) \doteq (uw + vz, zw)$ and $(u, z) \times (v, w) \doteq (uv, zw)$. This representation inherits the equality of the Cartesian product of integers: we do not need to form equivalence classes as representations are not required to be (internally) definitions. Instead we require a representation function. This takes the obvious form:

$$rep(u, v) = u/v$$

Of course, there maybe many different representations. In particular, there are many different ways of defining the integers, and this duplicity will be inherited in any representation of the rationals. But how do we proceed from a representation to the "actual" structure of the rationals? Obviously, we can just add the axiomatic account by brute force. But is there a uniform way of "abstracting" the structure from a representation?

5 Abstraction

> The real numbers should not be identified with the corresponding cuts because those cuts have "wrong properties"; namely, as sets they

contain elements, something that seems "foreign" to the real numbers themselves. Similarly, the natural numbers should not be ascribed set-theoretic or other "foreign" properties; they too should be conceived of "purely arithmetically". If one wishes to pursue your approach I should advise not to take the class itself (the system of mutually similar systems) as the number (Anzahl, cardinal number), but rather something new (corresponding to this class), something the mind creates. (Reck, 2020)

Dedekind links matters to the creation of new things, and here abstraction comes to the fore. The normal context adopted by contemporary abstractionists is second-order logic, but we develop an account of abstraction that is formulated within Feferman style theories.

5.1. Abstraction in T_0. Within the framework of T_0 the principle of abstraction is constituted by the following two axioms.

(Abst) $\quad EQ(R, A) \to \exists Z. (Z = a(A, R) \wedge \forall x, y \in A. [x] =_{a(A,R)} [y] \leftrightarrow R(x, y))$

(Com) $\quad \forall x \in a(A, R). \exists y \in A. [y] =_{a(A,R)} x$

Here $EQ(R, A)$ asserts that R is an equivalence relation on A. The first axiom introduces the new class $a(A, R)$ that contains elements of the form $[x]$ where $x \in A$. It does so by providing a relative equality relation ($=_{a(A,R)}$) for the new class in terms of the equivalence relation on the base class. We sometimes write $a(A, R)$ as A/R. The second axiom is a *completeness* axiom that insists that that there are no other objects in the new class except those produced by the abstraction. This ensures that the base class is a representation of the new one.

However, we do not assume that for a given structure all the operations preserve the equivalence. The "liftable" operations of an abstraction are those that are congruent. These operations maybe extended to the abstracted class as follows:

$$\text{If } o \in A_n \Rightarrow A \text{ then define } [o] \in (A/R)_n \Rightarrow A/R,$$
$$[o]([x_1], \ldots, [x_n]) \doteq [o(x_1, \ldots, x_n)]$$

For parameters other than those from A, [ ] acts as the identity.

5.2. Abstracting types. Under what circumstances is the abstracted structure a type? Presumably, when the relation of abstraction is expressible as a Boolean function, i.e.,

$$\exists f \in C \Rightarrow \mathbb{B}. \forall x, y \in C. f(x,y) = tt \leftrightarrow R(x,y)$$

5.3. Conservative extension. We shall get to a concrete example shortly, but first we observe a central property of such abstractions.

Definition 6. Let T_0^a be T_0 plus the axioms of abstraction.

Theorem 7. T_0^a *is a conservative extension of* T_0.

Proof. We have to show: if φ is a formula of the theory T_0, that is provable in the theory T_0^a, then it is provable in T_0. We set up a translation ς from T_0^a to T_0. The main clauses involving the new type are the following. The rest translate compositionally.

$$\varsigma(c =_{a(A,R)} d) \doteq R(\varsigma(c), \varsigma(d))$$
$$\varsigma(c \in a(A,R)) \doteq R(\varsigma(c), \varsigma(c))$$

where $\varsigma([c]) \doteq c$. The mapping ς is the identity on T_0. It is then routine to check that if T_0^a proves φ then T_0 proves $\varsigma(\varphi)$. □

Abstraction does not increase the logical strength the theory; it conservatively extends its ontology.

5.4. Rationals. We have seen how the rational numbers can be represented in terms of Cartesian products of the integers. The standard way in which this is turned into an internal definition of the rationals is via equivalence classes. In contrast, Fregean abstraction would have it that the actual rational numbers arise by abstraction from the integers, and that they form a new primitive type. This is furnished by the obvious principle of abstraction where we write $[u, z]$ as u/z.

$$\forall u, v \in \mathbb{Z}. \forall z, w \in \mathbb{Z}^+. (u/z =_\mathbb{Q} v/w) \leftrightarrow uw =_\mathbb{Z} vz$$

Addition and multiplication are congruent and so maybe lifted, e.g.,

$$\forall u, v \in \mathbb{Z}. \forall z, w \in \mathbb{Z}^+. u/z =_\mathbb{Q} v/w \doteq (vw + vz)/zw$$
$$\forall u, v \in \mathbb{Z}. \forall z, w \in \mathbb{Z}^+. u/z \times_\mathbb{Q} v/w \doteq uv/zw$$

These are now definitions rather than axioms. This is a type: the following is the internal equality function.

$$eq_{\mathbb{Q}}(u/z, v/w) \doteq e_{\mathbb{Z}}(uw, vz)$$

In this way we obtain a new type with new primitive operators together with a representation.

6 The natural numbers

We have treated the natural numbers as primitives that are built into the theory T_0. Here we discuss their representation and abstraction. This is made possible via the operations of the theory and the presence of inductive definitions together with the axioms of abstraction themselves.

6.1. Representation. A well-known definition of the natural numbers in the lambda calculus is the encoding of Alonzo Church. This representation is often called "iterative". Here 0 and $succ$ are represented as follows:

$$0' \doteq \lambda f. \lambda x. x$$
$$succ' \doteq \lambda n. \lambda f. \lambda x. f(nfx)$$

The actual type maybe formed using inductive generation by putting:

$$A = \{x. x = 0' \lor \exists y. x = succ'(y)\}$$
$$R = \{(y, x). x = succ'(y)\}$$

The type of numbers is then represented as $i(A, R)$, which we write as N_{Ch}. So, we have a structure:

$$\mathsf{N}_{Ch} = \langle N_{Ch}, =_{N_{Ch}}, 0', succ' \rangle$$

where $x =_{N_{Ch}} y \leftrightarrow eq_{N_{Ch}}(x, y) = tt$ where $eq_{N_{Ch}}(x, y)$ is defined by recursion. Consequently, this structure forms a type. The recursive representation function $rep \in N_{Ch} \Rightarrow N$ is defined as follows:

$$rep(0') = 0$$
$$rep(succ'(x)) = succ(rep(x))$$

Proposition 8. N_{Ch} *is a representation of* $\mathbb{N}$.

Proof. To show that *rep* is total and surjective we employ induction on N_{Ch}—derived from its definition in terms of inductive structures.

□

Actually, in this case *rep* is an isomorphism.

Definition 9. A numerical representation is *definitional* if *rep* an isomorphism.

There are of course many representations of numbers available in combinatorial logic and the lambda calculus. Indeed, it's possible to generate denumerable variations from the Church version. But because of some of its computational properties, the "recursive" one has attracted some attention (Parigot, 1990). This is given as follows:

$$0'' \doteq \lambda f. \lambda x. x$$
$$succ'' \doteq \lambda n. \lambda f. \lambda x. fn$$

The type is similarly formed using inductive definitions by putting:

$$A = \{x. x = 0'' \vee \exists y. x = succ''(y)\}$$
$$R = \{(y, x). x = succ''(y)\}$$

Write $i(A, R)$ as N_R. The corresponding structure is given as follows:

$$\mathsf{N}_R = \langle N_R, =_{N_R}, 0'', succ'' \rangle$$

Proposition 10. N_R *is a representation of* $\mathbb{N}$.

Moreover, the iterative and recursive representations are isomorphic.

Proposition. *The following is an isomorphism:*

$$\tau \in N_{Ch} \Rightarrow N_R$$
$$\tau(0') = 0''$$
$$\tau(succ'(n)) = succ''(\tau(n))$$

So, we have two quite different representations of the natural numbers. Indeed, in principle, any of the various lambda representations may be used.

Note that in the theory with stratified comprehension we can define the representation classes by comprehension:

$$N = \{z. \forall Y. ((0 \in Y \land (\forall x \in Y. succ(x) \in Y)) \to z \in Y)\}$$

6.2. Simple abstraction. Given this, we might abstract directly on the Church representation. This is an instance where the abstraction relation is representable as a Boolean function:

$$\forall x, y \in N_{Ch}. [x] =_{a(N_{Ch},=)} [y] \leftrightarrow eq_{N_{Ch}}(x, y) = tt$$

This is a special instance of abstraction where the relation is that of equality. Consequently, the basic operations of the structure are congruent. So, we may define:

$$[succ](x) \doteq [succ'(x)]$$

Alternatively, we might abstract on the recursive one:

$$\forall x, y \in N_{Rec}. [x] =_{a(N_{Ch},=)} [y] \leftrightarrow eq_{N_{Rec}}(x, y) = tt$$

Trivially, both the Church and recursive representations are not only isomorphic to each other, but isomorphic to their respective abstractions. In fact, the authors of (Linnebo & Pettigrew, 2014) argue for such an "instantiation" requirement on any legitimate abstraction.[2]

For instance, when we apply abstraction to a particular complete ordered field, such as the field of Dedekind cuts or equivalence classes of Cauchy sequences, we get back a complete ordered field (Linnebo & Pettigrew, 2014).

6.3. Foreign properties. But this does not exclude "foreign" properties. In this regard, note that we might also lift application to an operation of the abstraction:

$$[app]([n], [m]) = [app(n, m)]$$

---

[2] More exactly, they argue for such a property for any notion of abstraction that supports "non-eliminative structuralism" in mathematics. But that is not our motivation.

This is well-defined since application is congruent. In fact, because equality is that of the underlying representation, congruence plays no role in excluding "foreign" properties. And this is so for both representations.

However, there are obviously properties of the Church representation that are not properties of the recursive one, and visa versa. For example, the Church structure validates the equality $n(m) = n^m$. This is not so in the recursive representation. Moreover, in the Church representation a "one step" addition (in terms of reduction) is definable (Parigot, 1990). And while this is a more efficient way of representing addition (in terms of reduction) this is not available in the recursive representation N_{Rec}. In contrast, in N_{Rec}, there is an efficient predecessor (Parigot, 1990). And this is not available in the Church one. This is analogous to the differences between the Zermelo and von Neumann representations in set theory, but here application is replacing set membership. All this echoes Benacerraf's remarks about set-theoretic representations (Benacerraf, 1965).

The problem is, if we restrict abstraction to a single numerical representation that is isomorphic to the original, we cannot eliminate any foreign properties. And yet the foreign properties of one representation are not identical to those of the second. We have not abstracted away from these.

6.4. Family abstraction. The obvious way forward is to somehow abstract over families of representations with the goal of excluding these foreigners. For this purpose we require the following notion.

Definition 11. A family of structures $C[i]_{i \in I}$ with the same signature is an *Isomorphism Family* if $\tau_{ij} \in C[i] \Rightarrow C[j]$ are isomorphisms.

Abstraction over such families then takes the following form:

$$\forall x, y \in (\Sigma x \in I.C). [x] =_{a(\Sigma x \in I.C, \simeq)} [y] \leftrightarrow x \simeq y$$

where $x \simeq y \doteq \tau_{ij}(p_2 x) = p_2 y$.

We may "lift" the operations of the family to the abstracted domain as follows:

$$[o]([(x_1, i)], \ldots, [(x_n, i)]) \doteq [o^i(x_1, \ldots, x_n)]$$

where o^i is an operation in C[i]. Since τ_{ij} are isomorphisms, $\simeq$ is a congruence.

We can now apply this to our natural number representations.

Definition 12. A *numerical family* is a family of numerical representations $N[i]_{i \in I}$ with structure:

$$N[i] = \langle N[i], 0^i, succ^i \rangle$$

where $\tau_{ij} \in N[i] \Rightarrow N[j]$ is a family of isomorphisms between them.

Abstraction then proceeds as above:

$$\forall x, y \in (\Sigma x \in I. N). [x] =_{\Sigma x \in I. N/\simeq} [y] \leftrightarrow x \simeq y$$

where $x \simeq y \doteq \tau_{ij}(p_2 x) = p_2 y$. We are then able to abstract those operations of the structure that are congruent, e.g.,

$$[succ][x] \doteq [succ^i(x)]$$

Theorem 13. *Let $N[i]_{i \in I}$ be a numerical family. Then $a(\Sigma x \in I. N, \simeq)$ satisfies the axioms of the natural numbers given in T_0.*

But how exactly does this exclude any foreign material? In this regard, reconsider the addition of application to the lifted operations:

$$[app]([n], [m]) = [app(n, m)]$$

Assume the Church and recursive representations are members of the family. Then *app* is no longer well-defined: in the Church representation $app(n, m) = n^m$ but this is not so in the recursive one. This is now a foreign operator. Of course, we cannot claim to have excluded all foreign operators. What is excluded depends upon the choice of the indexed family of representations—and what representations are available in the theory of combinators/Feferman's theories.

However, under certain conditions we do maintain instantiation.

Definition 14. An *isomorphism family* is *rigid* if for every $i \in I$, the only automorphism is the identity.

Proposition 15. *Every definitional representation of the natural numbers given in T_0 is rigid.*

Proof. Let f be any isomorphism. Employ induction to prove $\forall n. f(n) = n$. The base case is the only concern. Argue by contradiction. Suppose that $f(0) \neq 0$. Because f is a homomorphism, $f(1) = succ(f(0))$. So $f(1) \neq 0$. And inductively, $\forall n. f(n) \neq 0$. Hence zero cannot be in the co-domain of f. This contradicts f being surjective. □

Indeed, any numerical representation that is obtained by inductive generation will satisfy induction, and so will be rigid. For rigid structures instantiation is maintained.

Proposition 16. *If the family $C[i]_{i \in I}$ is rigid then the abstraction satisfies for each i,*

$$C[i] \simeq a(\Sigma x \in I. C, \simeq)$$

Proof. By definition,

$$a(\Sigma x \in I. C, \simeq) = [C[i]]$$

where $[C[i]] \doteq \{x \in a(\Sigma x \in I. C, \simeq). \exists y \in C[i]. x = [y]\}$. Claim that the function $x \mapsto [x]$ is an isomorphism from $C[i]$ onto $[C[i]]$. Assume that $[x] = [y]$. It follows that $p_1 x = p_1 y$ and $\tau_{ii}(p_2 x) = p_2 y$. By rigidity, $x = y$. □

Corollary 17. *Let $N[i]_{i \in I}$ be an indexed family of definitional numerical representations. Then $N[i] \simeq a(\Sigma x \in I. N, \simeq)$.*

7 Containers

"Containers" allow the storage, insertion and access to objects. Examples of these abstract types include queues, stacks, lookup tables, lists and arrays. We shall illustrate matters with lists and queues as they represent two rather different structures in terms of the way they store and retrieve objects. They further illustrate the process of abstraction that parallels the natural numbers. However, our reason for selecting them as a case study in abstraction goes further: we employ them not only to further unpack the process of abstraction, but to illustrate how we might move to higher levels of abstraction.

7.1. Lists. The polymorphic version allows lists to be created for any given type. For instance, we might form lists of numbers $L[\mathbb{N}]$ or lists of Boolean values $L[\mathbb{B}]$, etc. This has the following structure:[3]

$$L[X] = \langle L[X], nil, cons, head, tail, =_X, =_{Q[X]} \rangle$$

The operations have the following functionalities:

$$nil \in L[X]$$
$$cons \in X \otimes L[X] \Rightarrow L[X]$$
$$head \in L[X] \Rightarrow X$$
$$tail \in L[X] \Rightarrow L[X]$$

Special equality conditions are given by the following axiom:

$$\forall x \in X. \forall y \in L[X]. (head(cons(x,y)) =_X x) \wedge (tail(cons(x,y)) =_{Q[X]} y)$$

Finally, we assume list induction:

$$\forall X. ((nil \in X \wedge \forall x \in Y. \forall y \in X. cons(x,y) \in X) \rightarrow \forall x \in L[Y]. x \in X)$$

Notice that in T_0 this only gives us induction for elementary formula. Recursion on lists is provided by an instance of the general recursion scheme:

$$lisrec \doteq \lambda f. \lambda z. \lambda x. cond(nil(z), x, f(head(x), listrec(f, tail(z)), x))$$

Given definition by cases, we are able to write these as recursion equations. For example, we have equality for the class as an instance:

$$eq(nil, nil) \doteq true$$
$$eq(cons(a,l), nil) \doteq false$$
$$eq(nil, cons(a,l)) \doteq false$$
$$eq(cons(a,l), cons(b,k)) \doteq eq(a,b) \wedge eq(l,k)$$

Here we assume that there is an equality function for the base type. The following is then clear.

---

[3] More formally, we should employ the mechanism of Feferman (2009) and Feferman (1991) to represent such structures. But we shall not fuss over this here.

Proposition 18. *Lists over a type form a type.*

7.2. Queues. These have the same signature as lists but have a different mechanism for selection: lists operate a "last-in" and "first-out" regime: queues reverse matters and employ a "last-in" and "last-out" one:

$$Q[X] = \langle Q[X], emp, enqueue, front, dequeue, =_X, =_{Q[X]} \rangle$$

The functions of the structure také the following form:

$$emp \in Q[X]$$
$$enqueue \in (Q[X] \otimes X) \Rightarrow Q[X]$$
$$dequeue \in Q[X] \Rightarrow Q[X]$$
$$front \in Q[X] \Rightarrow X$$

They are taken to satisfy the following equality axioms:

$$\forall x \in X.\, front(enqueue(emp, x)) =_X x \wedge dequeue(enqueue(emp, x)) =_{Q[X]} emp$$
$$\forall q \in Q'[X].\, \forall x \in X.\, front(enqueue(q, x)) =_X front(q)$$
$$\forall q \in Q'[X].\, dequeue(enqueue(q, x)) =_{Q[X]} enqueue(dequeue(q), x)$$

Here *isemp* is definable by recursion—where recursion on queues follows an analogous form to that for lists. And:

$$Q'[X] \doteq \{x \in Q[X].\, not(isemp(q)) = tt\}$$

In what follows we illustrate everything with lists, but it all applies to queues.

7.3. Representation. As in the case of numbers there are many ways of representing lists/queues in the theory of combinators, including the iterative and recursive ones (Parigot, 1990). To illustrate matters we again employ the Church representation, this time for lists:

$$nil \doteq p(tt)(tt)$$
$$cons \doteq \lambda h.\, \lambda x.\, pff(phx)$$
$$head \doteq \lambda z.\, p_1(p_2 z)$$
$$tail \doteq \lambda z.\, p_2(p_2 z)$$

Notice that these are implicitly polymorphic, i.e., work for all parameter types. The representation of the actual type is provided using inductive definitions:

$$List[X] = i(A[X], R[X])$$
$$A[X] = \{x = nil \vee \exists y \in X. x = cons(y, z)\}$$
$$R[X] = \{(y, x). \exists u \in X. x = cons(u, y)\}$$

Closure follows from the closure axiom for inductive generation, and we obtain induction for elementary formulae. Using recursion, we can define a recursive function from the representation to the actual type of lists:

$$rep \in List[X] \Rightarrow L[X]$$
$$rep(nil) = nil$$
$$rep(cons(x, y)) = cons(x, rep(y))$$

The following uses the same pattern of proof as that of the natural numbers.

Proposition 19. *Every definitional representation of lists is rigid.*

7.4. Abstraction. Our next objective is to abstract the type of lists. For this we again follow the lead of the natural numbers. As before we abstract over a family, $L[X, i]_{i \in I}$, of list representations with parameter X:

$$\forall x \in L[X, i]. \forall y \in L[X, j]. [x] =_{L[X]} [y] \leftrightarrow x \simeq y$$

where $x \simeq y \doteq \tau_{ij}(p_2 x) = p_2 y$, i.e., generated by the recursive isomorphisms between the list representations. From the general theory of family abstraction, the analogue of theorem 13 holds.

Theorem 20. *Let $L[X, i]_{i \in I}$ be an indexed family of definitional list representations. Then:*

(i) $L[X, i] \simeq a(\Sigma x \in I. L[X. i], \simeq)$
(ii) $a(\Sigma x \in I. L[X, i], \simeq)$ *satisfies the axioms for lists.*

But what are the fundamental properties that all such container types have in common? The problem is that there is no isomorphism that preserves all the operations.

7.5. Container types. Each member of the family of container structures $C[i]$, of which queues, stacks, lookup tables, lists and arrays are examples, has the following structure:

$$C[X,i] = \langle C[X,i], empty^i, insert^i, select^i, rest^i, =_X, =_{C[X,i]} \rangle$$

The *insert* operator inserts a single element of X, *select* selects an element, and *rest* returns the rest. They are governed by the same induction scheme as that of lists and queues, i.e., for elementary formula. For convenience later we state this as a schema with formula instead of classes:

$$((\varphi(empty^i) \wedge \forall x \in X. \forall y \in S[X,i]. \varphi(y) \to \varphi(insert^i(x,y)))) \to \forall x \in S[X,i]. \varphi(x)$$

Similarly, each is guaranteed to satisfy a parallel recursion scheme to lists and queues. Finally, each container type is governed by axioms that determine the relationships between insertion and selection/rest operators. From the perspective of their destructors they are a heterogeneous mix.

What do these things have in common? They are certainly not isomorphic.

7.6. Finite sets. Recursion on container types supports definitions of membership, quantification and extensional equivalence:

$$\in^i (empty^i, x) \doteq false$$
$$\in^i (insert^i(a,s), x) \doteq cond(eq(a,x), true, \in^i (x,s))$$
$$\forall x \in empty^i. g(x) \doteq true$$
$$\forall x \in^i insert^i(a,s). g(x) \doteq g(a) \wedge \forall x \in^i s. g(x)$$
$$u \equiv^i v \doteq \forall x \in^i u. x \in^i v) \leftrightarrow (\forall x \in^i v. x \in^i u)$$

These are local to each container type. However, extensional equivalence can be generalized to a notion that operates between container types:

$$u \equiv^{ij} v \doteq (\forall x \in^i u. x \in^j v) \leftrightarrow (\forall x \in^j v. x \in^i u)$$

This permits the abstraction of the notion of "finite set" from these container structures. Let $\langle S[X,i] \rangle_{i \in I}$ be an indexed family of container structures. Then we abstract as follows:

$$\forall x \in S[X,i]. \forall y \in S[X,j]. [x] =_{S[X]} [y] \leftrightarrow x \equiv^{ij} y$$

This gives the abstracted structure with the following signature:

$$S[X] = \langle S[X], \emptyset, \oplus, =_X, =_{S[X]} \rangle$$

where

$$\emptyset \doteq [empty^j]$$
$$\forall x \in X. \forall y \in S[X,j]. \oplus (x,[y]) \doteq [insert^i(x,y)]$$

Sets form a type. They also inherit a restricted induction principle, i.e., limited to elementary congruent formula, where an elementary formula φ is congruent exactly when the following holds:

$$\forall x \in S[X,i]. \forall y \in S[X,j]. x \equiv^{ij} y \rightarrow (\varphi^i(x) \leftrightarrow \varphi^j(y))$$

where φ^i is a formula that has the vocabulary of the structure $S[X,i]$.

For such formula we can lift to the abstract structure:

$$\forall x \in S[X,i]. [\varphi]([x]) \doteq \varphi^i(x)$$

For congruent elementary formula, we then have the following derived induction principle for sets:

$$([\varphi](\emptyset) \wedge (\forall x \in X. \forall y \in S[X]. [\varphi][y] \rightarrow [\varphi](\oplus (x,y))) \rightarrow \forall x \in S[X]. [\varphi](x)$$

Recursion can also be applied to sets. However, again some care is required in the formulation of permitted recursions. On the assumption that f is congruent, i.e.,

$$\forall u \in S[X,i]. \forall v \in S[X,j]. \forall x \in X. u \equiv^{ij} v \rightarrow (f^i(x,u) = f^j(x,v))$$

we may lift the operation to the abstract structure:

$$\forall u \in S[X,i]. \forall x \in X. [f]([u],x) \doteq [f^i(u,x)]$$

Then we have a derived recursion scheme where we put:

$$\forall x \in S[X]. [rec]_{g,[f]}([x]) \doteq [rec_{g,f}(x)]$$

Unpacked, and writing Rec for $[rec]$, this yields the following recursion:

$$Rec(\emptyset) = g$$

$$\forall u \in S[X]. \forall x \in X. Rec_{g,[f]}(\oplus (x,u)) = [f](x, Rec_{g,[f]}(u))$$

This form of recursion supports membership ($\in$) and quantification with the same definition format as for general containers. Given this, we may define Boolean equality for sets:

$$\forall u, v \in S[X]. u \equiv v \doteq (\forall x \in u. x \in v) \wedge (\forall x \in v. x \in u)$$

Finally, we can use induction on sets to show that extensional holds.

Proposition 21. *Sets are extensional, i.e.,*

$$\forall x, y \in S[X]. x =_{S[X]} y \leftrightarrow x \equiv y$$

So, sets behave as expected. Feferman (2009) provides an account of the abstract type of finite sets where a certain "density" requirement is added. This is replaced here by the completeness demand on abstractions. We shall see this connection more explicitly in the next case study.

8 Real numbers

The constructive real numbers provide an interesting case study in abstraction in that the representation function is an essential part of what it is to be a constructive real number: to be constructively given a real number r, we must be given a rule for calculating it from its rational base to any desired degree of accuracy. In this section we show that Feferman's account of the constructive reals given in Feferman (1992) and Feferman (1991) is exactly the notion obtained from Cauchy sequences by abstraction.

8.1. The Constructive real numbers. Feferman (1992) and Feferman (1991) provides such an account that is inspired by that of Bishop (Bishop & Bridges, 1985). In this account the Cauchy sequences are the "approximations" of the of reals. These are defined as pairs consisting of a sequence of rationals together with a function for uniform convergence:

$$Cauchy \doteq \{(x, g) \in (\mathbb{N}^+ \Rightarrow \mathbb{Q}) \otimes (\mathbb{N}^+ \Rightarrow \mathbb{N}^+). \forall k, n, m \in \mathbb{N}^+.$$
$$n, m > g(k) \rightarrow |x_n - x_m| < \frac{1}{k}\}$$

The identities, addition and multiplication on these proceed in the standard way. For example, for $(x, g), (x', g') \in Cauchy$,

$$(x, g) + (x', g') \doteq (\lambda u.\, xu + x'u,\, \lambda u.\, max(g(u), g'(u)))$$

The definitions of 0,×,< follows suite. These yield a new class:

$$\text{Cauchy} = \langle Cauchy, +, \times, 1, 0, <, = \rangle$$

But this is not the class of reals; it is the class of the "approximations" to them. We can take equality to be that of the following Cartesian product: $(\mathbb{N}^+ \Rightarrow \mathbb{Q}) \otimes (\mathbb{N}^+ \Rightarrow \mathbb{N}^+)$.

The reals themselves are to be taken as a new primitive structure:

$$\mathsf{R} = \langle \mathbb{R}, +_\mathbb{R}, \times_\mathbb{R}, 1_\mathbb{R}, 0_\mathbb{R}, <_\mathbb{R}, =_\mathbb{R}, lim \rangle$$

where the operations on the reals are taken as primitive, and obey the axioms of a "complete" ordered field where "completeness" is understood constructively, and is fixed by the presence of the *lim* operation:

$$lim \in Cauchy \Rightarrow \mathbb{R}$$

This is taken to be a homomorphism from the Cauchy sequences to the reals. Equality for the reals is given in terms of this homomorphism.

For $(x, g), (x', g') \in Cauchy$,

$$lim(x, g) =_\mathbb{R} lim(x', g') \leftrightarrow \forall k, n, m \in \mathbb{N}^+.\, n, m > g(k) \rightarrow |x_n - x'_m| < \frac{1}{k}$$

In Feferman's account there is a further "density" condition:

$$\forall x \in \mathbb{R}.\, \exists y \in Cauchy.\, lim(y) =_\mathbb{R} x$$

that guarantees that every constructive real is the limit of a sequence of approximations. In other words, *lim* has to be a surjective homomorphism.[4] Spelled out this may be taken as an axiomatization of the constructive real numbers. But now we see that the class of reals has a built in representation function, i.e.,

---

[4] Feferman puts this in terms of a new operation $\forall x \in \mathbb{R}.\, \forall k \in \mathbb{N}^+.\, |x_n - approx(x, k)| < \frac{1}{k}$ where $\mathbb{Q}$ is injected into $\mathbb{R}$. But this is a hidden operation that can be pulled out from "between the lines".

$$lim \in Cauchy \Rightarrow \mathbb{R}$$

i.e., the Cauchy class is a representation. But notice that the constructive reals are not an equivalence classes of Cauchy sequences. They are a new primitive class. Finally, observe that equality for the reals is not implementable; it is not a type.[5]

8.2. Abstracting the constructive real numbers. The correctness of this account is given additional credibility by the fact that it is exactly the account of the constructive reals that one obtains by Fregean abstraction. More explicitly, given the Cauchy structure:

$$\mathsf{Cauchy} = \langle Cauchy, +, \times, 1, 0, < \rangle$$

we may reverse engineer matters, and abstract the constructive reals via the standard Cauchy equality condition. For $(x, g), (x', g') \in Cauchy$,

$$lim(x, g) =_\mathbb{R} lim(x', g') \leftrightarrow \forall k, n, m \in \mathbb{N}^+. n, m > g(k) \rightarrow |x_n - x'_m| < \frac{1}{k}$$

By congruence, we define addition and multiplication for the reals by addition and multiplication on their approximations: for $(x, g), (x', g') \in Cauchy$,

$$lim(x, g) +_\mathbb{R} lim(x', g') \doteq lim((x, g) + (x', g'))$$

In addition, the principle of abstraction guarantees that every real number is the limit of such a sequence:

$$\forall r \in \mathbb{R}. \exists f \in \mathbb{R}. r =_\mathbb{R} lim(f)$$

i.e., density is guaranteed by surjectivity. So every regular sequence gives rise to a real number, and real numbers only arise as the result of such approximations. So, we have precisely Feferman's account.

The idea that computational abstraction, unlike mathematical abstraction, should leave behind an "implementation" trace is argued for as part of the informal discussion of computational abstraction given in Colburn & Shute (2007). They argue that

---

[5] However, the equality for the Cauchy sequences is that of the underlying types of operations. There are implementations of this in the typed lambda calculus where every term is strongly normalizing.

mathematical abstraction jettisons the concrete structure in favor of the more abstract one, and any hint of the representation is lost. In contrast, somehow computational abstraction holds on to it. However, matters appear to be more subtle than this crude division suggests. Fregean abstraction always leaves behind a representation trace. The very act of abstraction constructs a representation as part of the process of creating the more abstract structure. And this applies to all forms of abstraction. Of course, this still leaves the possibility of discarding the representation in the mathematical cases. But this does not occur in the case of the real numbers given above: the representation is actually part of the very notion of computational real number. Perhaps abstraction brings out the difference between classical and constructive analysis.

9 Representation and abstraction

We can go back and forth between equivalence relations and the representation functions; back and forth between abstraction and representation However, although they are mathematical cousins, there are significant conceptual differences. The "intentional stance" is different. The central issue concerns correctness. Under the representational stance, it is the concrete one that is "correct" relative to the abstract one. The correctness is evidenced by the existence of the surjective homomorphism from the concrete to the abstract. If matters go wrong we have to modify the concrete structure. Under abstraction, it is the abstraction that is correct relative to the more concrete object. The congruence relation placed upon the concrete structure determines what operations lift to the abstracted structure. The intentional stance dictates what is correct relative to what. It determines what governs what and what happens when things go awry. In the representational one we modify the encoding; in the abstraction scenario we abstract only in accord with the abstraction relation and the properties of the more concrete structure—if we get it wrong we modify the abstraction. In so far as the aim of the semantic enterprise is provide conditions of correctness (Boghossian, 1989), these differences of correctness are closely associated with the semantic ones. Under representation, the concrete structure is not intended as a semantic interpretation of abstract one. The abstract one provides the correctness conditions for the more concrete representation. The "intentional stance" dictates matters (Turner, 2020). In contrast, in abstraction, the concrete structure does provide the semantic grounding of the more abstract one.

Moreover, not only are there semantic differences between the two notions, but there are fundamental ontological ones. Representation assumes that the two structures are in place. On the face of it, abstraction is a process that creates new structures. There is a caveat here. The new structures arise by application of the axiom of abstraction. In a static sense, the axiom already sanctions all such structures. At least it has the potential to generate them. But in a dynamic sense it does not. But to express this difference we need to embed Feferman's theories in a form of modal logic (Linnebo, 2018) that can that articulate the difference between "actual" and "potential existence".

10 Further work

More cases studies are needed.

On the computational side topics to explore include "streams", "trees" and "inductive" data types in general. Indeed, Feferman's theory as a host theory is much richer than we have so far exploited.

On the mathematical side, and it is hard to keep these two completely apart, more cases studies internal to Bishop's mathematics need to be explored. It would be interesting to employ abstraction to design an actual programming language that is geared towards Bishop's mathematics. Indeed, the development of such a language is suggested in Feferman (1992). Finally, the step from the constructive real numbers to Euclidean geometry via invariant transformations is an instance of abstraction. It would be pleasant to unpack this in some detail. And there is more.

Bibliography

Abrial, J. R. (1988). The B tool (abstract). In R. E. Bloomfield, L. S. Marshall, & Roger B. Jones (Eds.), *VDM – The Way Ahead, Proc. 2nd VDM-Europe Symposium. Lecture Notes in Computer Science, 328* (pp. 86–87). Berlin: Springer.

Angius, N., & Primiero, G. (2018). The logic of identity and copy for computational artefacts. *Journal of Logic and Computation, 28*(6), 1293–1322.

Benacerraf, P. (1965). What numbers could not be. *The Philosophical Review, 74*(1), 47–73.

Beeson, M. (1985). *Foundations of Constructive Mathematics*. Berlin: Springer.

Bishop, E., & Bridges, D. S. (1985). *Constructive Analysis*. Berlin: Springer.

Bjørner, D., & Jones, C. B. (Eds.). (1978). *The Vienna Development Method: The Meta-Language. Lecture Notes in Computer Science, 61*. Berlin: Springer.

Boghossian, P. A. (1989). The rule-following considerations. *Mind, 98*(392), 507–549.

Colburn, T., & Shute, G. (2007). Abstraction in computer science. *Minds & Machines, 17*, 169–184.

Dale, N., & Walker, H. M. (1996). *Abstract data types: Specifications, implementations, and applications*. Boston: Jones & Bartlett Learning.

Ebert, P., & Rossberg, M. (2016). *Abstractionism. Essays in Philosophy of Mathematics.* Oxford: Oxford University Press.

Feferman, S. (1979). Constructive theories of functions and classes. In M. Boffa, D. van Dalen, K. McAloon (Eds.), *Logic Colloquium 78* (pp. 159–224). Amsterdam-New York: North-Holland.

Feferman, S. (1975). A language and axioms for explicit mathematics. In J. Crossley (Ed.), *Algebra and Logic, Lecture Notes in Mathematics, 450* (pp. 87–139). Berlin: Springer.

Feferman, S. (1992). A new approach to abstract data types. Informal development. *Mathematical Structures in Computers Science, 2*(2), 93–229.

Feferman, S. (1991). A new approach to abstract data types II computation on ADTs as ordinary computation. In E. Börger, G. Jäger, H. Kleine Büning, & M. M. Richter (Eds.), *Computer Science Logic (CSL). Lecture Notes in Computer Science, 626.* Berlin: Springer.

Feferman, S. (1995). Definedness. *Erkenntnis, 43*(3), 295–320.

Floridi, L. (2008). The method of levels of abstraction. *Minds & Machines, 18*(3), 303–329.

Fine, K. (2008). *The Limits of Abstraction.* Oxford: Oxford University Press.

Frege, G. (2003). *Posthumous writings.* Translated by P. Long & R. M. White. Oxford: Basil Blackwell.

Falguera, J. L., Vidal, C. M., & Rosen, G. (2021). Abstract objects. In E. N. Zalta (Ed.), *The Stanford Encyclopedia of Philosophy* (*Winter 2021 Edition*).

Gupta, A. (2021). Definitions. In E. N. Zalta (Ed.), *The Stanford Encyclopedia of Philosophy* (*Winter 2021 Edition*).

Hale, B., & Wright, C. (2001). *The Reason's Proper Study. Essays toward a Neo-Fregean Philosophy of Mathematics.* Oxford: Oxford University Press.

Heck, Jr. R. G. (1993). The development of arithmetic in Frege's Grundgesetze der Arithmetik. *Journal of Symbolic Logic, 58*(2), 579–601.

Hindley, R., & Seldin, J. P. (1986). *Lambda-Calculus and Combinators, An Introduction.* Cambridge: Cambridge University Press.

Jones, C. B. (1980). *Software Development. A Rigorous Approach*: Oxford: Prentice Hall.

Linnebo, Ø. (2018). *Thin Objects: An Abstractionist Account.* Oxford: Oxford University Press.

Linnebo, Ø., & Pettigrew, R. (2014). Two types of abstraction for structuralism. *The Philosophical Quarterly, 64*(255), 267–283.

Liskov, B., & Zilles, S. (1974). Programming with abstract data types. In B. Leavenworth (Ed.), *Proceedings of the ACM SIGPLAN Symposium on Very High Level Languages* (pp. 50–59). New York, NY: Association for Computing Machinery.

Mancosu, P. (2016). *Abstraction and Infinity.* Oxford: Oxford University Press.

Parigot, M. (1990). On the representation of data in lambda-calculus. In E. Börger, H. K. Büning, M. M. Richter (Eds.), *CSL '89. CSL 1989. Lecture Notes in Computer Science, 440* (pp. 309–321). Berlin: Springer.

Reck, Erich. (2020). Dedekind's Contributions to the Foundations of Mathematics. In E. N. Zalta (Ed.), *The Stanford Encyclopedia of Philosophy (Winter 2020 Edition)*.

Spivey, J. M. (1992). *The Z Notation: A Reference Manual* (2nd ed.). Oxford: Prentice Hall.

Thomas, P., Robinson, H., & Emms, J. (1988). *Abstract Data Types: Their Specification, Representation, and Use*. Oxford: Clarendon Press.

Turner, R. (2018). *Computational Artifacts. Towards a Philosophy of Computer Science*. Berlin: Springer.

Turner, R. (2021). Computational abstraction. *Entropy, 23*(2), 213.

Turner, R. (2020). Computational intention. *Studies in Logic, Grammar and Rhetoric, 63*(1), 19–30.

Wright, C. (1983). *Frege's Conception of Numbers as Objects*. Aberdeen: Aberdeen University Press.

Author biography. Raymond Turner, Emeritus Professor of Logic and Computation, University of Essex, is an English logician, philosopher, and theoretical computer scientist. He is best known for his work on logic in computer science and for his pioneering work in the philosophy of computer science.

Symposia and Thematic Panels

Symposia and Thematic Panels Organized at the Congress

(a) Summaries of Symposia and Panels[1]

(i) DLMPST/IUHPST Commissions' Symposia

Special session for the IUHPST Essay Prize: What Is the Value of History of Science for Philosophy of Science?
Symposium of the Joint Commission of the IUHPST 1
Organiser: Hasok Chang (University of Cambridge)

This special symposium, chaired by Hasok Chang (University of Cambridge, UK) as the Chair of the DLMPST-DHST Joint Commission, featured the winning entry for the second IUHPST Essay Prize in History and Philosophy of Science. This prize competition seeks to encourage fresh methodological thinking on the history and philosophy of science as an integrated discipline. The winner of the 2019 prize was the essay entitled 'Negotiating History: Contingency, Canonicity, and Case Studies', by Dr. Agnes Bolinska and Dr. Joseph D. Martin of the University of Cambridge. The runner-up was the essay 'History and Philosophy of Science after the Practice-Turn: From Inherent Tension to Local Integration' by Mr. Max W. Dresow of the University of Minnesota.

Agnes Bolinska and Joseph Martin's winning essay begins with an insightful and systematic typology of the many difficulties faced by the case-study method in the history and philosophy of science. In discussing remedies to these difficulties, Bolinska and Martin focus on the 'metaphysical' type of worry: "what if history itself is just inherently unsuited to providing evidential support for philosophical claims?" The core of their response is a consideration of 'canonicity': a case study can be philosophically informative if it is canonical with respect to a particular philosophical aim. A historical case is canonical when its philosophically salient features provide a good causal account of the scientific process in question. And the appreciation of relevant historical contingencies is crucial for the identification of a canonical case. Bolinska and Martin offer an illuminating analysis of the concept of contingency, disambiguating it from the notion of chance and showing how it is crucially implicated in the historical-cum-philosophical explanation of past scientific episodes. Their essay addresses the prize

---

[1] Unless indicated otherwise, the summaries were authored by the organisers of the symposia.

question directly: what can the history of science do for the philosophy of science? Bolinska and Martin offer a subtle and original answer: working through a canonical case helps philosophers reach a clearer understanding of the philosophical issues in question and provide evidence for or against particular epistemological claims about science. The adeptness with which both historical and philosophical concerns are handled in this essay is a clear sign of a productive collaboration between the co-authors across the philosophy/history boundary.

The runner-up essay by Max Dresow conveys an excellent sense of the history of the methodological debates, demonstrating a firm command of the relevant literature and providing an insightful overall perspective on the problem of history-philosophy interaction. Dresow goes beyond the delightful diagnosis of the problem that he provides, by articulating how history is used in three distinct modes of practice-based philosophy of science: the functional-analytic approach, the integrative history of the recent, and the phylogenetic approach. All three approaches are characterized in a way that is suggestive and instructive for both historians and philosophers. An important general message emerges from Dresow's discussion: the history-philosophy relation looks inherently problematic only if we are trying to discern an overall relationship at the level of whole disciplines; these worries largely dissipate when we consider how historical sources and facts are used 'locally' in pursuit of specific philosophical aims.

At the symposium Dr. Bolinska and Dr. Martin presented the content of their winning entry, highlighting the following points. Recent work on the use of historical case studies as evidence for philosophical claims has advanced several objections to this practice. Their two-fold goal was first to systemize these objections, showing how an appropriate typology can light the path toward a resolution, and second, to show how some of these objections can be recast as advantages for the historically sophisticated philosopher, specifically by describing how attention to contingency in the historical process can ground responsible canonicity practices.

Systematizing objections to the use of historical case studies for philosophical ends shows that they fall largely into two categories: methodological objections and metaphysical objections. The methodological objections are not unique to the use of historical cases—they would also apply to other forms of philosophical reasoning. Case studies demand responsible handling, but this is unsurprising. History is messy and philosophy is difficult. But the need for care is hardly the mark of a hopeless endeavor. Rather, attention to the ways in which history is messy and in which philosophy is difficult can be resources for developing better historiographical and philosophical practices.

Metaphysical objections do, however, raise special problems for the use of historical case studies. Attention to what makes for a canonical case can address these problems. A case study is canonical with respect to a particular philosophical aim when the philosophically salient features of the historical system provide a reasonably complete causal account of the results of the scientific process under investigation. Dr. Bolinska and Dr. Martin showed how to establish canonicity by evaluating relevant contingencies using two prominent examples from the history of science: Eddington's confirmation of Einstein's theory of general relativity using his data from the 1919 eclipse and Watson and Crick's determination of the structure of DNA. These examples suggest that the analogy between philosophical inquiry and the natural sciences, although imperfect, has important elements that make it worth retaining. This is not to say that we should think of philosophy as modeled on scientific practice, but rather that both succeed by virtue of something more general: their reliance on shared principles of sound reasoning.

Taking seriously the practices necessary to establish the canonicity of case studies makes clear that some examples of the historical process of science are more representative of its general ethos than others. With historiographical sense, we can pick these examples out. Doing so requires attention to the contingencies of history. Rather than undermining the use of historical cases, philosophical attention to contingency aids the development of case studies as resources by making explicit otherwise tacit assumptions about which features of them are most salient and why.

These considerations help us address the question of the value of history of science for the philosophy of science. It is possible, even easy, to use the rich resources that history provides irresponsibly to make a predetermined point. But that is not a genuine case of history of science informing philosophy of science—in part because it proceeds in the absence of historiographical sense. By outlining the practices that render particular cases canonical for certain philosophical aims, the authors offered a route by which such sense can be integrated into standard philosophical practices.

Can the History of Science Be Used to Test Philosophy?
Symposium of the Joint Commission of the IUHPST 2
Organiser: Hasok Chang (University of Cambridge)

This symposium, chaired by Jouni-Matti Kuukkanen (University of Oulu, Finland) examined the evidential relations between history and philosophy from various angles. Can the history of science show evidential support and falsifications for the philosophical theories about science? Or is it always a case of stalemate in which each reconstruction of history is only *one* possible reconstruction amongst several others? One suggestion has naturally been that the whole approach aimed at testing and comparing alternative philosophical models by recourse to historical data is misguided at worst, or in need of serious reformulation at best.

The tradition that looms large over this discussion is the attempt to turn philosophy of science into an empirically testable discipline. History and philosophy of science is then understood as a science of science in a close analogy to the natural sciences. One view is that philosophers provide theories to test and historians produce data by which these theories are tested. The most vocal and well-known representative of this approach is the VPI (Virginia Polytechnic Institute) project. The two most notable publications of this endeavour are 'Scientific Change: Philosophical Models and Historical Research' and *Scrutinizing Science: Empirical Studies of Scientific Change*.[2] A conference organised in 1986 preceded the latter publication. The key idea is testability; that historical case studies perform the role of empirical validation or falsification of the philosophical models of science. In this way, case studies were meant to provide 'a reality check for philosophy of science.'[3]

It is the role and status of case studies, and the rationale using case studies, that is brought back to the table and in the locus of this symposium. More generally, the authors are probing the appropriate evidential relationship between history and philosophy. The symposium makes evident a new sticking point in the debate regarding the empirical accountability of philosophical theories: Should very recent science rather than the history of science function as a source of empirical information? Or should we rather focus on finding more sophisticated evidential modes for the history of science?

Four papers were presented. First Raphael Scholl (University of Cambridge, UK) spoke on 'Scenes from a Marriage: On the Confrontation Model of History and Philosophy of Science'. According to the 'confrontation model', integrated history and philosophy of science operates like an empirical science. It tests philosophical accounts of science against historical case studies much like other sciences test theory against data. However, the confrontation model's critics object that historical facts can neither support generalizations nor genuinely test philosophical theories. According to Dr. Scholl, most of the model's defects can be traced back to its usual framing in terms of two problematic

---

[2] Larry Laudan et al. (1986), Scientific change: Philosophical models and historical research, *Synthese*, *69*, 141–223; Arthur Donovan & Larry Laudan, *Scrutinizing Science: Empirical Studies of Scientific Change* (Dordrecht: Kluwer Academic Publishers, 1988).
[3] Jutta Schickore (2018), Explication work for science and philosophy, *Journal of the Philosophy of History*, *12*, 1, 4.

accounts of empirical inference: the hypothetico-deductive method and enumerative induction. This framing can be taken to suggest an unprofitable one-off confrontation between particular historical facts and general philosophical theories. He outlined more recent accounts of empirical inquiry, which describe an iterative back-and-forth movement between concrete (rather than particular) empirical exemplars to their abstract (rather than general) descriptions. Reframed along similar lines, the confrontation model continues to offer both conceptual insight and practical guidance for a naturalized philosophy of science.

Luca Tambolo (Independent scholar, Italy) gave a presentation on 'The Problem of Rule-Choice Redux'. This paper tackled the contribution that history of science can make to the *problem of rule-choice*, i.e., the choice from among competing methodological rules. Taking his cue from Larry Laudan's writings, Dr. Tambolo discussed what he calls *historicist naturalism*, i.e., the view that history of science plays a pivotal role in the justification of rules, since it is one source of the evidence required to settle methodological controversies. He presented cases of rule-choice that depend on conceptual considerations alone, and in which history of science does not factor. Moreover, there are cases in which methodological change is prompted—and explained—by empirical information that is not historical in nature: as suggested by what we call *scientific naturalism*, the justification of methodological choices comes from our knowledge of the structure of the world, as expressed by our currently accepted scientific theories. Due to its backward-looking character, historicist naturalism does not satisfactorily deal with the case of newly introduced rules, for which no evidence concerning their past performance is available. In sum, the contribution that history of science can make to rule-choice is more modest than Laudan suggests.

Veli Virmajoki (University of Turku, Finland) gave a presentation entitled 'The Science We Never Had'. He presented arguments to the effect that there are historiographical and philosophical reasons to resist the idea that there have been sciences in the past, drawing on the insights from the historians of science. If there were no sciences in the past, it is difficult to see how the history of science could provide evidential support (or falsifications) for the philosophical theories of science. He examined different ways of understanding the relationship between the history and philosophy of science in the situation where the practices of the past cannot be judged as sciences. Among the alternatives there are three main lines along which the philosophy of science may proceed. 1. We can study how science would have been different, had its history been different. 2. We can test philosophical accounts using counterfactual scenarios. The question is not whether an account captures what actually happened but what would have happened, had science proceeded in accordance with the account. 3. We can estimate the possible future developments of science by studying what factors behind the development of science could change either due to a human intervention or due to a change in other area of society. As he pointed out, each of the lines 1–3 requires that counterfactual scenarios are built. Luckily, each of the lines can be shown to be a variation of the structure that is implicit in the explanations in the historiography of science. Moreover, this general structure is often implicit in more traditional case studies in the philosophy of science, and therefore the lines 1–3 are not too exotic despite the first impression. He concluded that the value of history of science is that it provides the materials to build the counterfactual scenarios.

Jouni-Matti Kuukkanen (University of Oulu, Finland) spoke on 'Truth, Incoherence and the Evolution of Science'. He began by recalling Thomas Kuhn's argument that scientific development should be understood as an ever-continuing evolutionary process of speciation and specialization of scientific disciplines. This view was first time expressed explicitly in *The Structure of Scientific Revolutions*.[4] Kuhn kept on returning to it until the end of his life. In his last published interview, Kuhn laments that "I would now argue very strongly that the Darwinian metaphor at the end of the book [*SSR*] is right and should have been taken more seriously than it was".[5] However, in this paper, rather than focusing on the evolution of Kuhn's notion of evolutionary development of science as such, Dr. Kuukkanen addressed two of its significant consequences regarding scientific progress. The one is the resulting incoherence of science as a global cognitive venture. The other is the relation of incoherence with truth as an aim of science. Kuhn remarked that "[S]pecialization and the narrowing of the range of expertise now look to me like the *necessary price* of increasingly powerful cognitive tools [...]. [T]o anyone who values the unity of knowledge, this aspect of specialization [...] is a condition to be deplored".[6] These words imply that the evolution of science gradually decreases the unity of science. Further, the more disunified science is, the more incoherent in total it is. Kuhn rejected the idea that science converges on the truth of the world, or the teleological view of scientific development, in part because he saw it as a historically unsubstantiated claim. But is truth as an aim of science also conceptually incoherent in Kuhn? It seems evident that the evolutionary view of scientific development makes the goal of progressing towards the singular Truth with the capital T impossible. As Nicholas Rescher, for example, has argued, the true description of the world should form a maximally coherent whole or a manifestation of ideal coherence.[7] But what if truth is seen as local, applicable

---

[4] Thomas S. Kuhn, *The Structure of Scientific Revolutions*, 2nd enlarged ed. (Chicago: University of Chicago Press, 1970).
[5] Thomas S. Kuhn, *The Road Since Structure* (Chicago: University of Chicago Press, 2000), p. 307.
[6] Ibid., p. 98.
[7] Nicholas Rescher, *The Coherence Theory of Truth* (Oxford: Oxford University Press, 1973); Nicholas Rescher (1985), Truth as ideal coherence, *Review of Metaphysics, 38*, 795–806.

in the specialized disciplines, so that they produce truths of the matters they are specialized in describing? Could science aim at producing a large set of truths about the world without the requirement of their systematic coherence? Science would be a collective of true beliefs without directionality or unity. This paper offered valuable insights on relations between truth, incoherence and the evolution of science within the Kuhnian philosophical framework.

Messy Science
Symposium of the Joint Commission of the IUHPST 3
Organiser: Hasok Chang (University of Cambridge)

This symposium, chaired by Catherine Kendig (Michigan State University, USA) examined the currently fashionable idea that science is 'messy' and, because of this messiness, abstract philosophical thinking is only of limited use in analysing science. But in what ways is science messy, and how and why does this messiness surface? Is it an accidental or an integral feature of scientific practice? In this symposium, the presenters showed their attempts to understand some of the ways in which science is messy and draw out some of the philosophical consequences of taking seriously the notion that science is messy. Four presentations were made in this session.

Jutta Schickore (Indiana University Bloomington, USA) spoke on 'Scientists' Reflections on Messy Science', focusing on what *scientists themselves* say about messy science, and whether they see its messiness as a problem for its functioning. Examining scientists' reflections about 'messy science' can fulfill two complementary purposes. Such an analysis helps to clarify in what ways science can be considered 'messy' and thus improves philosophical understanding of everyday research practice. The analysis also points to specific pragmatic challenges in current research that philosophers of science can help address. It has become a commonplace in recent discussions about scientific practice to point out that science is 'messy', so messy in fact that philosophical concepts and arguments are not very useful for the analysis of science. However, the claim that science is messy is rarely spelled out. What does it mean for science to be messy? Does it mean that scientific concepts are intricate or confused? Or that procedures are dirty and often unreliable? Or that methodological criteria are often quite sloppily applied? Is it really such a novel insight that actual science is messy? Moreover, it is not entirely clear what such messiness means for philosophy and philosophy's role for understanding or improving science. Philosophical analysis aims at clarifying concepts and arguments, at making distinctions, and at deriving insights that transcend the particulars of concrete situations. Should we be worried that philosophical concepts and arguments have become too far removed of actual scientific practice to capture how science really works or to provide any guidance to scientists?

Dr. Schickore addressed these sets of questions in an indirect way, shifting the focus from analyzing scientific concepts, methods, and practices to analyzing scientists' reflections on scientific practice. Thereby she sought to carve out a new niche for philosophical thinking about science. She began with a brief survey of recent philosophical debates about scientific practice, trying to clarify what it is that analysts of science have in mind when they are referring to the 'messiness' of research practices, and also what, in their view, the messiness entails for philosophical analysis. In the main part of the talk, she examined what scientists themselves have said about messy science. Do they acknowledge that science is messy? If so, what aspects of scientific research do they highlight? Do they see the messiness of science as a problem for its functioning, and if so, why? To answer these questions, she drew on a diverse set of materials—among other things, methods sections in experimental reports, articles and editorials in general science journals, as well as interviews with scientists. Analyzing scientists' own conceptualizations of scientific research practice proves illuminating in a number of ways. Today, scientists themselves are often reluctant to admit that science is messy— much more reluctant than they were a century or two ago. She stressed that it matters— and why it matters—whether scientists themselves are right or wrong about how science really works. In conclusion, she suggested that examining scientists' reflections about 'messy science' can fulfill two complementary purposes. On the one hand, such an analysis helps to clarify in what ways science can be considered 'messy' and thus improves philosophical understanding of everyday research practice. On the other hand, this analysis points to specific pragmatic challenges in current research that philosophers of science can help address.

Next, Jordi Cat (Indiana University Bloomington, USA) presented a paper entitled 'Blur Science through Blurred Images. What the Diversity of Fuzzy Pictures Can Do for Epistemic, Methodological and Clinical Goals'. Different kinds of images include hand-drawings, analogical and digital photographs, and computer visualizations. Historically these have been introduced in an ongoing project of simulation of blurred vision that began with so-called artificial models, artificial visual aberrations and photographic simulations, and experiments. Computer simulations followed suit, each with their own specific conditions. Different kinds of pictures, like the roles and goals they serve, do not always arise to replace others, but instead develop different relations to others and introduce new uses. In the new pictorial regime, research and clinical practice rely on a combination of drawings, different kinds of photographs, and computer visualizations.

Dr. Cat also showed how the simulations and the pictures play a number of roles: providing illustration and classification, prediction, potential explanations (a deeper level of classification), exploration, testing, evidence for or against explanatory hypotheses, evidence for or against the effectiveness of research tests and techniques, evidence for or against the reliability of diagnostic tests and the effectiveness of corrective treatments, and tracking the evolution of conditions and treatments. Fuzziness or blur in images deserves critical attention as a subject and resource in scientific research practices and clinical interventions. He discussed how the project of engaging blur in vision optics is embedded in a constellation of different mathematical and pictorial tools with different standards and purposes—both investigative and clinical—which are often inseparable. An expression of this is the variety of kinds of pictures of blurred vision, many of which do appear blurred, and their different and shifting roles and uses. Their use runs against the commitment to sharpness as an ideal of, for instance, scientific representation, reasoning, and decision making. Dr. Cat's analysis contradicts and supplements a number of other accounts of the significance of images in terms of their content and use. A central issue is how the central interest in the phenomenon of blur in visual experience prompts pervasive and endemic considerations of subjectivity and objectivity. Different relations and tensions between standards of subjectivity and objectivity play a key role in the evolution of research and clinical intervention. This aspect finds expression in the interpretation, production, and use of pictures.

Bettina Dietz (Hong Kong Baptist University, Hong Kong), in her paper 'Tinkering with Nomenclature. Textual Engineering, Co-authorship, and Collaborative Publishing in Eighteenth-Century Botany', explored how the messiness of eighteenth-century botanical practice, resulting from a constant lack of information, generated a culture of collaborative publishing. Given the amount of information required for an accurate plant description let alone a taxonomic attribution, eighteenth-century botanists and their readers were fully aware of the preliminary nature of their publications. They openly acknowledged the necessity of updating and correcting them, and developed collaborative strategies for doing so efficiently. Authors updated their own writings in cycles of iterative publishing, most famously Carl Linnaeus, but this could also be done by others, such as the consecutive editors of the unpublished manuscripts of the German botanist Paul Hermann (1646–1695), who became his co-authors in the process.

Hermann had spent several years in Ceylon as a medical officer of the Dutch VOC (United East India Company) before he returned to the Netherlands in 1680 with an abundant collection of plants and notes. When he died almost all of this material, eagerly awaited by the botanical community, was still unpublished. As the information economy of botany, by then a discipline aiming for the global registration and classification of plants, tried to prevent the loss of precious data, two botanists—William Sherard (1650–1728) and Johannes Burman (1706–1779)—consecutively took on the task of ordering, updating, and publishing Hermann's manuscripts. The main goal of these cycles of iterative publishing was, on the one hand, to add relevant plants and, on the other, to identify, augment, and correct synonyms—different names that various authors had given to the same plant over time. As synonyms often could not be identified unambiguously, they had to be adjusted repeatedly, and additional synonyms, which would, in turn, require revision in the course of time, had to be inserted. The process of posthumously publishing botanical manuscripts provides insights into the successive cycles of accumulating and re-organizing information that had to be gone through. As a result, synonyms were networked names that were co-authored by the botanical community. Co-authorship and a culture of collaborative publishing compensated for the messiness of botanical practice.

The last presenter in this symposium was Catherine Kendig (Michigan State University, USA), who presented a paper on 'Messy Metaphysics: The Individuation of Parts in Lichenology', an investigation of how biological classification can sometimes rely on messy metaphysics. Focusing on the lichen symbiont, she explored what grounds we might have for relying on overlapping and conflicting ontologies. Lichens have long been studied and defined as two-part systems composed of a fungus (mycobiont) and a photosynthetic partner (photobiont), such as algae or cyanobacteria. This bipartite metaphysics underpins classificatory practices and determines the criteria for stability that rely on the fungus to name lichens despite the fact that some lichens are composed of three or more parts. She examined how reliable taxonomic information can be gleaned from metaphysics that makes it problematic to even count biological individuals or track lineages.

The standard view of lichens has been that they are systems that have one fungus—typically an Ascomycete or Basidiomycete. Although other fungi are known to be parts of the lichen (in a less functional or evolutionarily impactful role), the classical view of lichen composition of mycobiont-photobiont has been widely accepted. This bipartite view suggests that the criteria for lichen stability is the presence of the same mycobiont in the lichen system and underpins classificatory practices that rely on the fungus to name lichens. But this one-lichen, one-fungus metaphysics ignores relevant alternatives. Recent discoveries show that some lichens are composed of three rather than two symbiotic parts.[8] The metaphysical concept of the lichen and what are considered to be its parts

---

[8] P.-L. Chagnon, J. M. U'Ren, J. Miadlikowska, F. Lutzoni & A. E. Arnold (2016), Interaction type influences ecological network structure more than local abiotic conditions: Evidence from endophytic and endolichenic fungi at a continental scale, *Oecologia*, *180*(1), 181–191.

determines how lichens are individuated and how they are named and tracked over time. Naming the lichen symbiont relies on capturing its parts but also on the means by which we attribute parthood. If we say that something is a part of something else, reference to its parthood is typically thought to be metaphysically grounded (e.g., its parthood is due to a particular relationship of composition, kind membership, or inheritance), or, saying that something is a part might be indicative of our understanding of its role in a process (e.g., which entities are involved in a pathway's functioning over time). Brett Calcott suggests that parts may play different epistemic roles depending on how they are used in lineage explanations and for what purpose parthood is attributed to them.[9] A part may be identified by the functional role it plays as a component of a biological process. Or, parts-talk may serve to indicate continuity of a phenomenon over time, despite changes between stages, such that one can identify it as the same at time T1 as at time T2. Dr. Kendig employed Calcott's account of the dual role of parts in order to shed light on the messy individuation activities, partitioning of the lichen symbiont, and criteria of identity used in lichenology. She used this case to explore what grounds we have for relying on different ontologies, what commitments we rely upon for our classifying practices, and how reliable taxonomic information can be gleaned from these messy individuation practices. Ontological messiness may be both problematic in making it difficult to count biological individuals or track lineages, and useful in capturing the divergent modes of persistence and routes of inheritance in symbionts.

The History and Ontology of Chemistry
Symposium of the Joint Commission of the IUHPST 4
Organiser: Hasok Chang (University of Cambridge)

This symposium, chaired by Hasok Chang (University of Cambridge, UK), made a historical-philosophical examination of chemical ontology. Philosophers thinking about the metaphysics of science would do well to scrutinize the history of the concepts involved carefully. The idea of 'cutting nature at its joints' does not offer much practical help to the scientists, who have to seek and craft the taxonomic and ontological notions according to the usual messy procedures of scientific investigation. And we philosophers of science need to understand the nature of such procedures. This session showcased various attempts to do such historical-philosophical work, with a focus on chemistry.

---

[9] Brett Calcott (2009), Lineage explanations: Explaining how biological mechanisms change, *British Journal for the Philosophy of Science*, *60*, 51–78.

Robin Hendry (Durham University, UK), in his presentation entitled 'The History of Science and the Metaphysics of Chemistry', provided a general framing of the issue. Any scientific discipline is shaped by its history, by the people within it and by the cultures within which they work. But it is also shaped by the world it investigates: the things and processes it studies, and the ways in which it studies them. The International Union of Pure and Applied Chemistry (IUPAC) has developed different systems of nomenclature for inorganic and organic substances, based systematically on their structure at the molecular scale. These systems reflect both chemistry's historical development and particular metaphysical views about the reality of chemical substances. Thus, for instance, IUPAC names many inorganic substances on the basis of a system which is the recognisable descendant of the scheme of binomial nomenclature proposed by Antoine Lavoisier and his associates in the 1780s as part of their anti-phlogistic campaign. IUPAC's nomenclature for organic substances is based on a theory of structure that was developed in the 1860s and 1870s to provide an account of various kinds of isomerism. Both of these were reforming developments: attempts to introduce order, clarity and precision into an otherwise chaotic and confused scene, based on a particular foundational conception of the field (or rather sub-field). But order, clarity and precision may come at a cost: by tidying things up in one way, laying bare one set of patterns and structures, chemists might have obscured or even buried other patterns and structures. Looking back into the history, we recognize the contingent decisions taken by past chemists that led to our present conceptions, and the possible paths-not-taken that might have led to different ontological conceptions. Such decisions were, and will continue to be, influenced by various types of forces that shape science.

If the history of chemistry is a garden of forking paths, then so is the metaphysics of chemistry. Suppose that one is primarily engaged in developing an account of what the world is like, according to chemistry, in the respects in which chemistry studies it. One might start with modern chemistry, studying its currently accepted theories and the implicit assumptions underlying its practices, and think about how the world would be, in the respects in which chemistry studies it, if those theories and assumptions were broadly true. In this kind of project the particular metaphysical views about the reality of chemical substances that underlie modern chemistry are of central interest, as is the story of how modern chemistry came to be the way it is. That story also includes the options not taken up: alternative systems of nomenclature based on different ways of thinking about chemical reality. That story is an indispensable part of understanding, in both historical and epistemic terms, why modern chemistry is the way it is. How do we know that modern chemistry will present us with a coherent set of metaphysical views about the reality of chemical substances, something that can be regarded as, or can perhaps be shaped into, a metaphysics of chemistry? Of course, we don't. Chemistry and its history might present two kinds of difficulty: metaphysical disunity in modern chemistry, and historical options not taken up, but which demand to be taken seriously. There are different ways to respond to these difficulties, not all of them being pluralist (for disunity may reflect disagreement). And we don't know a priori that modern chemistry cannot be understood on the basis of a coherent metaphysical view of chemical reality. Any philosopher who wishes to bring their study of science into contact with metaphysics must acknowledge the different forces that have shaped science as we find it, but acknowledging them brings with it the recognition that there are different ways to proceed. If the history of chemistry is a garden of forking paths, then so is the metaphysics of chemistry.

This presentation was followed by three concrete studies. Marina Paola Banchetti-Robino (Florida Atlantic University, USA), in her presentation on 'Early Modern Chemical Ontologies and the Shift from Vitalism to Mechanicism', discussed the shift from vitalism to mechanicism that took place in early modern investigations of matter.[10] This was a gradual and complex process, with corpuscularianism as an important commonality shared by the competing perspectives. From a philosophical point of view, one of the more significant changes that occurred in chemical philosophy from the late 16th to the 17th century is the shift from the vitalistic metaphysics that had dominated Renaissance natural philosophy to the mechanistic theory of matter championed by the Cartesians and Newtonians. The shift away from vitalism and toward mechanicism was gradual rather than abrupt, and aspects of vitalism and of mechanicism coexisted in interesting ways within the chemical ontologies of many early modern chymists. The gradual demise of vitalism resulted not from the victory of reductionistic mechanicism but, rather, from the physicalistic and naturalistic rationalization of chemical qualities and processes that opened the door for Lavoisier to articulate his quantitative and operational conception of simple substances.

In spite of the tensions between these two opposing metaphysical paradigms, one important thread that connects early modern chymical theories, whether vitalistic or mechanistic, is their ontological commitment to corpuscular theories of matter. The

---

[10] Selected references: Bernadette Bensaude-Vincent & Isabelle Stengers, *A History of Chemistry* (Cambridge, MA: Harvard University Press, 1996); Antonio Clericuzio, *Elements, Principles and Corpuscles: A Study of Atomism and Chemistry in the Seventeenth Century* (Dordrecht: Kluwer Academic Publishers, 2000); Allen G. Debus, *The Chemical Philosophy: Paracelsian Science and Medicine in the Sixteenth and Seventeenth Centuries* (New York: Dover Publications, 2002); Trevor H. Levere, *Transforming Matter: A History of Chemistry from Alchemy to the Buckyball* (Baltimore: Johns Hopkins University Press, 2001); William R. Newman, *Atoms and Alchemy: Chymistry & the Experimental Origins of the Scientific Revolution* (Chicago: University of Chicago Press, 2006).

historical process whereby ancient Democritean atomism was revived in the 16th century is quite complex, but it would be a mistake to assume that particulate theories of matter need imply a commitment to physicalism and mechanicism. In fact, although the atomism of such natural philosophers as Gassendi and Charleton was indeed mechanistic, one finds many examples of medieval, Renaissance, and early modern atomism that embraced vitalistic metaphysics while endorsing a corpuscularian theory of matter. As it happens, there is strong evidence to show that, for much of the 17th century, chemical philosophers adopted a view of matter that was both ontologically corpuscularian and metaphysically vitalistic. In other words, these chemical philosophers adhered to a particulate matter theory while also embracing the idea that chemical qualities and operations involved the action of vital spirits and ferments. Dr. Banchetti-Robino examined these ideas by focusing on some of the more significant transitional chemical philosophies of the 16th and 17th centuries, in order to establish how chymists at this time adhered to complex corpuscularian ontologies that could not be subsumed under either a purely vitalistic or a purely mechanistic metaphysical framework. To this end, she focused on the chemical philosophies of Jan Baptista van Helmont, Daniel Sennert, Sebastian Basso, and Pierre Gassendi and the contributions that each of these important figures made to the subtle and graduate shift from vitalism to mechanicism.

Sarah Hijmans (Université Paris-Diderot, France), in her presentation entitled 'The Building Blocks of Matter: The Chemical Element in 18th- and 19th-century Views of Composition', addressed the history of the concept of chemical element. She started by noting that the IUPAC holds a double definition of chemical element.[11] These definitions loosely correspond to Lavoisier's and Mendeleev's respective definitions of the element: whereas Lavoisier (1743–1794) defined the element as a simple body, thus provisionally identifying all indecomposable substances as the chemical elements, Mendeleev (1834–1907) distinguished between elements and simple bodies. He reserved the term 'element' for the invisible material ingredient of matter, detectable only through its atomic weight, and not isolable in itself. Today, philosophers of chemistry generally agree that two meanings of the term 'element' co-exist, and that this leads to confusion. In order to study the nature of the chemical element, philosophers often refer to Lavoisier's and Mendeleev's views as illustrations of the two meanings. Thus, their definitions are analysed individually as well as compared to each other, independently of their historical context. This reinforces the idea that Mendeleev's definition marks a rupture in the

---

[11] IUPAC, *Compendium of Chemical Terminology*, 2nd ed. (the Gold Book), compiled by A. D. McNaught and A. Wilkinson (Oxford: Blackwell Scientific Publications, 1997). XML on-line corrected version: http://goldbook.iupac.org (2006–).

history of the chemical element: it is presented as the return to a pre-existing metaphysical view[12] or the establishment of a new concept of element.[13]

Ms. Hijmans argued that the change in the conception of the element was part of a broader evolution of chemical practice. A view very similar to Mendeleev's was already present in early 19th-century chemical atomism, and developed in a rather continuous way through the century. However, little is known about the evolution of the concept of chemical element during the early 19th century: where did the change in definition between Lavoisier and Mendeleev come from?

The aim of Hijmans' paper was to historicise the notion of chemical element, and study its development in the context of 18th- and 19th-century chemistry. Based on the works of Hasok Chang, Ursula Klein and Robert Siegfried,[14] she argued that the change in definition does not in itself constitute a rupture in the history of the chemical element; rather, it is part of a broader evolution of chemical practice which connects the two definitions through a continuous transfer of ideas. Indeed, a view very similar to Mendeleev's was already present in early 19th-century chemical atomism. The 'theory of chemical portions', identified by Klein,[15] transformed the stoichiometric proportions in which elements combined into an intrinsic quality of the elements: it "identified invisible portions of chemical elements [...] as carriers of the theoretical combining weights".[16] This theory in turn overlaps with Daltonian atomism, which constituted the height of 'compositionism'.[17] Compositionism was based on the assumption that chemical composition consisted of a rearrangement of stable building blocks of matter. This view was dominant in the 18th century and played a crucial role in Lavoisier's Chemical Revolution.[18] Thus, through a historical analysis this paper identified the continuity

---

[12] Eric Scerri, *The Periodic Table: Its Story and Its Significance* (Oxford: Oxford University Press, 2007), 114–116; Elena Ghibaudi, Alberto Regis and Ezio Roletto (2013), What do chemists mean when they talk about elements?, *Journal of Chemical Education*, *90*, 1626–1631, on p. 1627.
[13] Bernadette Bensaude-Vincent (1986), Mendeleev's periodic system of chemical elements, *The British Journal for the History of Science*, *19*(1), 3–17, on p. 12.
[14] Hasok Chang (2011), Compositionism as a dominant way of knowing in modern chemistry, *History of Science*, *49*, 247–268; Hasok Chang, *Is Water H$_2$O? Evidence, Realism and Pluralism* (Dordrecht: Springer, 2012); Ursula Klein (1994), Origin of the concept of chemical compound, *Science in Context*, *7*(2), 163–204; Ursula Klein (2001), Berzelian formulas as paper tools in early nineteenth-century chemistry, *Foundations of Chemistry*, *3*, 7–32; Ursula Klein, *Experiments, Models, Paper Tools: Cultures of Organic Chemistry in the Nineteenth Century* (Stanford, CA: Stanford University Press, 2003); Robert Siegfried, *From Elements to Atoms: a History of Chemical Composition* (Philadelphia: American Philosophical Society, 2002).
[15] Klein, Berzelian formulas, pp. 15–17; Klein, *Experiments, Models, Paper Tools*, ch. 1.
[16] Klein, Berzelian formulas, p. 15.
[17] Chang, Compositionism.
[18] Chang, Compositionism; Chang, *Is Water H$_2$O?*, pp. 37–41, 135; Siegfried, *From Elements to Atoms*; Klein, Origin of the concept.

between the views of Lavoisier and Mendeleev. This provides an example of how historical thinking can shed a new light on chemical ontology. Perhaps, a better understanding of the historical constitution of the chemical element will show the contingency of the current double definition, and thus help resolve the question of the nature of chemical element today.

Karoliina Pulkkinen (University of Cambridge, UK) examined the history of the late 19th-century attempts to find periodic regularities among the chemical elements in her presentation on 'Some Sixty or More Primordial Matters: Chemical Ontology and the Periodicity of the Chemical Elements'. Accounts on the periodic system often draw attention to how two of its main discoverers had contrasting views on the nature of chemical elements. Where Julius Lothar Meyer saw it likely that the elements were comprised of the same primordial matter,[19] Dmitrii Ivanovich Mendeleev opposed to this view. Instead, Mendeleev argued that each element was its distinct, individual, autonomous entity, and he discouraged from making representations of periodicity that suggested otherwise.[20] While Meyer saw it likely that all elements were comprised of the same primordial matter, Mendeleev saw each element as a distinct, individual, autonomous entity and refrained from making representations of periodicity that suggested otherwise. Following Andrea Woody's rich article on the law of periodicity as a theoretical practice,[21] Ms. Pulkkinen explored how Meyer's and Mendeleev's ontological views on primordial matter shaped their ideas on how to represent periodicity. She started by showing how Meyer's views on the nature of the elements were not an endorsement of the truth of the hypothesis on the primordial matter. Instead, for Meyer, taking the view on board was needed for conducting further investigations on the relationship between atomic weight and other properties of elements. With respect to Mendeleev, she showed how his metaphysical views on nature of elements influenced his evaluation of other investigators' representations of periodicity. Especially Mendeleev's rejection of graphs[22] and equations for representing periodicity is in part explained by his views on the nature of the elements. Among the many attempts of rendering periodicity

---

[19] Lothar Meyer (1870), Die Natur der chemischen Elemente als Function ihrer Atomgewichte, *Annalen der Chemie und Pharmacie*, VII, *Supplementband*, 354–363, on p. 358; Lothar Meyer, *Modern Theories of Chemistry* (London: Longmans, Green, and Co., 1888), p. 133.
[20] Dmitri I. Mendeléeff, *Principles of Chemistry*, 7th ed. (London: Longmans, Green, and Co., 1905), pp. 22–24.
[21] Andrea Woody, Chemistry's periodic law: Rethinking representation and explanation after the turn to practice, in L. Soler, S. Zwart, M. Lynch & V. Israel-Jost (Eds.), *Science After the Practice Turn in the Philosophy, History, and Social Studies of Science* (New York: Routledge, 2014), pp. 123–150.
[22] Bernadette Bensaude-Vincent, Graphic representations of the periodic system of chemical elements, in U. Klein (Ed.), *Tools and Modes of Representation in the Laboratory Sciences* (Dordrecht: Kluwer Academic Publishers, 2001), pp. 117–132.

to more mathematical language, she especially focused on the equations created by the Russian political philosopher and lawyer Boris N. Chicherin. Mendeleev's ontological views influenced his rejection of Chicherin's equations. The examples of Meyer and Mendeleev show that their ontological commitments directed both their own representations of periodicity and their evaluations of other investigators' representations. Even though we are warned not to confuse means of representation with what is being represented,[23] the case of Meyer and Mendeleev suggests that ontological views on the nature of elements influenced representing periodicity.

Denial of Facts: Instrumentation of Science, Criticism, and Fake News
Organisers: Benedikt Löwe (University of Hamburg, University of Amsterdam, and University of Cambridge) and Daya Reddy (University of Cape Town)

For scientists and rational thinkers, the increasing acceptance of positions that constitute outright denial of established scientific consensus is disconcerting. In recent years, science denial movements have become more vocal and widespread, from climate change deniers via vaccination opponents to politicians whose statements are directly and openly in contradiction with established facts. The phenomenon of denial of (scientific) facts used to be confined to the fringes of our societies, but now transformed into a phenomenon with relevant policy effects and long-term consequences for everyone on the entire globe. Both logic and philosophy of science can contribute to our understanding of this phenomenon and possibly show paths to react to it and counter it.

The *Division for Logic, Methodology and Philosophy of Science and Technology* of the *International Union of History and Philosophy of Science and Technology* (DLMPST/IUHPST) and the *International Science Council* (ISC), the global umbrella organisation for all of the natural and social sciences, decided to tackle this important topic in a symposium during the 2019 congress, bringing together logicians and philosophers of science to discuss both the philosophical theories underlying the phenomenon of denial of facts and their potential consequences for science policy makers, science communicators, and other stakeholders. The two symposium organisers and authors of this report represented the two involved institutions: the first author was Secretary General of DLMPST/IUHPST at the time of the symposium; the second author was the President of the ISC.

---

[23] Steven French (2010), Keeping quiet on the ontology of models, *Synthese*, 2(172), 231–249.

Already at the time of the symposium in the summer of 2019, there was a sense of urgency: participants felt that positions of science denialism were mainstreamed in the general political discussion and that quick action was needed; and yet, none of us could foresee how this topic developed into a matter of life and death in the global Covid pandemic that started in 2020. In the global discourse on how to react to the pandemic, science denialists used the full force of their deceptive communication techniques in order to cast doubt on scientifically accepted interventions and mitigation measures saving the health and lives of uncounted people. Many of the themes of our symposium were in full display during the public debates of the Covid pandemic: e.g., it was a huge surprise for those of us who have worked on the epistemology of peer review in science to witness discussions in the general public of the epistemic difference between a preprint and a paper published in a peer-reviewed journal. Some attitudes exhibited when journalists got a glimpse of the internal scientific vetting process reminded participants of the 2019 symposium of Sven Ove Hansson's warning in Prague that there is a close link between emphasising science exceptionalism and nurturing science denialism: when the observation of actual science does not meet the expectations of the idealised picture of exceptional epistemic access, doubts are created or reinforced. We firmly believe that analysing the connections between philosophy of science, the public image of science, and the rise of science denialism and fake news is an urgent task for the immediate future.

Our symposium featured four presentations, two of which focused on specifying the meaning of terms such as 'fake news' and the other two on the relationship of science denialism, science communication, and philosophy of science.

In the opening presentation entitled 'Fake News, Pseudoscience, & Public Engagement', Daya Reddy aimed at delineating the boundaries between 'fake news' and 'pseudoscience'. Reddy used the term 'fake news' to refer to information that is deliberately fabricated, and often distributed in ways that mimic the formats of news media, thus lending it, at least superficially, a semblance of credibility; in contrast, he characterised 'pseudoscience' by scientific claims that are characterised by a lack of supporting evidence, erroneous arguments, and a general incompatibility with the scientific method. Reddy gave a number of examples of the devastating effects of policy-makers under the influence of fake news or pseudoscience and encouraged scientists to embrace their role as communicators. At the time of the symposium, the *International Science Council* that had been created just a year earlier (July 2018) by the merger of the *International Council for Science* (ICSU) and the *International Social Science Council* (ISSC) in the Founding General Assembly in Paris. The central strategic document for this merger was the position paper 'Science as a Global Public Good' (Boulton, 2021). Reddy outlined that the view of science as a public good implies (via an implicit social contract) a responsibility for the scientific community not only to disseminate scientific knowledge within its ranks, but to ensure that it is made accessible to the public. In light of the rise of fake news and science denialism where traditional methods of science communication falter, this responsibility entails the duty to engage with new ideas about how to reach the goal of a scientifically educated citizenry. Such an engagement would require leadership at the nexus of science education, science communication, public outreach, sociology of science, and the behavioural sciences.

This line of thought was continued by a two-part joint presentation by Romy Jaster (Berlin) and David Lanius (Karlsruhe) entitled 'Truth and Truthfulness'. Its two parts were subtitled 'Part I: What Fake News Is and What It's Not' and 'Part II: Why They Matter'. Jaster and Lanius started by providing a systematic account of fake news contrasting it with related phenomena, such as journalistic errors, selective and grossly negligent reporting, satire, propaganda, and conspiracy theories. They conceptual analysis was based on the assumption that fake news are news reports lacking in truth (they are false or misleading) and truthfulness (they are circulated by people with an intention to deceive or a bullshit attitude). Jaster and Lanius then argued that their conceptual analysis is likely to contribute to improve the public debate. The results reported on in this presentation are published in Jaster & Lanius (2021) and relate to Gelfert (2018) and (2021).

The other two presentations focused on a different angle: the question of whether philosophers of science or science communicators are (partially) to be blamed for science denialism.

The presentation entitled 'Unwitting Complicity: When Science Communication Breeds Science Denialism' by Alex Gelfert (Berlin) started by discussing the psychological motivations for holding denialist positions. Gelfert argued that science denialism seems to find an audience especially amongst those who consider certain scientific facts a threat to their deeply held convictions or self-image. Dismissing such scientific findings, then, may be an (epistemologically flawed) attempt at reducing cognitive dissonance. In this situation, science communicators cannot change the mind of such individuals by merely reiterating the scientific facts.

Gelfert argued that certain quite common modes of science communication ignore this psychological situation and fail to regain trust and make the science denialist willing to consider scientific evidence afresh; this form of science communication might worsen the denialist attitudes.

The presentation entitled 'The Philosophical Roots of Science Denialism' by Sven Ove Hansson (Stockholm) went even further and addressed the direct role of certain philosophical traditions in the development of science denialism. Hansson's analysis was that some of the most important thought patterns of science denialism are based on the methodology of radical doubt that was developed in philosophical scepticism. Even worse, Hansson argued that the common assumption of *science exceptionalism* (i.e., the idea that the epistemological foundations of science are different from those of our other forms of knowledge) that is usually assumed in many discussions in philosophy of science is one of of necessary requirements for science denialism. Based on his analysis of the philosophical roots of science denialism, Hansson made recommendations what philosophers could and should do to defend science against the current onslaught of science denialism and other forms of pseudoscience.

References

Boulton, G. S. (2021). Science as a global public good. *International Science Council Position Paper*. 2nd edition. International Science Council.
Gelfert, A. (2018). Fake news: A definition. *Informal Logic, 38*(1), 84–117.
Gelfert A. (2021). What is Fake News? In M. Hannon & J. de Ridder (Eds.), *The Routledge Handbook of Political Epistemology* (pp. 171–180). Routledge.
Jaster, R. & Lanius, D. (2021). Speaking of fake news: Definitions and dimensions. In S. Bernecker, A. Floweree, & T. Grundmann (Eds.), *The Epistemology of Fake News* (pp. 19–45). Oxford University Press.

The Gender Gap in the Sciences and Philosophy of Science

Organisers: Benedikt Löwe (University of Hamburg, University of Amsterdam, and University of Cambridge) and Helena Mihaljević (Hochschule für Technik und Wirtschaft, Berlin)

We use the term *the gender gap in the sciences* to refer to the phenomenon that in some scientific disciplines, women are statistically underrepresented among the researchers and in particular among those with research leadership positions.

Most of us are familiar with this phenomenon, but if we think about this description of the gender gap more carefully, we realise that it raises several methodological questions: Which disciplines? Does the gender gap manifest differently in different disciplines? Is it a global phenomenon or does it differ from country to country? Do differences between disciplines differ between countries? Are the social constructs that occur in the description ('scientific disciplines', 'researchers', 'research leadership positions') sufficiently stable that we can compare data across countries? How do we determine whether the gender gap grows or shrinks? What are the reasons for the gender gap? Do these reasons differ from discipline to discipline or from country to country? What can we do to close the gender gap? How do we measure success or failure of interventions intended to close the gender gap?

The project *A Global Approach to the Gender Gap in Mathematical, Computing, and Natural Sciences: How to Measure It, How to Reduce It?* was an international and interdisciplinary effort funded by the *International Council for Science* (ICSU) from 2017 to 2019 to better understand the manifestation of the gender gap and to provide solution concepts and ways to evaluate them. It was a collaboration of eleven institutions, co-ordinated by the *International Mathematical Union* (IMU) through its *Committee for Women in Mathematics* and the *International Union of Pure and Applied Chemistry* (IUPAC). The other partner organisations were the *International Union of Pure and Applied Physics* (IUPAP), the *International Astronomical Union* (IAU), the *International Union of Biological Sciences* (IUBS), the *International Council for Industrial and Applied Mathematics* (ICIAM), the *International Union of History and Philosophy of Science and Technology* (IUHPST), the *United Nations Educational, Scientific and Cultural Organization* (UNESCO) through its project *STEM and Gender Advancement* (SAGA), *Gender in Science, Innovation, Technology and Engineering* (GenderInSITE), the *Organization of Women in Science for the Developing World* (OWSD), and the *Association for Computing Machinery* (ACM), through ACM-W. The involvement of so many different disciplines gave an opportunity to elaborate common grounds as well as discipline-specific differences.

Philosophy of science had an interesting and curious dual role in this project: while the gender gap phenomenon definitely applies to the discipline of philosophy of science, it is at the same time a meta-discipline that reflects on disciplinary practices and provides concepts and tools to analyse methods and techniques.

Since IUHPST was one of the partner institutions and officers of the *Division for Logic, Methodology and Philosophy of Science and Technology* (DLMPST) of the IUHPST were involved in several of the project's tasks and events, it was very appropriate to present the project and its findings in a symposium at CLMPST 2019 in Prague. The symposium was organised by the two authors of this paper and had two presentations.

The first presentation, representing the meta-theoretical approach of philosophy of science, was by Helen Longino (Stanford), at the time of the symposium the First Vice

President of DLMPST, whose talk entitled 'How Science Loses by Failing to Address the Gender (and Other) Gaps' presented the importance of thinking of the gender gap (and other participation gaps) not only as harmful to the underrepresented groups, but as universally harmful to the scientific endeavour. This argument follows Longino's work on communities as the locus of knowledge production and her argument that diverse communities have epistemic advantages over homogeneous communities (Longino, 1990, 2002).

The second presentation was a report from the work of the project itself: Helena Mihaljević gave a talk entitled 'What Can Publication Records Tell about the Gender Gap in STEM?' in which she presented technical and conceptual problems provided by the large data sources used to study publication patterns and how these problems were handled in the project (cf. in particular Mihaljević & Santamaría, 2021). She presented results of the project and discussed what differences in publication behaviour can tell us about the current state and possible future developments of the gender gap in the respective fields. More details can be found in Mihaljević et al. (2019) and Gledhill et al. (2019).

From its inception, the *Gender Gap in Science* project has placed a strong focus on dissemination, expansion of networks, and making women visible in various scientific disciplines. The *Database of Good Practices*, for instance, whose conception, development and provision was one of the work packages, is a means of structured assessment of existing approaches and initiatives to reduce the gender gap, especially with regard to young women (the database can be found on the website of the IMU).

The results of the project have been continuously communicated to diverse audiences; a collection of project-related publications can be found on the project website. Among those publications are also detailed analyses of particular questions concerning publication patterns such as publications in top ranked journals (Mihaljević & Santamaría 2020) and patterns in invitation to talks at the *International Congresses of Mathematicians* (ICM; Mihaljević & Roy, 2019).

Particularly noteworthy is the final report of the project, which has been made available as an open access book (Roy et al., 2020). It includes the central results of the three work packages as well as a collection of recommendations addressed to different stakeholders.

The main lasting legacy of the project is the creation of the *Standing Committee for Gender Equality in Science* (SCGES), an association of numerous of the international organizations that collaborated in the project. SCGES was established in September 2020 with the goal to promote gender equity and help reduce the gender gap in science. The IUHPST has a voice on SCGES via a representative and a deputy representative; the IUHPST representative is Catherine Jami (Paris) who is also the chair of SCGES. The deputy representative is provided by DLMPST: from 2020 to 2021, this was Delia Kesner (Paris); since the end of 2021, Hanne Andersen (Copenhagen) has been in this role.

References

Gledhill, I. M. A., Roy, M.-F., Chiu, M.-H., Ivie, R., Ponce-Dawson, S., & Mihaljević, H. (2019). A global approach to the gender gap in mathematical, computing and natural sciences: How to measure it, how to reduce it? *South African Journal of Science, 115*(3/4).

Longino, H. (1990). *Science as Social Knowledge*. Princeton University Press.

Longino, H. (2002). *The Fate of Knowledge*. Princeton University Press.

Mihaljević, H. & Roy, M.-F. (2019). A data analysis of women's trails among ICM speakers. In C. Araujo, G. Benkart, C. Praeger, & B. Tanbay (Eds.), *World Women in Mathematics 2018* (pp. 111–128). Association for Women in Mathematics Series, Vol. 20. Springer.

Mihaljević, H. & Santamaría, L. (2020). Authorship in top-ranked mathematical and physical journals: Role of gender on self-perceptions and bibliographic evidence. *Quantitative Science Studies, 1*(4), 1468–1492.

Mihaljević, H. & Santamaría, L. (2021). Disambiguation of author entities in ADS using supervised learning and graph theory methods. *Scientometrics, 126*, 3893–3917.

Mihaljević, H., Tullney, M., Santamaría, L., & Steinfeldt, C. (2019). Reflections on gender analyses of bibliographic corpora. *Frontiers in Big Data*, Section Data Mining and Management.

Roy, M.-F., Guillopé, C., Cesa, M., Ivie, R., White, S., Mihaljević, H., Santamaría, L., Kelly, R., Goos, M., Ponce Dawson, S., Gledhill, I., Chiu, M.-H. (2020). A global approach to the gender gap in mathematical, computing, and natural sciences: How to measure it, how to reduce it? *International Science Council*.

Philosophy of Science and the Periodic Table. A Symposium on the Occasion of the International Year of the Periodic Table (IYPT) 2019

Organisers: Gisela Boeck (University of Rostock) and Benedikt Löwe (University of Hamburg, University of Amsterdam, and University of Cambridge)

The periodic table of chemical elements is one of the most recognizable icons in the entire history of science. Its ubiquitous presence in all kinds of scientific environments and even popular culture worldwide is a clear testimony to its usefulness and informativeness. The year 1869, in particular the publication of the first table by Dmitri Mendeleev in the *Journal of the Russian Chemical Society* in May of that year (cf. Figure 1) is usually considered as the year of discovery of the periodic system; thus, we celebrated its 150th anniversary in the year 2019. On the initiative of the *International Union for Pure and Applied Chemistry* (IUPAC), the United Nations General Assembly proclaimed the year 2019 to be the *International Year of the Periodic Table* which was celebrated by academic and public events around the world. Many scientific unions joined IUPAC in this initiative, among them the *International Union for History and Philosophy of Science and Technology* (IUHPST). It was therefore very appropriate to mark this particular celebration with a symposium at our 2019 congress where we had the opportunity to reflect on the interplay between philosophy of science and the periodic table. Our symposium involved four presentations, of which two were personal reflections by junior researchers in history and philosophy of science (the second and fourth author) who gave an account on how the periodic table influenced their own research. We print reports of these two presentations, entitled 'Understanding the Chemical Element: A Personal

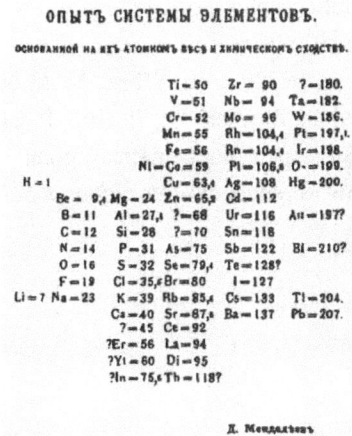

Figure 1. Mendeleev's first periodic table, published in May 1869 (image: public domain)

Reflection on the Periodic Table' and 'Values in Science and Early Periodic Tables', authored by Sarah Hijmans and Karoliina Pulkkinen, respectively, below.

The symposium opened with an overview presentation entitled 'Why Should Philosophers Care about the Periodic Table' by Hasok Chang. Chang gave a brief overview of the history of attempts to create a convenient and informative ordering of the chemical elements in the 19th century, presented some debates concerning the epistemic merits of Mendeleev's system, and showed how the history of the periodic table can be used to make effective illustrations of epistemic values in action, focusing especially on explanation and prediction.

A local angle was explored in the presentation entitled 'Mendeleev's Dedicated Supporter and Friend, the Czech Chemist Bohuslav Brauner and the Worldwide Reception of the Periodic System' by Soňa Štrbáňová (Prague) in which she discussed Bohuslav Brauner (1855–1935) who became an enthusiastic supporter of Mendeleev's system and contributed greatly to its improvement and dissemination in the entire world.

'Understanding the Chemical Element: A personal Reflection on the Periodic Table' (S. Hijmans, CNRS). At our symposium, I presented a personal reflection on the relevance of this table for my doctoral research. Although it is not directly aimed at studying the periodic table itself, my research focuses on the history and philosophy of that which is classified in it: chemical elements. While the concept of element remains difficult to grasp, it is clear that classifications have played an important role in identifying chemical elements, at least since at least the nineteenth century.

In order to illustrate why the chemical element might be philosophically interesting, let us consider an example. The transition metal vanadium, the 23rd element of the periodic table, forms a soft, silvery white metal when pure. When one shakes a solution of vanadium (V) ions with a mercury-zinc amalgam, the ions give off their electrons one by one and the solution slowly changes color, from yellow to purple via blue and green (a video of this experiment is available on the YouTube channel *Periodic Videos*.) If the solution is then left in contact with air, it will oxidize over time and slowly turn yellow again. This is a fairly standard chemical experiment that can teach students about the oxidation states of transition metals; but it also gives rise to philosophical questions: What is the relation between the soft metal and the colored solutions? How come a single element can form substances with such varying properties? Which part of vanadium is conserved in this process?

There is no clear answer to this type of question. As Elena Ghibaudi and her colleagues (2013, 1628) state: "the problem of unambiguously defining what is conserved in a chemical transformation has been a matter of discussion for a long time within the chemistry community". There is no consensus regarding a single definition of the concept of element, and the *International Union of Pure and Applied Chemistry* (IUPAC, 1997) proposes two definitions. Whereas professional chemists generally know how to navigate this ambiguity, it can be difficult for students to connect the chemical formulae in their textbooks to the experiments they observe in the classroom, and to switch between the microscopic and macroscopic meanings of the notion of element (Ghibaudi, Regis & Roletto, 2013, 1628–1629).

Historical research can help in improving our understanding of the concept of element, for example by studying the historical origin and development of definitions of this concept. Mendeleev himself is often credited with formulating a new definition of the element on the basis of its atomic weight, a view that played an important role in the development of his periodic table (see Bensaude-Vincent, 2019). His definition contradicted the dominant conception of much of the eighteenth and nineteenth centuries, according to which chemical elements could only be provisionally identified through the operations of chemical decomposition. According to that definition, promoted most famously by the French chemist Antoine-Laurent de Lavoisier, any substance that could not be decomposed in the laboratory should be seen as a chemical element (at least until further experiments proved otherwise).

However, the study of historical definitions only provides us with a partial view of the development of the concept of element. Whereas definitions can provide criteria for the identification of elements, such as their existence in the form of indecomposable substances in the case of Lavoisier, these criteria may not always be applied in practice. For instance, even Lavoisier himself omitted certain indecomposable substances from his list of elements because they behaved like compounds (Perrin, 1973). My doctoral project therefore aims at complementing analyses of definitions with a study of identifications of chemical elements in practice. Following a case-study based approach of integrated History and Philosophy of Science, I argue that the actual identification of elements happened on the basis of a complex argumentation which required abstract inferences as well as laboratory decompositions.

There are multiple ways in which the periodic table relates to this project. Firstly, it provides a philosophical motivation: the tremendous success of this classification accentuates the need for an understanding of the nature of that which is classified in it. Furthermore, the success of Mendeleev's version of the table, and the importance of his philosophical views in developing it, motivate historical research into conceptions of the element that preceded his work. Even though it has been argued that he broke with his predecessors' definition of the concept, it is interesting to see whether any aspects of his view may have been implicitly present in his predecessors' chemical practice. Moreover, Mendeleev's periodic table is only one example of a classification of chemical elements. Chemical classifications of course much preceded Mendeleev's publication, and through my research I have found that classification practices were important in the identification of chemical elements long before 1869. Rather than individually, elements were almost always considered as part of a family of substances.

Various families of chemical elements that form the columns of the periodic table today date back to the late eighteenth and early nineteenth century. In fact, the histories of some chemical elements are so closely intertwined that they can hardly be told separately. One relatively little-known example is that of tantalum and niobium. These metals were initially discovered separately (under the names of tantalum and columbium) in 1801, but then argued to be identical in 1809, leading to the retraction of columbium. In the 1840s, the (re)discovery of niobium sparked a renewed interest in tantalum minerals, leading to the announcement of at least four other 'tantalum metals'. The existence of these metals was the subject of a debate that lasted twenty years, which resulted in the retraction of all tantalum metals except for tantalum and niobium. Throughout the debate, various authors reflected on how the tantalum metals would be classified, internally ordering the metals according to the type of compounds they formed, their crystal structure, valence and atomic weight, and proposing placements of the group relative to other chemical families.

More importantly however, the question of classification was not just something to be considered once elements were identified: the place (or lack thereof) of an element in a family of elements could even determine its identification itself. The most striking examples of this kind of reasoning can be found within the family of the halogens. In 1810, the English chemist Humphry Davy argued that chlorine could not be decomposed, not even using the most powerful instruments available, and that it should therefore be seen as a chemical element. Yet, despite this empirical evidence, most chemists refused to accept the new element for over six years. Depending on how one viewed the formation of acids and salts, the reaction that produced hydrochloric acid from chlorine and hydrogen could be viewed as either a chemical combination or a decomposition. As a result, the same experiment could be taken as proof for two different interpretations of the nature of chlorine. It was only after the discovery of iodine that chlorine came to be seen as more than a strange exception, and the existing theories were gradually revised. Once chlorine and iodine were commonly accepted as chemical elements, fluorine was added to this family on the basis of an analogical inference: since hydrochloric acid was composed of hydrogen and chlorine, the analogous substance hydrofluoric acid was likely composed of a similar constituent (namely fluorine) and hydrogen. By the 1820s, this was a widely accepted view, despite the fact that fluorine could not be produced in the form of a simple substance until 1886.

In short, substances that could not be decomposed in the laboratory could only be accepted as elements in case this identification was coherent with chemical classifications. If their chemical behavior resembled that of existing compounds, they would be classified as compounds and predicted to contain previously unknown elements. These are only a few examples among many that show the importance of classifications of elements, often even determining the identity of elements themselves. These examples therefore place the work of Mendeleev (and the other discoverers of the periodic table) in the context of a much larger collective endeavor of chemical classification. Rather than as a break with his predecessors, Mendeleev's table could be seen as the matured product of a period of collaborative research on the elements. In my opinion, the collective nature of this achievement only gives us more reason to celebrate it today.

'Values in science and early periodic tables' (K. Pulkkinen, University of Helsinki). In successfully encapsulating much of chemical knowledge, the periodic system of chemical elements serves as a fruitful ground for philosophical investigations. It invites questions regarding the role of theory and empiricism in chemistry, laws of nature, chemical practice, classification, and scientific representation. Although the modern periodic system has been looked at from all the above angles—no doubt facilitated by its persistence as a scientific representation—also the process for its discovery and development warrants philosophical attention. When examining its development, a notable feature is that it was discovered multiple times. Comparing different versions of the early periodic systems highlights the distinctness of chemists' design-choices and their explanations of what they had discovered. In conducting a comparison between different early periodic systems, I argued that many of the contrasts between the systems of three of its discoverers—the Russian chemist Dmitri Mendeleev, the German Julius Lothar Meyer, and English John Newlands—can be explained by their differing emphasis on different epistemic values. Newlands highlighted the simplicity of his arrangements; Meyer was more careful about the quality of data that gave rise to his system of elements; and Mendeleev sought to make his system more complete. By doing so, the talk illustrated that the periodic system is also of interest from the point of view of values in science. My study was subsequently published as (Pulkkinen, 2020).

References

Bensaude-Vincent, B. (2019). Reconceptualizing chemical elements through the construction of the periodic system. *Centaurus, 61*(4), 299–310.

Ghibaudi, E., Regis, A., & Roletto., E. (2013). What do chemists mean when they talk about elements? *Journal of Chemical Education, 90*, 1626–31.

IUPAC. (1997). Chemical element. In A. D. McNaught & A. Wilkinson (Eds.), *Compendium of Chemical Terminology*, 2nd ed. Oxford: Blackwell.

Perrin, C. E. (1973). Lavoisier's table of the elements: A reappraisal. *Ambix, 20*(2), 95–105.

Pulkkinen, K. (2020). Values in the development of early periodic tables. *Ambix, 67*(2), 174–198.

Philosophy of Big Data

Symposium of the DLMPST Commission for the History and Philosophy of Computing
Organiser: Paula Quinon (Warsaw University of Technology)

The symposium 'Philosophy of Big Data' was devoted to a discussion of philosophical problems related to Big Data, an increasingly important topic within philosophy of computing. Big Data is worth studying from an academic perspective for several reasons. First of all, ontological questions are central: what Big Data is, whether we can speak of its components as separate ontological entities, and what their mereological status is. Second, epistemological questions: what kind of knowledge does Big Data induce, and what methods are required for accessing valuable information. These general questions have also very specific counterparts which raise a series of methodological questions. Should data accumulation and analysis follow the same general patterns for all Sciences, or should those be relativised to particular domains? For instance, shall medical doctors and businessmen focus on the same issues related to gathering of information? Is the quality of information similarly important in all the contexts? Can one community be inspired by the experience of another? To what extent do human factors influence information that we derive from Big Data? In addition to these theoretical academic issues, Big Data also represents a social phenomenon. 'Big Data' is nowadays a fancy business buzzword, which—together with 'AI' and 'Machine Learning'—shapes business projects and the R&D job market, with data analysts among the most attractive job titles. It is believed that 'Big Data' analysis reveals opportunities and generates additional profits. However, it is not clear what counts as Big Data in the industry, and critical reflection upon it seems necessary. The symposium has gathered philosophers, scientists and experts in commercial Big Data analysis to reflect on these questions. The opportunity to exchange ideas, methodologies and experiences from different perspectives and with divergent goals fostered this exchange and certainly enriched academic philosophical reflection; we believe that in the future it will also prove useful from practical perspectives in science and business.

Symposium Chair: Giuseppe Primiero (University of Milan)

Presented papers:

Jens Ulrik Hansen (Roskilde University), **'Philosophizing on Big Data, Data Science, and AI'**

To whom does the concept of 'Big Data' belong? Is Big Data a scientific discipline? Does it refer to datasets of a certain size? Or does Big Data more precisely refer to a collection of information technologies? Or is it a revolution within modern business? Certainly, 'Big Data' is a buzzword used by many different people, businesses, and organizations to mean many different things.

Similar consideration can be offered to the concepts of 'Data Science' and 'AI'. Within academia, Data Science has on several occasions been used to refer to a 'new' science that mixes statistics and computer science. Another use of the term pertains to what 'Data Scientists' (mainly in industry) are doing. (These might differ). Likewise, the term 'AI' has been used to refer to the study of Artificial Intelligence as a scientific discipline. However, AI is also the new buzzword within industry. Here, AI might better be translated as 'Automated Intelligence'. Within industry, AI is essentially the same as what 'Big Data' used to refer to, however, the focus has moved towards how models can be embedded in applications that automatically make decisions, instead of simply deriving insights from data.

Why are the different usages of these concepts relevant? On the one hand, if we want our science and philosophy to have relevance beyond academia, it does matter how the concepts are used outside academia in the mainstream public and business world. On the other hand, there is a much stronger sense in which the usages and different meanings of these concepts matter. It matters to our philosophy. For instance, if we want 'Philosophy of Big Data' to be about the ethics of automatic profiling and fraud detections used in welfare, health and insurance decisions, the dataset sizes and information technologies used do not really matter. Instead, it is how data about individuals are collected and shared, how biases in data transfer to biases in machine learning model predictions, how predictive models are embedded in services and application, and how these technologies are implemented in private and public organizations. Furthermore, if by 'Philosophy of Big Data' we are interested in the epistemological consequences of Big Data, it is again other aspects that are central.

In this talk I will therefore argue for the abandonment of terms like 'Philosophy of Big Data', 'Philosophy of Data Science', 'Philosophy of AI', etc. Instead, I suggest that we as philosophers paint a much more nuanced picture of a wide family of related concepts and technologies related to Big Data, Data Science, AI and their cousins, such as 'Cognitive Computing', 'Robotics', 'Digitalization', and 'IoT'.

Helena Kossowska (University of Warsaw), **'Big Data in Life Science'**

In this talk, I am going to discuss differences between the ways in which data for Big Data Analysis are gathered in the context of business and Life Sciences, especially in medical biology projects. Since both the size and complexity of experimental projects in life sciences is varied, I would like to focus on big interdisciplinary projects that usually combine different testing methods.

In business the process usually starts with collecting as much information as possible. Only then do people try to determine what can be inferred from the data, forming assumptions upon which the subsequent analysis is carried out. In Life Sciences the operating model is different: it starts with planning what information a scientist needs to collect in order to get the answer to the scientific question. Moreover, scientists usually have a limited budget for their broad experimental projects, and collection of each and every information has its cost. For that reason, the scope of collected information, as well as type and size of the study group, should be carefully planned and described. Furthermore, in medical sciences, the cooperation between various medical and scientific units is crucial. Therefore, one often has to deal with data collected by different teams, using different methods and storage formats (not all of them being digital). Thus, data in life sciences is not only big, varied and valuable, but also tends to occupy substantial space in laboratories and archives.

It is only recently that scientists have at their disposal high-throughput genomic technologies that enable the analysis of whole genomes or transcriptomes originating from multiple samples. Now they are able to correlate these data with phenotypic data such as biochemical marks, imaging, medical histories, etc.

Some of the challenges in that endeavor are choosing the best measurement methods that can be used by different people or teams, and collecting the most reliable data. Later there comes the problem of digitising the results of measurements and combining them with the other data. Furthermore, genomic experiments tend to yield huge files of raw data that need to be analysed using specific algorithms. It is not obvious what should be done with those raw data after the analysis. Should they be saved, because there is a chance for a better analysing algorithm in the future? Should they be deleted, to make room for future data? Should they be shared in some commonly accessible databases?

Life Science is developing rapidly, bringing about spectacular discoveries. Yet scientists are often afraid of Big Data, even though they deal with it very often. In my opinion there is a need for discussion resulting in development of guidelines and standards for collecting diverse types of scientific data, combining and analysing them in a way that maximises the reliability of results.

Sabina Leonelli (University of Exeter), **'Semantic Interoperability: The Oldest Challenge and Newest Frontier of Big Data'**

A key task for contemporary data science is to develop classification systems through which diverse types of Big Data can be aligned to provide common ground for data mining and discovery. These systems determine how data are mined and incorporated into machine learning algorithms; which claims—and about what—data are taken as evidence for; whose knowledge is legitimised or excluded by data infrastructures and related algorithms; and whose perspective is incorporated within data-driven knowledge systems. They thus inform three key aspects of data science: the choice of expertise and domains regarded as relevant to shaping data mining procedures and their results; the development and technical specifications of data infrastructures, including what is viewed as essential knowledge base for data mining; and the governance of data dissemination and re-use through such infrastructures. The challenge of creating semantically interoperable data systems is well-known and has long plagued the biological, biomedical, social and environmental sciences, where the methods and vocabulary used to classify data are often finely tailored to the target systems, and thus tend to vary across groups working on different organisms and ecosystems. A well-established approach to this challenge is to identify and develop one centralised system, which may serve as a common standard regardless of the specific type of data, mining tools, learning algorithms, research goals and target systems in question. However, this has repeatedly proved problematic for two main reasons: (1) agreement on widely applicable standards unavoidably involves loss of system-specific information that often turns out to be of crucial importance to data interpretation; and (2) the variety of stakeholders, data sources and locations at play inevitably results in a proliferation of classification systems and increased tensions among different interest groups around what system to adopt and impose on others. Taking these lessons into account, this paper takes some steps towards developing a conceptual framework through which different data types and related infrastructures can be linked globally and reliably for a variety of purposes, while at the same time preserving as much as possible the domain- and system-specific properties of the data and related metadata. This enterprise is a test case for the scientific benefits of epistemic pluralism, as advocated by philosophers such as John Dupré, Hasok Chang, Ken Waters and Helen Longino. I argue that 'intelligent data linkage' consists of finding ways to mine diverse perspectives and methods of inquiry, rather than to overcome and control such diversity.

Domenico Napoletani, Marco Panza and Daniele C. Struppa (Chapman University),
'Finding a Way Back: Philosophy of Data Science on Its Practice'

Because of the bewildering proliferation of data science algorithms, it is difficult to assess the potential of individual techniques, beyond their obvious ability to solve the problems that have been tested on them, or to evaluate their relevance to specific datasets. In response to these difficulties, an effective philosophy of data science should be able not only to describe and synthesise the methodological outline of this field, but also to project back on the practice of data science a discerning frame that can guide, as well as be guided by, the development of algorithmic methods. In this talk we attempt some first steps in this latter direction. In particular we will explore the appropriateness of data science methods for large classes of phenomena described by processes mirroring those found in developmental biology. Our analysis will rely on our previous work (Napoletani, Panza & Struppa, 2011, 2014, 2017) on the motifs of mathematisation in data science: the principle of forcing, which emphasises how large data sets allow mathematical structures to be used in solving problems irrespective of any heuristic motivation for their usefulness; and Brandt's principle (Napoletani, Panza & Struppa, 2017), which synthesises the way forcing local optimization methods can be used in general to build effective data-driven algorithms. We will then show how this methodological frame can provide useful broad indications on key questions of stability and accuracy for two of the most successful methods in data science, deep learning and boosting.

References

Napoletani, D., Panza, M., & Struppa, D. C. (2011). Agnostic science. Towards a philosophy of data analysis. *Foundations of Science*, *16*, 1–20.
Napoletani, D., Panza, M., & Struppa, D. C. (2014). Is big data enough? A reflection on the changing role of mathematics in applications. *Notices of the American Mathematical Society*, *61*(5), 485–490.

Napoletani, D., Panza, M., & Struppa, D. C. (2017). Forcing optimality and Brandt's principle. In J. Lenhard & M. Carrier (Eds.), *Mathematics as a Tool. Boston Studies in the Philosophy and History of Science 327*. Cham: Springer.

Wolfgang Pietsch (Technical University of Munich), **'On the Epistemology of Data Science—the Rise of a New Inductivism'**

Data science, here understood as the application of machine learning methods to large data sets, is an inductivist approach, which starts using facts to infer predictions and general laws. This basic assessment is illustrated by a case study of successful scientific practice from the field of machine translation, and also by a brief analysis of recent developments in statistics, in particular the shift from so-called data modelling to algorithmic modelling as described by the statistician Leo Breiman. The inductivist nature of data science is then explored by discussing a number of interrelated theses. First, data science leads to the increasing predictability of complex phenomena, especially to more reliable short-term predictions. This essentially follows from the improved ways of storing and processing data by means of modern information technology, in combination with the inductive methodology provided by machine learning algorithms. Second, the nature of modelling changes from heavily theory-laden approaches with little data to simple models using a lot of data. This change in modelling can be observed in the mentioned shift from data to algorithmic models. The latter are in general not reducible to a relatively small number of theoretical assumptions, and must therefore be developed or trained with a lot of data. Third, there are strong analogies between exploratory experimentation, as characterised by Friedrich Steinle and Richard Burian, and data science. Most importantly, a substantial theory-independence characterises both scientific practices. They also share a common aim, namely to infer causal relationships by a method of variable variation, as will be elaborated in more detail in the following theses. Fourth, causality is the central concept for understanding why data-intensive approaches can be scientifically relevant, in particular why they can establish reliable predictions or allow for effective interventions. This thesis states the complete opposite of the popular conception that with big data correlation replaces causation. In a nutshell, the argument for the fourth thesis is contained in Nancy Cartwright's point that causation is needed to ground the distinction between effective strategies and ineffective ones. Because data science aims at effectively manipulating or reliably predicting phenomena, correlations are not sufficient, but rather causal connections must be established. Sixth, the conceptual core of causality in data science consists in difference-making rather than constant conjunction. In other words, variations of circumstances are much more important than mere regularities of events. This is corroborated by an analysis of a wide range of machine learning algorithms, from random trees or forests to deep neural networks. Seventh, the fundamental epistemological problem of data science as defined above is the justification of inductivism. This is remarkable, since inductivism is by many considered a failed methodology. However, the epistemological argument against inductivism is in stark contrast to the various success stories of the inductivist practice of data science, so a re-evaluation of inductivism may be needed in view of data science.

Gregory Wheeler (Frankfurt School of Finance & Management), **'Prolegomena to Machine Epistemology'**

Until very recently, artificial intelligence was an interesting if mildly disreputable academic endeavor. There were logical approaches emphasizing intelligent thinking and statistical approaches emphasizing intelligent action, but neither really worked that well. And the philosophy of computer science was as speculative and detached from practice as the philosophy of mind once was. Those days are over. AI works now, and it runs on statistical machine learning methods. The philosophy of machine learning engages two types of topics (i) those that concern the foundations of machine learning, and (ii) those that contrast the performance of common machine learning algorithms to the performance of common programs in traditional and formal epistemology. This talk presents examples of each.

Symposium of the DLMPST Commission on Arabic Logic
Organiser: Wilfrid Hodges

The Arabic Logic Commission offered a symposium of three half-hour talks. The first was to be a tutorial introduction to Arabic Hypothetical Logic, by Saloua Chatti. The second was to be a report by Wilfrid Hodges on the—apparently newly discovered—twelfth-century logic diagrams of Abū al-Barakāt al-Baghdādī. The third was to be a commentary on Barakāt's diagrams from the point of view of modern logic diagrams.

Fate has decreed that at every Arabic Logic Commission symposium at least one speaker will fail to appear. In Prague Fate went into overdrive and we lost two of the three speakers. Saloua Chatti was taken to hospital with appendicitis shortly before the Congress; she sent her slides and they were read by Wilfrid Hodges. The third speaker never appeared, thanks to a communication breakdown. So, we had two half-hour lectures, and most of the audience stayed into the next hour for further discussion of the material. In fact, sixteen people attended the symposium. None of the audience were specialists in Arabic Logic, so the symposium proved to be purely educational.

Tutorial, Saloua Chatti (Professor of Philosophy at the University of Tunis, Tunisia), **'An Introduction to Arabic Hypothetical Logic'**

At least from the tenth century AD, Arabic logicians have recognised a branch of logic which they called 'conditional' (*shartī*). This branch of logic has a complicated history going back to ancient Greece. Within Arabic logic its history can be divided into three periods: (1) Al-Fārābī's *istithnā'ī* ('exceptive') logic in the early tenth century; (2) several new kinds of *shartī* logic introduced by Ibn Sīnā (known in the West as Avicenna) in the early eleventh century; (3) revisions and developments of *shartī* logic by various logicians from the twelfth century to modern times.

Al-Fārābī's 'exceptive' logic is very similar to the 'hypothetical' logic described in Latin by Boethius in the sixth century; the two logics must have a common origin. Modern scholars have tended to borrow Boethius's name 'hypothetical' to describe Arabic *shartī* logic; the name is less misleading than 'conditional'. But al-Fārābī's 'exceptive' logic can also be read as a kind of propositional logic. For example, two of his main sentence types are 'If A is B then C is D' and 'Either A is B or C is D'. The first type allows the inference rule *modus ponens*, in the form:

> If A is B then C is D.
> But A is B.
> Therefore C is D.

The second type allows the inference rule:

> Either A is B or C is D.
> But A is B.
> Therefore C is not D.

In other words, al-Fārābī reads 'Either ... or' as an exclusive disjunction. He interprets 'if ... then' sometimes as an intensional conditional and sometimes as an intensional (and potential) biconditional.

Avicenna introduced several new hypothetical logics at various stages in his career. Very early he used and defined explicitly the inclusive disjunction as in the inference below:

> Either A is B or C is D.
> But A is not B.
> Therefore C is D.

Earlier logicians had used inclusive disjunction, but Avicenna seems to have been the first to make it the main reading of 'Either ... or'. In his later treatise *al-Ishārāt wa at-tanbīhāt*, he distinguished between three main kinds of disjunction. These are: the exclusive disjunction, the negated conjunction, and the inclusive disjunction. This distinction will be widely used by his followers such as al-Khūnajī and al-Tūsī.

But his most original contribution was a new form of hypothetical logic based on the introduction of quantification inside the hypothetical sentences, which gave rise to quantified hypothetical syllogisms close to Aristotle's categorical syllogisms. Its sentence forms included universal or existential quantification over times or situations, as in

> Whenever A is B, C is D.
> Never when A is B, C is D.
> Sometimes when A is B, C is D.
> Not whenever A is B, C is D.

The hypothetical syllogistic moods containing these propositions are different from the *istithnā'ī* syllogistic moods provided by al-Fārābī, which are also provided by Avicenna but given much less importance. They are called *iqtirānī* or 'recombinant', because of their closeness to the usual categorical syllogisms. He provided several hypothetical systems containing 1) only quantified conditional sentences, 2) quantified conditional plus disjunctive sentences and 3) quantified conditional plus disjunctive plus usual predicative sentences.

Avicenna rejected al-Fārābī's biconditional and distinguished between two main kinds of conditional: 1) a strong conditional where the consequent really follows from the antecedent, and 2) a conditional called '*ittifāq*' where there is no such relation of following from.

His followers such as Al-Khūnajī in the twelfth century recombined the components of Avicenna's hypothetical logics in a new way, concentrating on the first kind of conditional. Al-Khūnajī introduced a range of more complicated sentence forms, developing in his own way the quantified system.

But it is through his brief treatise *al-Jumal*, wherein al-Khūnajī develops a kind of propositional logic by clarifying Avicenna's three kinds of disjunction and his two kinds of conditional, and by doing so arrives at many new laws of propositional logic that were influential in later North African logic. This influence can be seen through Ibn Arafa's (fourteenth century) commentary of *al-Jumal*, and al-Sanūsī's (fifteenth century) commentary of Ibn Arafa's treatise. Both scholars 'extensionalised' al-Khūnajī's definitions of both the disjunction and the conditional and arrived at more complex propositions such as the distributivity laws involving the disjunction and the conjunction and other laws of propositional logic.

The slides of this lecture are online at http://wilfridhodges.co.uk/chattiprague.pdf.

For the development from al-Khūnajī onwards, see also:
Khaled El-Rouayheb. *The Development of Arabic Logic (1200–1800)*. Basel: Schwabe Verlag, 2019.

Lecture, Wilfrid Hodges (British Academy), **'Abū al-Barakāt and His 12th Century Logic Diagrams'**

Abū al-Barakāt bin Malkā al-Baghdādī, Barakāt for short, was a leading philosopher and physicist in Iraq in the twelfth century. He had a reputation for originality and insight, and among other things he contributed to making acceleration and inertia basic notions of physics. He invented a way of doing the most basic logic—Aristotle's logic of categorical syllogisms—by using diagrams consisting of labelled horizontal lines. Unlike other methods before modern times, his method would show both when a pair of formal premises has a logical conclusion, and when it doesn't. (Aristotle could show both these things but using different methods in the two cases).

By 'formal' we mean using letters for terms, as in 'Every A is a B', 'No A is a B', 'Some A is a B', 'Some A is not a B'. Leibniz in the 17th century used picture diagrams to validate syllogisms. Leibniz's idea, followed later by Euler and Venn, was to use pictures to represent Aristotle's *formal sentences*, so as to copy Aristotle's reasoning but in pictures. Barakāt's idea was different: he used diagrams to represent *interpretations* of the letters by showing the set-theoretic relationships between the class of As and the class of Bs, etc. A syllogism is valid if and only if every interpretation making its premises true, makes its conclusion true too (as in Tarski's model-theoretic consequence).

We give two examples. First, here is Barakāt's proof that 'Every A is a B' and 'Every B is a C' together entail 'Every A is a C'. He considers all the four ways in which the two premises can be true:

```
C_____    C_____    C_____    C_____
 B_____     B_____     B_____     B_____
  A____      A____      A____      A____
```

We check by inspection that in each of these four cases 'Every A is a C' is true. (Aristotle gave no proof of this syllogism; he regarded it as self-evident).

Second, here is Barakāt's proof that 'No A is a B' and 'Every B is a C' don't entail any relationship from A to C. He illustrates three interpretations that make these two premises true:

```
                    C  rational
                    B  human___
         A  horse__
```

```
         C  animal__
         B  horse___
                    A  human___
```

```
         C  black___
         B  crow____
                    A  human___
```

In the first interpretation 'Some A is a C' is false; in the second, 'Some A is not a C' is false; and in the third both 'Every A is a C' and 'No A is a C' are false. So, there is no valid syllogistic conclusion.

Although Barakāt's method has been hiding in plain sight in his book *Kitāb al-mu'tabar*, it seems that the method was not deciphered before 2017. It raises a wide range of questions, which are gradually being answered.

Relevant papers are:

Wilfrid Hodges (to appear), A correctness proof for al-Barakāt's logical diagrams, *Review of Symbolic Logic*.

Jules Janssens (2016), Abū al-Barakāt al-Baghdādī and his use of Ibn Sīnā's *al-Ḥikma al-'Arūḍiyya* (or another work closely related to it) in the logical part of his *Kitāb al Mu'tabar*, *Nazariyat—Journal for the History of Islamic Philosophy and Sciences*, *3*(1), 2016, 1–22.

The slides of this lecture are online at http://wilfridhodges.co.uk/arabic66.pdf.

A YouTube tutorial on Barakāt's logical diagram method is available through the Arabic Logic Commission webpage at http://arabiclogic.com.

Symposium summary written by Saloua Chatti and Wilfrid Hodges.

Panels

Teaching panel
Organiser: Joeri Witteveen (University of Copenhagen)

ERC panel
Organiser: Hanne Andersen (University of Copenhagen)

Journal panel
Organiser: Hanne Andersen

In addition to many regular sessions, CLMPST 2019 also contained panel sessions on how to teach philosophy of science, on how to acquire research funding, and on current developments in scholarly publishing.

The session on teaching history and philosophy of science was organized by Joeri Witteveen (University of Copenhagen/Utrecht University) as a panel with contributions from Mieke Boon (University of Twente), Hasok Chang (University of Cambridge), Hans Halvorson (Princeton University), Mikkel Willum Johannsen (University of Copenhagen), Alan Love (University of Minnesota) and Roy Wagner (ETH Zürich). The session focused specifically on the role of philosophy of science in the education of scientists. In presentations of their own individual experiences and in conversations with the audience, panel members addressed how philosophy of science courses for science students differ from the traditional philosophy of science courses offered in philosophy programs; what kind of teaching material is useful when teaching history, philosophy and sociology to science students rather than to students in the fields of history, philosophy and sociology of science themselves; how the teaching of philosophy of science can benefit from current research in integrated history and philosophy of science, philosophy of science in practice, and socially relevant philosophy of science; and how philosophers of science can collaborate with scientists in promoting philosophical reflection in traditional science programs.

The session on research funding was organized by the PC chair, Hanne Andersen (University of Copenhagen). Focusing primarily on large-scale grants, the session drew its examples from the ERC funding scheme. As part of the session, previous ERC panel member Atocha Aliseda (National Autonomous University of Mexico) together with grantees Tarja Knuuttila (Univesity of Vienna) and Barbara Osimani (Marche Polytechnic University) presented their experiences as applicants and as evaluator, gave advice on how to develop and present a large-scale research endeavor, and answered questions from the audience.

The panel on scholarly publishing consisted of five editors from prominent journals in history and philosophy of science: Rachel Ankeny (University of Adelaide, editor of *Studies in History and Philosophy of Biological and Biomedical Sciences*), Otávio Bueno (University of Miami, editor of *Synthese*), Sabina Leonelli (University of Exeter, editor of *History and Philosophy of the Life Sciences*), Thomas Reydon (Leibniz University Hannover, editor of *Journal for General Philosophy of Science*), and K. Brad Wray (University of Aarhus; editor of *Metascience*). They reflected on recent developments in scholarly publishing, such as the Open Science movement, gave advice to especially novices in the field on good publication and reviewing practices, and answered questions from the audience.

(ii) Contributed Symposia

Approaching Probabilistic Truths, in Comparison with Approaching Deterministic Truths

Organisers: Theo Kuipers (University of Groningen) and Ilkka Niiniluoto (University of Helsinki)

The symposium 'Approaching Probabilistic Truths, in Comparison with Approaching Deterministic Truths' (APT) was held on August 6th, 2019, at the 16th International Congress of Logic, Methodology and Philosophy of Science and Technology (CLMPST 2019) organised in Prague. It was initially proposed by Theo Kuipers and then co-organised and co-chaired also by Ilkka Niiniluoto. According to the original plan as it still appears on the official program of the conference, six talks should have been presented at the APT symposium, in presentation order: by Ilkka Niiniluoto (University of Helsinki), by Gustavo Cevolani (IMT School for Advanced Studies Lucca) and Roberto Festa (University of Trieste), by Gerhard Schurz (University of Düsseldorf), by Theo Kuipers (University of Groningen), by Igor Douven (Pantheon-Sorbonne University), and by Graham Oddie (University of Colorado Boulder). However, two speakers (Kuipers and Oddie) could not participate for personal reasons; thus, only five presentations were given, with Niiniluoto giving both his own talk and an overview of Kuipers'.

The goal of the APT symposium was to explore the issue of probabilistic truth approximation from the viewpoint of current theories of 'truthlikeness' (or verisimilitude), and related approaches. The notion of truthlikeness was originally introduced by Karl Popper to make sense of the widespread idea that science, and human knowledge more generally, aim to approach the truth about the world. According to Popper, and many realist philosophers of science after him, scientific theories are always conjectural and corrigible, but later theories may still be 'closer to the truth' than earlier ones; thus, scientific progress consists of approaching truth or increasing verisimilitude. In the last decades, truthlikeness theorists developed different accounts of truth approximation, dealing both with the 'logical' problem (i.e., defining when a theory is closer than another to the given truth), and the 'epistemic' problem of verisimilitude (i.e., evaluating claims of truth approximation in the light of empirical evidence and non-empirical features of relevant theories, even when the truth is unknown). The main results of this thriving research program, both philosophical and technical, are summarised in such works as Niiniluoto (1987, 1998, 1999, 2018), Kuipers (1987, 2000, 2019), Oddie (1987, 2016), and Zwart (2001).

Traditionally, almost all such accounts assumed that 'the truth' to be approached is 'deterministic', i.e., the descriptive or factual truth about some domain of reality or the 'nomic' truth about what is physically or biologically possible. The APT symposium aimed at exploring the prospects of relaxing such an assumption, i.e., of extending the theory of truth approximation to the case where the truth is 'probabilistic'. Here, the target to be approached may be a collection of statistical facts, or the objective probability distribution of some process, or a fully probabilistic law. Given the widespread use of probabilistic and statistical methods in all branches of both theoretical and applied science, it seems clear that adequate theories of truth approximation should also be able to deal with the problem of approaching probabilistic truths. To this purpose, one needs to tackle again, on a new level, both the logical and the epistemic problems: the task becomes to find appropriate measures for the closeness of theories to probabilistic truths, and to evaluate claims about such distances on the basis of empirical evidence.

The speakers at the symposium addressed one or both of such problems, laying down the foundations for a theory of probabilistic truth approximation. A quick overview of the talks will give an idea of the main topics discussed.

The first session of the APT symposium featured three presentations, given by Niiniluoto, Cevolani, and Schurz. In his talk on 'Approaching Probabilistic Laws', Niiniluoto suggested addressing the problem of probabilistic truth approximation as a problem of probabilistic 'legisimilitude', i.e., to treat 'the truth' as defined by some relevant probabilistic law. By applying his favorite similarity approach to verisimilitude, he showed how to employ mathematical measures of the distance between probability distributions (like the Kullback-Leibler divergence) to address both the logical and the epistemic problem of truthlikeness in a probabilistic context, discussing both pros and cons of this approach. In the second talk 'Approaching Deterministic and Probabilistic Truth: A Unified Account', Cevolani showed how to extend the 'basic feature' approach to measuring truthlikeness (Cevolani & Festa, 2020) to cover both deterministic and probabilistic truth approximation; and compared the resulting unified account to other accounts of truthlikeness, revealing interesting differences and similarities. The third talk by Schurz, 'Approaching Objective Probabilities by Meta-Inductive Probability Aggregation', dealt with the logical and the epistemic problems from the point of view of his relevant consequence approach to truthlikeness (initially developed with Paul Weingartner). Schurz linked the issue of probabilistic truthlikeness both with the discussion of different 'scoring rules' used in various settings, and with the formal learning theory as developed within his recent work on optimal meta-induction (Schurz, 2019).

In the second session of the APT symposium, only two talks were presented, since Oddie was unable to give his presentation 'Credal Accuracy in an Indeterministic Universe'. In the first talk, read by Niiniluoto, Kuipers discussed the problem of 'Inductively Approaching a Probabilistic Truth and a Deterministic truth, the latter in comparison with Approaching it in a Qualitative Sense'. Building on his revised theory of nomic truth approximation (Kuipers, 2019), Kuipers discussed both deterministic and probabilistic legisimilitude from the perspective of the theory of inductive probabilities developed in the Carnap-Hintikka tradition, modelling the latter as convergence to the true probabilistic distribution in a multinomial context (the typical example being random sampling with replacement in an urn with colored balls). Finally, in his talk 'Optimizing Group Learning of Probabilistic Truths', Douven discussed truth approximation in a social setting. In particular, he studied the evolution of the collective opinion of a set of (not necessarily human) agents who update their beliefs in a version of the well-known Hegselmann-Krause model, exploring how effectively different updating methods can track the underlying truth.

The APT symposium was very successful and promoted a lively debate among the attendees on how best to tackle the issue of probabilistic truth approximation. Proposals based on existing theories of truthlikeness were compared with different approaches, and new ideas for further exploration of this topic emerged during the discussion. Given that many open problems were left on the table, and foreseeing an interest in contributing from other scholars not attending the symposium in Prague, the organisers decided to promote an open call for papers in order to reach a wider audience and collect new proposals from the community. In the end, an agreement was reached with editor-in-chief Wiebe van der Hoek for publishing a Topical Collection in the *Synthese* journal, titled 'Approaching Probabilistic Truths', and edited by Niiniluoto, Cevolani, and Kuipers. At the moment, ten papers have been published 'online first' in the collection after the usual review process.[24] These include both the six papers originally scheduled at the Prague symposium, and four other papers contributed via the call, as listed below:

1. *Ilkka Niiniluoto,* **'Approaching Probabilistic Laws'**
2. *Alfonso García-Lapeña,* **'Truthlikeness for Probabilistic Laws'**
3. *Gustavo Cevolani and Roberto Festa,* **'Approaching Deterministic and Probabilistic Truth: A Unified Account'**

---

[24] See the topical collection "Approaching Probabilistic Truths" of the journal *Synthese*, published in volume 199 (2021; pages 4195–4216, 6009–6037, 8001–8028, 8281–8298, 9041–9087, 9359–9389, 9391–9410, 10499–10519, 11465–11489, and 11729–11764) and volume 200 (2022; article 113).

4. Gerhard Schurz, '**Probabilistic Truthlikeness, Content Elements, and Meta Inductive Probability Optimization**'
5. Theo Kuipers, '**Approaching Probabilistic and Deterministic Nomic Truths in an Inductive Probabilistic Way**'
6. Alexandru Baltag, Soroush Rafiee Rad, and Sonja Smets, '**Tracking Probabilistic Truths: A Logic for Statistical Learning**'
7. Graham Oddie, '**Propositional and Credal Accuracy in an Indeterministic World**'
8. Igor Douven, '**Scoring, Truthlikeness and Value**'
9. David Atkinson and Jeanne Peijnenburg, '**Probabilistic Truth Approximation and Fixed Points**'
10. Leander Vignero and Sylvia Wenmackers, '**Degree of Riskiness, Falsifiability, and Truthlikeness: A Neo-Popperian Account Applicable to Probabilistic Theories**'

Overall, the APT Topical Collection provides the first systematic exploration of probabilistic truth approximation, bringing together approaches, methods and perspectives from philosophy of science, formal epistemology, and other related disciplines. We are confident that these preliminary results, the multiplicity of analytical methods employed, and the diversity of the topics discussed during the symposium and in the published papers will be instrumental in promoting further developments and new ideas on the issue of probabilistic truth approximation.

References

Cevolani, G., & Festa, R. (2020). A partial consequence account of truthlikeness. *Synthese, 197*, 1627–1646.
Kuipers, T. (Eds.), (1987). *What is Closer-to-the-Truth?* Rodopi.
Kuipers, T. (2000). *From Instrumentalism to Constructive Realism.* Dordrecht: Kluwer.
Kuipers, T. (2019). *Nomic Truth Approximation Revisited.* Cham: Springer.
Niiniluoto, I. (1987). *Truthlikeness.* D. Reidel.
Niiniluoto, I. (1998). Verisimilitude: the third period. *The British Journal for the Philosophy of Science, 49*, 1–29.
Niiniluoto, I. (1999). *Critical Scientific Realism.* Oxford: Oxford University Press.
Niiniluoto, I. (2018). *Truth-Seeking by Abduction.* Berlin: Springer.
Oddie, G. (1986). *Likeness to Truth.* Dordrecht: Reidel.
Oddie, G. (2016). Truthlikeness. In E. N. Zalta (Ed.), *The Stanford Encyclopedia of Philosophy* (Winter 2016 Edition).
Schurz, G. (2019). *Hume's Problem Solved: The Optimality of Meta-Induction.* Cambridge: MIT Press.

Zwart, S. D. (2001). *Refined Verisimilitude.* Dordrecht: Springer.

Authors of the symposium summary: Theo Kuipers (University of Groningen), Ilkka Niiniluoto (University of Helsinki) and Gustavo Cevolani (IMT School for Advanced Studies Lucca).

Association for the Philosophy of Mathematical Practice (APMP) symposium

Organisers: Andrew Arana (Université de Lorraine) Silvia De Toffoli (Princeton University)

The philosophy of mathematics has experienced a significant resurgence of activity during the last 20 years, much of it falling under the widely used label 'philosophy of mathematical practice'. This is a general term for a gamut of philosophical approaches that can also include interdisciplinary work. In order to give focus to this new research community, in 2009 the Association for the Philosophy of Mathematical Practice (APMP) was founded—for more information, see: http://philmathpractice.org.

APMP members promote a broad, outward-looking approach to the philosophy of mathematics, which engages with mathematics in practice, including issues in history of mathematics, the applications of mathematics, cognitive science, etc. The APMP aims to become a common forum that will stimulate research in philosophy of mathematics related to mathematical activity, past and present. It also aims to reach out to the wider community of philosophers of science and stimulate renewed attention to the very significant, and philosophically challenging, interactions between mathematics and science. Therefore, a symposium organised on behalf of the APMP fits well with the aims of this Congress.

To organise this symposium, we asked the members of APMP to submit proposals for taking part in this meeting, and we made an appropriate selection of submission so as to shape a one-day program. The aim of the meeting is to manifest the presence and activity of APMP within the larger community of philosophers of science and logicians. We sought contributions that put into focus different aspects of the philosophy of mathematical practice—both in term of topics and methods—and in grouping them together we aimed at promoting dialogue between them. In order to reach this, we opted for the format of twelve presentations that showcase the diversity of philosophical work done under the umbrella of APMP.

Program:

1. *Michael Friedman*, '**Heterogeneous Mathematical Practices: Complementing or Translation?**'
2. *Bernd Buldt*, '**Abstraction by Parametrization and Embedding. A Contribution to Concept Formation in Modern and Contemporary Mathematics**'
3. *Andrew Aberdein*, '**Virtues, Arguments, and Mathematical Practice**'
4. *Brendan Larvor and Gila Hanna*, '**As Thurston Says**'
5. *Markus Pantsar*, '**Complexity of Mathematical Cognitive Tasks**'
6. *Henrik Kragh Sørensen and Mikkel Willum Johansen*, '**Employing Computers in Posing and Attacking Mathematical Problems: Human Mathematical Practice, Experimental Mathematics, and Proof Assistants**'
7. *Gisele Secco*, '**The Interaction between Diagrams and Computers in the First Proof of the Four-Color Theorem**'
8. *John Mumma*, '**The Computational Effectiveness of Geometric Diagrams**'
9. *Ladislav Kvasz*, '**On the Relations between Visual Thinking and Instrumental Practice in Mathematics**'
10. *Arezoo Islami*, '**Who Discovered Imaginaries? On the Historical Nature of Mathematical Discovery**'
11. *Janet Folina*, '**The Philosophy and Mathematical Practice of Colin Maclaurin**'
12. *Marlena Fila*, '**On Continuity in Bolzano's 1817 Rein Analytischer Beweis**'

Bolzano's Mathematics and the General Methodology of the Sciences
Organiser: Steve Russ (University of Warwick)

This Symposium, which consisted of ten talks, was effectively the third meeting in a series of meetings in Prague on the work of Bernard Bolzano initiated by Steve Russ and Arianna Betti (in 2010 and 2014 respectively). Some details of previous meetings, as well as further material related to this Symposium, can be found from the links on the webpage at www.bernardbolzano.org.

It was a particular pleasure this year (2019) to benefit both from the organisational infrastructure of the CLMPST and from being an integral part of a much broader community. The venue of the Congress was especially appropriate for this meeting, since Bolzano spent much of his life and scientific activity in Prague and his work belongs to the cultural heritage of Bohemia. The talks, which were well-attended and generally provoked good discussions, fell naturally into the following four groups.

Concepts and Methodology

The Symposium began with a talk by Michael Otte, entitled 'Bolzano, Kant and the evolution of the concept of concept'. Bolzano, as well as Quine, derived their conception of the analytic/synthetic distinction from Kant's *Kritik der reinen Vernunft*. But while Quine thinks (as in analytical philosophy in general) that a theory of linguistic meaning is what comes out of empirical research into linguistic behaviour, Bolzano understands that the analytic/synthetic distinction reflects the complementarity of sense and reference of our semiotic representations (say, linguistic or mathematical). From Otte's point of view, Bolzano took into account 'the pragmatic aspects of language', with analytical propositions being 'particularly useful to the pragmatics of mathematical discourse'. This was followed by a talk from Annapaola Ginammi and Arianna Betti on the relationship between concepts and grounding. The topic of grounding, in various senses, has attracted much attention in recent years and is one to which Bolzano scholars have also contributed (work by Betti, Roski, Rumberg, Šebestík, among others), because of the introduction by him, as early as 1810, of the *Abfolge* relationship—often translated as a 'ground-consequence' relationship—which is rightly regarded as a major contribution to semantics. The work of Ginammi and Betti here was a careful and very clear analysis of the mature work by Bolzano found in disparate parts of his *Wissenschaftslehre*. They suggest their account does better justice than previous ones to all Bolzano's claims; it may also offer an improved understanding of the explanatory role of grounding. Lastly in this opening group of talks, was Paola Cantù's task to describe Bolzano's further constraints on concepts through notions of their 'Correct Ordering'. She described the natural applications of this idea to building both complex concepts out of simpler ones, and to developing proofs. She began by using Dale Johnson's 'Prelude to Dimension Theory' (1977) paper to illustrate and introduce a 'theory of definition', and made the fascinating and provocative claim that in Bolzano's work we saw 'mathematics leading the philosophy' (rather than the converse). Another interesting topic raised, perhaps more in the abstract than in the talk, was the potential for Bolzano's ideas about the ordering of concepts in definitions and proofs (in so far as they were known) to influence the emergence of axiomatics later in the century.

Annapaola Ginammi. Photo by Romana Kovácsová

The Mathematical Infinite and other Problems

Johan Blok was unfortunately not able to attend to give his talk, which was replaced by discussion of his abstract and framed around four or five questions. His key suggestion that Bolzano's later interest in entertaining 'objectless' concepts might have contributed to the resolution of difficult problem areas really needed his prior thinking and command of sources.

The next two talks referred to very familiar problem areas: comparing the sizes of infinite collections on the one hand and reacting to the continued widespread use of the language (at least) of infinitesimals on the other hand. But each speaker, in their own way, introduced new perspectives. The first talk, from Kateřina Trlifajová, 'Bolzano and the Part-Whole Principle for Infinite Collections', supports the intuitive idea that the whole is greater than the part even for infinite collections (contradicted by the Dedekind/Cantor criterion of 1-1 correspondence for equality of size). She has shown that we can make a consistent system preserving the part-whole principle using ideas from non-standard analysis, and that such a system may be related to more recent work on a theory of numerosities (see the first reference of her abstract). Also interesting here is Bolzano's mysterious reference to the 'determining ground' of a collection: this might enable some reconciliation between the part-whole principle and 1-1 correspondence.

Thanks to the second talk in this group, we now know a good deal more about the change of mind that occurred for Bolzano between his early published work (up to 1817), and his later unpublished work from the 1830s. In a talk a year earlier in Prague, Elías Fuentes Guillén had shown us passages in the notebooks (from around 1817), in which Bolzano is already becoming less dogmatic about rejecting the infinitely small than he had been in the publications of that time. In his Symposium talk, Fuentes Guillén brought forward evidence from the notebooks that the 'variable quantities ω' used in Bolzano's published works from 1816–1817 have been carefully chosen, and that there is an 'intrinsic difference' between these quantities and the quantities ε made famous by Weierstrass. This challenges the conventional historiography that these quantities are 'equivalent' and that the uses of them for definitions such as that of the continuity of function are also 'equivalent'.

Mathematical Manuscripts

The two afternoon sessions, each of which consisted of two talks, addressed Bolzano's mathematics in the light of a number of writings which he did not publish during his lifetime, but which were either nearly ready for publication (e.g., his *Paradoxien des Unendlichen)*, were drafts of works (e.g., the volumes comprising his project of a *Grössenlehre)*, or were personal documents (e.g., his mathematical notebooks written from 1799). The vast majority of this material was only rediscovered during the 20th century and was—or will be—published for the first time in the *Bernard Bolzano Gesamtausgabe (BBGA)*, which began to be published in 1969 by Frommann-Holzboog Verlag and is currently edited by Edgar Morscher.

Jan Šebestík. Photo by Romana Kovácsová

The first afternoon session was opened by a talk on the *Concursprüfung* for the Chair of Elementary Mathematics at the University of Prague in 1804, in which Bolzano took part. This talk, by Davide Crippa and Elías Fuentes Guillén, and with the participation of Jan

Makovský, presented the ongoing findings on such a *Concursprüfung*. As they showed, the study of the mathematical practices which prevailed in Prague at the turn of the 19th century, as well as the context in which the examination took place, helps to understand better the content of the examination and the criteria in its assessment. Moreover, they showed that Bolzano's examination not only reveals his familiarity with some of the more novel mathematical practices of the time, such as Lagrange's 'functional calculus', but also incipient foundational and conceptual concerns.

The second talk of this session, entitled 'Looking at Bolzano's Manuscripts', was delivered by Jan Šebestík, who guided the audience on a journey through the material that forms part of the *BBGA*. Šebestík noted that among these manuscripts are extracts and commentaries on the works of authors such as Lagrange, Gauss and Carnot, which constitute a valuable source of information on the development of Bolzano's mathematical thinking, as well as manuscripts which cast light on key notions in his late works. In particular, Šebestík highlighted Bolzano's 'preliminary sketches or auxiliary notes' on the notions of *Grösse* and *Inbegriff* (collection, i.e., a whole with parts), which are central to his *Grössenlehre* and *Paradoxien*, and urged the new generations interested in Bolzano's mathematics to pay special attention to his manuscripts on geometry and mechanics.

Kinds of Numbers

After a short pause, the second afternoon session, headed 'Kinds of Numbers', began with a talk by Anna Bellomo entitled 'Bolzano's Measurable Numbers: Sets or Sums?' Bolzano developed the notion of measurable numbers within his *Reine Zahlenlehre*, an unfinished work which was part of his *Grössenlehre,* and the most complete version of which was not published until 1976 as part of the *BBGA*. In her talk, Bellomo challenged the set-theoretic interpretation of Bolzano's measurable numbers. Instead, she argued for an interpretation in line with his theory of collections, and by carefully examining §107 of the 7th section of that work (which can be considered equivalent to Cauchy's convergence criterion), she provided compelling evidence that Bolzano's proposal would be 'best expressed in terms of parts and wholes'.

The Symposium closed with a talk by Peter Simons, 'On the Several Kinds of Number in Bolzano'. Simons discussed Bolzano's 'revolutionary' treatment of natural numbers in the light of the theory of collections (*Mengen, Reihen, Summen* and *Vielheiten* being different kinds of collection). He reconstructed such a treatment from passages of the *Wissenschaftslehre, Paradoxien* and *Grössenlehre,* and, on the basis of Bolzano's distinction between 'concrete and abstract units of a given kind A', he analysed Bolzano's account of numbers as members of a series which would not admit repetitions (unlike sequences in the modern sense). In the last part of his talk, Simons attempted to 'simulate sequences using Bolzano's methods', in particular by resorting to 'collections of collections'.

The talk 'On Continuity in Bolzano's 1817 *Rein Analytischer Beweis...*', by Marlena Fila, was presented at the Symposium of the Association for the Philosophy of Mathematical Practice (APMP). We include it here because its content clearly belongs to the BMMS of which we accounted it as an 'honorary' component. Fila argued that the division of the meaning of continuity into continuity of real numbers and that of functions first appeared in Bolzano's 1817 work. According to this, Bolzano's 'most insightful contribution' in that work would be the formulation of, on the one hand, the greatest lower bound principle; and, on the other hand, the definition of continuous function. Her presentation included several helpful visualisations of the proof structure of Bolzano's work.

Congress dinner. Photo by Romana Kovácsová

The Symposium represented a valuable occasion for junior and senior researchers coming from different backgrounds and working on Bolzano from different perspectives (history of philosophy, history of logic, history and philosophy of mathematical practice), to meet and learn from each other. Several contributors were members of the Internationale Bernard Bolzano-Gesellschaft, or the APMP. We also acknowledge here welcome support for the meeting from the publisher Frommann-Holzboog Verlag. Finally, we are pleased to list some recent publications by contributors, closely related to their presentations at the Symposium.

References

Fuentes Guillén, E. (2022). Bolzano's theory of *meßbare Zahlen*: Insights and uncertainties regarding the number continuum. In B. Sriraman (Ed.), *Handbook of the History and Philosophy of Mathematical Practice*. Springer.

Ginammi, A., Koopman, R., Wang, S., Bloem, J., & Betti, A. (2022). Bolzano, Kant and the traditional theory of concepts: A computational investigation. In A. de Block & G. Ramsey (Eds.), *The Dynamics of Science: Computational Frontiers in History and Philosophy of Science* (pp. 186–203). Pittsburgh: Pittsburgh University Press.

Fuentes Guillén, E. & Crippa, D. (2021). The 1804 examination for the chair of elementary mathematics at the University of Prague. *Historia Mathematica, 56*, 24–56.e18.

Bellomo, A. & Massas, G. (2021). Bolzano's mathematical infinite. *The Review of Symbolic Logic*, first view, 1–55.

Bair, J., Błaszczyk, P., Fuentes Guillén, E., Heinig, P., Kanovei, V., & Katz, M. G. (2020). Continuity between Cauchy and Bolzano: Issues of antecedents and priority. *British Journal for the History of Mathematics, 35*(3), 207–224.

Fuentes Guillén, E. & Martínez Adame, C. (2020). The notion of variable quantities ω in Bolzano's early works. *Historia Mathematica, 50*, 25–49.

Rusnock, P. & Šebestík, J. (2019). *Bernard Bolzano: His Life and Work*. Oxford: Oxford University Press.

Trlifajová, K. (2018). Bolzano's infinite quantities. *Foundations of Science, 23*, 681–704.

This report was compiled in Prague by Steve Russ with generous and substantial assistance from Davide Crippa and Elías Fuentes Guillén.

Epistemic and Ethical Innovations in Biomedical Sciences
Organiser: David Casacuberta (Autonomous University of Barcelona)

About 90% of the biomedical data accessible to researchers was created in the last two years. This certainly implies complex technical problems on how to store, analyse and distribute data, but it also brings relevant epistemological issues. In this symposium we will present some of such problems and discuss how epistemic innovation is key in order to tackle such issues.

Databases implied in biomedical research are so huge that they rise relevant questions about how scientific method is applied, such as what counts as evidence of a hypothesis when data cannot be directly apprehended by humans, how to distinguish correlation from causation, or in which cases the provider of a database can be considered co-author of a research paper. Current characterizations of hypothesis formation, causal link, or authorship are not sufficient for analysis of this issue, and we need some innovation in the methodological and epistemic fields in order to revise these and other relevant concepts.

At the same time, due to the fact that a relevant quantity of such biomedical data is linked to individual people, and that some knowledge from biomedical sciences can be used to predict and transform human behavior, there are ethical questions which are difficult to solve, as they imply new challenges. Some of these are in the awareness field, so patients and citizens understand these new ethical problems that did not arise before the development of Big Data; others relate to the way in which scientists can and cannot store, analyse and distribute information; and some others relate to the limits of which technologies are ethically safe, and which bring an erosion of basic human rights.

During the symposium we will present a coherent understanding on what epistemic innovation is, and some of logical tools necessary for its development; then we will discuss several cases on how epistemic innovation applies to different aspects of the biomedical sciences, while also commenting on its relevance when tackling ethical problems that arise in contemporary biomedical sciences.

This symposium was organised around the following talks:

David Casacuberta (Universitat Autònoma de Barcelona, Spain), **'Innovative Tools for Reaching Agreements in Ethical and Epistemic Problems in Biosciences'**

This talk will present several innovative methodological tools that are being used in biomedical sciences when epistemic and/or ethical problems arise, and when there are different stakeholders with different values, priorities, and aims who need to reach an agreement. Biomedical sciences may include scientists and technologists from very different fields, who therefore have different languages, aims, methodologies, and techniques. Reaching an agreement when there are so many differences among them can become very complex. Besides, biomedical sciences, either when applied or when gathering information about human subjects, can generate complex ethical problems which imply reaching agreements among very different agents, such as scientists, doctors, nurses, politicians, citizens or animal rights activists.

After a brief presentation of the state of the art in the subject, we will discuss two main methodological tools:

The Ethical Matrix. First developed to discuss when it is ethically admissible to introduce GMO foods in a specific environment, this is a very powerful tool to find agreement in lots of different ethical problems in biomedical sciences, and it can be helpful also when analysing epistemically complex situations where agreements among very different disciplines have to be made.

Value Maps. Built in a collaborative manner, these maps can help researchers to realise ethical implications of their work of which they were not previously aware, and also to discover non-epistemic values that, nonetheless, can be helpful to improve innovate processes in scientific research.

Alger Sans (Universitat Autònoma de Barcelona), **'The Incompleteness of Explanatory Models of Abduction in Diagnosis: The Case of Mental Disorders'**

Abduction is known as 'procedure in which something that lacks classical explanatory epistemic virtue can be accepted because it has virtue of another kind' (Gabbay & Woods, 2005; Magnani, 2017). In classical explanations this lack implies that the specific explanation model of abduction should be considered as special case of abduction. That is because to be an explanation implies something more than to be an abduction: a conclusion, in the sense that the burden of proof falls on the abductive product. To have a conclusive form means that explanation and the theory that needed it are already attuned and, of course, this case eliminates the possibility of accepting something because it has virtue of another kind. It is interesting to note that this causal transformation is the cause of the confusion between the explanation model of abduction and inference to the best available inference, which is also known as IB(A)E. On the other hand, the difference between each is the role of the conclusion.

This last point is important because the special case of explanatory abduction is also suitable to conceptualise medical diagnosis, while IB(A)E not. The reason is that medical diagnosis is only possible if the relation with the medical knowledge of the doctor is tentative. That is, only if there is the lack that abduction implies. In other words, the causality form of abduction is substantially different than IB(A)E, because diagnosis needs a virtue of another kind for to be accepted (Aristotle, Rh, I, 1355b10-22).

However, the other face of this situation is that the specific and causal form of explanatory abduction is only useful in specific medical diagnosis: in cases where it is possible to draw a causal course of facts, as in Neurology (Rodríguez, Aliseda & Arauz, 2008). I want to use this last medical discipline as example because it is possible to see in it one mechanism for diagnosis of brain problems. I want to contrast this medical idea with another medical discipline, which studies the brain too, but from a different point of view: psychiatry. When trying to explain psychiatric diagnosis through classical explanatory abduction, it is possible to see that there is something wrong. One the one hand, the generalisation from enumeration is more difficult than in other medical disciplines; and on the other hand, it is more visible here that a difference between simple diagnosis, and diagnosis plus prescription, is needed in the characterization of abduction.

The reason is that abduction is one form of human reasoning, and if there is one area in which diagnosis does not have causal dependency, then it is possible that classical explanatory model of abduction: a) is a more specific kind of diagnosis (some part of general abduction), or b) diagnosis needs something more for their good conceptualization. I want to try to defend *b* from an analysis of EC-Model of abduction in which I try to defend the necessity to infer moral values.

Anna Estany (Universitat Autònoma de Barcelona, Spain), **'Design Epistemology as Innovation in Biomedical Research'**

The idea of design has reached our theories of epistemology: a field that, at first glance, seems to be quite far removed from the analysis of practical situations. However, we should bear in mind that epistemology has shifted from an *a priori* perspective to a naturalised one, in the sense that we cannot engage in epistemology without accounting for the empirical results of science when it comes to configuring methodological models. In addition, the philosophy of science has expanded its field of analysis beyond pure science, and this has made it necessary to consider the epistemology of applied science. At this point it is relevant design epistemology, as an alternative to analytic epistemology.

The objective is to explore just how far design epistemology (DE) can be adopted as a methodological framework for research in biomedical sciences and in some of their applications as public health. To this end, we will analyse different approaches to DE and to related terms and expressions such as 'design thinking', 'design theory', and 'designedly ways of knowing'. One of the issues that we need to address is precisely the polysemy that exists in the field of design, relating to many different concepts. Thus, it seems impossible not to engage in a certain amount of conceptual analysis before we can embark on the study of the role of DE in public health research.

Another of the questions that we will consider here is where to place biomedical sciences within the fields of academic knowledge and research. The disciplines involved range from biology to medicine, and also to the applications of these bodies of knowledge, as in the case of public health. We will examine some of the definitions provided by international organizations and we will locate public health within the framework of healthcare services and their organization. Finally, we will see how DE can offer proposals and solutions to the challenges that a phenomenon as complex as public health currently faces. That is, we will measure DE proposals against public health research needs. Design epistemology asks a whole series of questions which, at one and the same time, constitute different perspectives and proposals concerning how to understand the subject of DE itself. On the one hand, we have DE as an alternative to classic epistemology, which is often described as 'analytic' and juxtaposed with 'synthetic', which is how DE would be described, as it would also cover the applied sciences. On the other hand, DE is said to have a series of defining characteristics, among which we can highlight 'interdisciplinary' as a means of addressing dynamic and complex problems; and a prominent element of social concern expressed through 'design thinking' that revolves around human-scale design. Around these principal axes, we are going to examine a series of proposals and considerations relevant to biomedical sciences.

Angel Puyol (Universitat Autònoma de Barcelona, Spain), **'Solidarity and Regulatory Frameworks in (Medical) Big Data'**

The use we make of digital technologies produces a huge amount of data, known as Big Data, whose management is becoming increasingly more difficult. One of the problems in this regard is the possibility of data controllers abusing their position of power and using the available information against the data subject. This abuse can have several faces. Prainsack (2015) identifies at least three types of abuse: hypercollection, harm, and humiliation. Hypercollection means that just because institutions can collect information about customers or citizens for purposes other than the ones for which it was collected in the first place, they do so. Harm occurs when the information obtained is used against the interests or rights of the data subject. This damage is accompanied by humiliating effects when making people partake in their own surveillance.

In the face of this new reality, the question of how to govern data use has become more important than ever. The traditional way of governing the use of data is through data protection. Recently, the European Union has published a new General Data Protection Regulation (GDPR) that follows this approach. However, authors such as Prainsack and Buyx (2016) rightly point out that the strictly regulatory approach is insufficient for dealing with all abuses related to the use of Big Data. On the one hand, excessive control can curb the opportunities and benefits of digital technologies for users and for society as a whole. On the other hand, control and regulation may be insufficient in controlling all risks associated with the use of Big Data.

In opposition to the strict regulatory approach, Prainsack and Buyx propose a new one, based on solidarity. The solidarity approach entails the acceptance of the impossibility of eliminating the risks of modern data usage. The authors base their proposal on a solidarity formula whose objective is to compensate those affected by possible abuses: harm mitigation funds. Such funds would help to ensure that people who accept those risks and are harmed as a result have appropriate support.

The paper does not question the adequacy of harm mitigation funds, but rather the conception of solidarity that Prainsack and Buyx choose to justify them. I would argue that this conception of solidarity, based on psychology and moral sociology, has less normative force than exists in the strict regulatory approach, which is based on the defence of fundamental rights. If we want the policy of harm mitigation funds to have a normative force similar to that of the strict regulatory approach, then we must choose a conception of solidarity based on respect for fundamental rights.

In this paper, I first present the context in which it makes sense to oppose a solidarity-based perspective to a strictly regulatory one. Then I review what I believe are the weak points in Prainsack and Buyx's ideas regarding solidarity. And finally, I introduce an alternative conception of solidarity that normatively better justifies any public solidarity policy addressing the risks of Big Data, including harm mitigation funds.

Factivity of Understanding: Moving beyond the Current Debates
Symposium of the East European Network for the Philosophy of Science (EENPS)
Organiser: Lilia Gurova (New Bulgarian University)

The symposium on the factivity of understanding was organised at the invitation of the *Program Committee of the 16th CLMPST*. Its main objective was to show the ongoing research of some of the members of the recently established *East European Network for the Philosophy of Science* (2015). The original symposium project was developed by Lilia Gurova, Richard David-Rus, Daniel Kostic, Insa Lawler, Stefan Petkov and Martin Zach but given some more fortunate and some unfortunate circumstances, at the congress it was presented by Richard David-Rus, Daniel Kostic, Stefan Petkov, and Martin Zach. The following is a general summary of the symposium, with brief descriptions of the contributions of the individual participants.

General Summary

There are several camps in the recent debates on the nature of scientific understanding. There are factivists and quasi-factivists who argue that scientific representations provide understanding insofar as they capture some important aspects of the objects they represent. Representations, the (quasi-)factivists say, yield understanding only if they are at least partially or approximately true. The factivist position has been opposed by the non-factivists who insist that greatly inaccurate representations can provide understanding given that these representations are *effective* or *exemplify* the features of interest. Both camps face some serious challenges. The factivists need to say more about how exactly partially or approximately true representations, as well as non-propositional representations, provide understanding. The non-factivists are expected to put more effort into the demonstration of the alleged independence of effectiveness and exemplification from the factivity condition. The aim of the proposed symposium is to discuss in detail some of these challenges and to ultimately defend the factivist camp.

Individual Contributions

'The Factivity of Model-Based Explanations' (*Insa Lawler*) defends a factive account of model-based explanations (ME). The *explananda* of MEs are argued to be 'relaxed' approximate descriptions of the explanandum-phenomenon. The *explanantia* of MEs involve correct propositions that are *extracted* from the model. On this account, the indispensable idealisations, which many successful models contain, can contribute to factive understanding by enabling the extraction of correct explanatory information.

A different argument for the factivity of scientific understanding provided by models containing idealisations is presented in **'Understanding Metabolic Regulation: A Case for the Factivists'** (*Marin Zach*). The central claim of this paper is that such models bring understanding if they correctly capture the causal relationships between the entities, which these models represent.

What happens, however, when understanding is provided by explanations, which do not refer to any causal facts? This question is addressed in **'Factivity of Understanding in Non-causal Explanations'** (*Daniel Kostic*). The author argues that the factivity of understanding can be analysed and evaluated by using some modal concepts that capture 'vertical' and 'horizontal' counterfactual dependency relations which the explanation describes.

'**Scientific Explanation and Partial Understanding**' (*Stefan Petkov*) focuses on cases where the explanations consist of propositions which are only partially true (in the sense of da Costa's notion of partial truth). The author argues that such explanations bring partial understanding insofar as they allow for an inferential transfer of information from the *explanans* to the *explanandum*.

One of the biggest challenges to factivisim, the existence of non-explanatory representations, which do not possess propositional content but nevertheless provide understanding, is addressed in '**Considering the Factivity of Non-explanatory Understanding**' (*Richard David-Rus*). This paper argues against the opposition between effectiveness and veridicality. Building on some cases of non-explanatory understanding, the author shows that effectiveness and veridicality are compatible, and that we need both.

'**Effectiveness, Exemplification, and Factivity**' (*Lilia Gurova*) further explores the relation between the factivity condition and its suggested alternatives—effectiveness and exemplification. The author's main claim is that the latter are not alternatives to factivity, strictly speaking, insofar as they could not be construed without any reference to truth conditions.

Formalism, Formalization, Intuition and Understanding in Mathematics: From Informal Practice to Formal Systems and Back Again

Organiser: Máté Szabó (AHP Université Lorraine, IHPST Paris)

This symposium is the outcome of a German-French research project (ANR/DFG) between the Archives Henri-Poincaré, University of Lorraine/CNRS (Nancy), the Institute for History and Philosophy of Science and Technology, Université Paris 1 Panthéon-Sorbonne/CNRS (Paris) and the Munich Center of Mathematical Philosophy (Munich).

The aim of the symposium is to investigate the interplay between informal mathematical theories and their formalization, and argue that this dynamism generates three different forms of understanding:

1. Different kinds of formalizations fix the boundaries and conceptual dependences between concepts in different ways, thus contributing to our understanding of the content of an informal mathematical theory. We argue that this form of understanding

of an informal theory is achieved by recasting it as a formal theory, i.e., by transforming its expressive means.
2. Once a formal theory is available, it becomes an object of understanding. An essential contribution to this understanding is made by our recognition of the theory in question as a formalization of a particular corpus of informal mathematics. This form of understanding will be clarified by studying both singular intended models, and classes of models that reveal the underlying conceptual commonalities between objects in different areas of mathematics.
3. The third level concerns how the study of different formalizations of the same area of mathematics can lead to a transformation of the content of those areas, and a change in the geography of informal mathematics itself.

In investigating these forms of mathematical understanding, the project will draw on philosophical and logical analyses of case studies from the history of mathematical practice, in order to construct a compelling new picture of the relationship of formalization to informal mathematical practice. One of the main consequences of this investigation will be to show that the process of acquiring mathematical understanding is far more complex than current philosophical views allow us to account for.

While formalization is often thought to be negligible in terms of its impact on mathematical practice, we will defend the view that formalization is an epistemic tool, which not only enforces limits on the problems studied in the practice, but also produces new modes of reasoning that can augment the standard methods of proof in different areas of mathematics. Reflecting on the interplay between informal mathematical theories and their formalization means reflecting on mathematical practice and on what makes it rigorous, and how this dynamism generates different forms of understanding.

We therefore also aim to investigate the connection between the three levels of understanding described above, and the notion of rigor in mathematics. The notion of formal rigor (in the proof theoretic sense) has been extensively investigated in philosophy and logic, though an account of the epistemic role of the process of formalization is currently missing. We argue that formal rigor is best understood as a dynamic abstraction from informally rigorous mathematical arguments. Such informally rigorous arguments will be studied by critically analysing case studies from different subfields of mathematics, in order to identify patterns of rigorous reasoning.

Talks:

1. *Marco Buzzon (University of Macerata, Italy),* **'Mathematical VS Empirical Thought Experiments: Between Informal Mathematics and Formalization'**
2. *Michael Andrew Moshier (Chapman University, United States),* **'The Independence of Excluded Middle from Double Negation via Topological Duality'**
3. *Marco Panza (CNRS, France),* **'Formalisation and Understanding in Mathematics'**
4. *Alberto Naibo (IHPST (UMR 8590, University Paris 1 Panthéon-Sorbonne, France),* **'A Formalization of Logic and Proofs in Euclid's Geometry'**
5. *Máté Szabó (AHP Univ Lorraine, IHPST Paris 1, France) and Patrick Walsh (Carnegie Mellon University, United States),* **'Gödel's and Post's Proofs of Incompleteness'**
6. *Pierre Wagner (Institut d'histoire et de philosophie des sciences et des techniques, France),* **'Gödel's and Carnap on the Impact of Incompleteness on Formalization and Understanding'**
7. *Silvia De Toffoli (Princeton University, United States),* **'The Epistemic Basing Relation in Mathematics'**
8. *Benedict Eastaugh (Munich Center for Mathematical Philosophy, LMU Munich, Germany) and Marianna Antonutti Marfori (Munich Center for Mathematical Philosophy, LMU Munich, Germany),* **'Epistemic Aspects of Reverse Mathematics'**

Authors of the symposium summary: Máté Szabó, Gerhard Heinzmann (University of Lorraine at Nancy), Marco Panza (CNRS), Marianna Antonutti Marfori (Ludwig Maximilian University, Munich).

From Contradiction to Defectiveness to Pluralism

Organisers: María Del Rosario Martínez Ordaz (Federal University of Rio de Janeiro) and Otávio Bueno (University of Miami)

In their day-to-day practice, scientists make constant use of defective information (false, imprecise, conflicting, incomplete, inconsistent, etc). The philosophical explanations of the tolerance of defective information in the sciences are extremely varied, making philosophers struggle at identifying a single correct approach to this phenomenon. Thus, we adopt a pluralist perspective on this issue in order to achieve a broader understanding of the different roles that defective information plays (and could play) in the sciences.

This symposium was devoted to exploring the connections between scientific pluralism and the handling of inconsistent as well as other types of defective information in the sciences. The main objectives of this symposium are (a) to discuss the different ways in which defective information could be tolerated (or handled) in the different sciences (formal, empirical, social, health sciences, etc). as well as (b) to analyze the different methodological tools that could be used to explain and handle such type of information.

The symposium was divided into two parts: the first addressed the issue of inconsistency and scientific pluralism. This part included discussions of the possible connections between the different ways in which scientists tolerate contradictions in the sciences and particular kinds of scientific pluralism. This analysis is extremely interesting, as the phenomenon of inconsistency toleration in the sciences has often been linked to the development of a plurality of formal approaches, but not necessarily to logical or scientific pluralism. In fact, scientific pluralism is independent of inconsistency toleration.

The second part of the symposium was concerned with a pluralistic view on contradictions and other defects. This part was devoted to exploring under which circumstances (if any) it is possible to use the same mechanisms for tolerating inconsistencies and for dealing with other types of defective information. This part included reflections on the scope of different formal methodologies for handling defectiveness in the sciences, as well as considerations on scientific communicative practices and their connections with the use of defective information, and reflections on the different epistemic commitments that scientists have towards defective information.

Contributions:

1. *Carolin Antos (University of Konstanz) and Daniel Kuby (University of Konstanz),* **'Mutually Inconsistent Set-Theoretic Universes: An Analysis of Universist and Multiversist Strategies'**
2. *Jody Azzouni (Tufts University),* **'Informal Rigorous Mathematics and Its Logic'**
3. *Diderik Batens (Ghent University),* **'Handling of Defectiveness in a Content-Guided Manner'**
4. *Jonas Becker Arenhart (Federal University of Santa Catarina) and Décio Krause (Federal University of Santa Catarina),* **'Quasi-Truth and Defective Situations in Science'**
5. *Xavier de Donato-Rodriguez (University of Santiago de Compostela),* **'Inconsistency and Belief Revision in Cases of Approximative Reduction and Idealization'**

6. *Michèle Friend (The George Washington University)*, '**Disturbing Truth**'
7. *Moisés Macías Bustos (University of Massachusetts-Amherst and National Autonomous University of Mexico)*, '**Lewis, Stalnaker and the Problem of Assertion & Defective Information in the Sciences**'
8. *María Del Rosario Martínez-Ordaz (National Autonomous University of Mexico) and Otávio Bueno (University of Miami)*, '**Making Sense of Defective Information: Partiality and Big Data in Astrophysics**'
9. *Joke Meheus (Ghent University)*, '**Logic-based O in the Biomedical Domain: From Defects to Explicit Contradictions**'

Karl Popper: His Science and His Philosophy
Organiser: Zuzana Parusniková (Czech Academy of Sciences)

Karl Popper is one of the few philosophers of science who is well-known to scientists and respected by them. Apart from the direct influence of his views on science, it is his methodology that most appeals to scientists. He earned the highest accolades for his emphasis on criticism as the essence of progress in science. Not surprisingly, Popper inspired scientists (especially the Nobel Prize winners Peter Medawar, Jacques Monod and John Eccles, in addition to the biologist Donald Campbell, the biochemist Günter Wächtershäuser and the mathematician Hermann Bondi), and he won recognition from the scientific establishment (he was elected a Fellow of the Royal Society in 1976).

Unlike Popper's methodology and epistemology that have been widely and vividly discussed, his impact on scientific research and his contributions to it have received less attention. The aim of this symposium was thus to evaluate the impact that Popper has had in the natural and mathematical sciences. The structure of the symposium provided a unique opportunity to open a debate between scientists and philosophers of science. The topics selected were quantum mechanics, evolutionary biology, cosmology, mathematical logic, statistics, and cognitive science.

The first talk, *Popper and the Quantum Controversy,* was delivered by Olival Freire Junior from the Universidade Federal da Bahia, Brazil. Freire argued that Popper fully accepted the probabilistic descriptions and suggested his propensity interpretation in quantum theory, yet without attachment to determinism; simultaneously Popper criticised the introduction of subjectivist approaches in this scientific domain, thus aligning himself with the realist position in the quantum controversy. Freire also pointed out Popper's collaboration with physicists such as Jean-Pierre Vigier and Franco Selleri, who were hard critics of the standard interpretation of quantum physics. From this collaboration emerged a proposal of an experiment to test the validity of some presumptions of quantum theory. Initially conceived as an idealised experiment, it spurred a debate which survived Popper himself.

This topic was further elaborated by Flavio del Santo (The Institute for Quantum Optics and Quantum Information of the Austrian Academy of Sciences, Vienna, and the University of Vienna, Austria). For Popper, he stressed, quantum formalism could and should be interpreted realistically: in this regard Popper invented a thought experiment in which he intended to show that a particle can have both precise position and momentum at the same time through the correlation measurement of an entangled two-particle system. Thus Popper systematised his critique of the Copenhagen interpretation of quantum mechanics, proposing an alternative interpretation based on the concept of ontologically real probabilities (propensities).

The field of cognitive science was opened by Peter Århem from the Karolinska Institutet, Sweden in his talk *Popper on the Mind-Brain Problem*. He focused on two aspects of Popper's philosophy of mind. One aspect related to the ontology of mind and the theory of biological evolution in which Popper found support for his interactionist view on the mind-brain problem. The second aspect concerned Popper's observation that the mind has similarities with forces and fields of forces. Århem further addressed Popper's hypothesis that consciousness acts on fields of probability amplitudes rather than on electromagnetic fields. As an illustration a case from the field of neuroscience was discussed, relating to the development of theories about mechanisms underlying the nerve impulse.

These arguments were further developed by David Miller (University of Warwick, UK). He explored some similarities between the theories of minds as force fields and the proposal that the propensities that are fundamental to Popper's propensity interpretation of probability should be likened to forces. This latter proposal was made indirectly in one of Popper's earliest publications on the propensity interpretation, but never decisively pursued. Instead, Popper adopted the idea that propensities (which are measured by probabilities) be likened to partial or indeterministic causes. Miller maintained that this was a wrong turn, and that propensities are better seen as indeterministic forces.

Denis Noble (University of Oxford, UK) introduced the theme of evolutionary biology in his contribution *The Rehabilitation of Karl Popper's Views of Evolutionary Biology and the Agency of Organisms*. He observed that Popper contrasted what he called 'passive Darwinism' (essentially the neo-Darwinist Modern Synthesis) with 'active Darwinism' (based on the active agency of organisms). This was a classic clash between reductionist views of biology that exclude teleology and intentionality, and those that see these features of the behaviour of organisms as central in the adaptability driver. In their investigations of how organisms can harness stochasticity in generating functional responses to environmental challenges, Denis Noble and Raymond Noble developed a theory of choice that reconciles the unpredictability of a free choice with its subsequent rational explanation. Here, stochasticity is seen as the clay from which the active behaviour of organisms develops and therefore influences the direction of evolution.

In a critical response to this contribution, Philip Madgwick from the Milner Centre for Evolution, University of Bath, UK, defended the 'Neo-Darwinism' or 'the Modern Synthesis'. Evolutionary biology, Madgwick argued, has tended to understand the 'choices' underlying form and behaviour of organisms as deterministic links in the chain between genotypic causes and phenotypic effects. As selection acts on phenotypes, there is little room for concepts like 'free will' or 'meaningful choice'—instead, agency becomes a useful 'thinking tool' rather than a 'fact of nature'. Critics like Karl Popper have suggested that evolutionary theory has gone further in (unscientifically) denying the existence of what it cannot explain (namely, agency). Madgwick evaluated this line of criticism, highlighting four different aspects of arguments against the concept of agency within modern evolutionary theory.

The first afternoon session began with the theme of cosmology. Helge Kragh (University of Copenhagen, Denmark) presented his paper on *Popper and Modern Cosmology: His Views and His Influence*. He analysed the interaction between Popper's philosophy of science and developments in physical cosmology in the post-World War II era. The impact of Popper's philosophical views, and of his demarcation criterion in particular, is still highly visible in the current debate concerning the so-called multiverse hypothesis. Popper's views, however, changed somewhat over time. While he had some sympathy for the now defunct steady-state theory, he much disliked the big bang theory, which since the mid-1960s has been the generally accepted framework for cosmology. According to Popper, the concept of a big bang as the beginning of the universe did not belong to science proper.

In response to Kragh, Anastasiia Lazutkina (Leipzig University, Germany) defended the view that a methodological analysis of contemporary cosmological models that is in line with Popper's demarcation criterion between scientific and non-scientific cosmology can greatly benefit from the use of formal methods. The application of these formal criteria reveals that there are two contrasting approaches in cosmology: one focusing on small scale phenomena (e.g., galaxies, clusters), the other forming a model of the universe as a whole and then working the way toward smaller scales. The applicability of the Popperian methodology in these cases was discussed.

Statistics and logic were the last topics of the Symposium. Stephen Senn from the Luxembourg Institute of Health, UK, confronted the positions of Bayesianism and falsificationism in his paper *De Finetti Meets Popper or Should Bayesians Care about Falsificationism?* He discussed the classical De Finetti's theory of learning and the formal frequentist Neyman-Pearson approach to hypothesis testing, and their attitudes towards proving scientific laws. He concluded that falsificationism is important for Bayesians also, although it is an open question as to whether it is enough for frequentists.

In response to this paper, Timothy Childers (Institute of Philosophy, Czech Academy of Sciences, Czechia) argued that there are forms of Bayesianism that are closer to Popper's methodology (Howson, Urbach). This form of Bayesianism is motivated by the acceptance of a negative solution to the problem of induction and a deep scepticism towards mechanical approaches to scientific methodology. He further discussed the potential and limits of falsificationism in testing.

The Symposium closed with a talk *Karl R. Popper: Logical Writings* by Thomas Piecha (University of Tübingen, Department of Computer Science, Germany), a co-author of a critical edition *The Logical Writings of Karl Popper* (Springer, in print). He highlighted Popper's position, in which logic is a metalinguistic theory of deducibility relations that are based on certain purely structural rules. Logical constants are then characterised in terms of deducibility relations. He discussed in detail several later developments and discussions in philosophical logic that Popper's works on logic anticipate.

The Symposium proved the fruitfulness of a multidisciplinary approach, opening a dialogue across scientific disciplines and illuminating past and present interactions between philosophy and science. It was also an incentive for the Organiser to extend the field of topics, and to involve other scholars in this project, resulting in a publication *Karl Popper's Science and Philosophy*, edited by *Parusniková*, Zuzana and Merritt, David (Springer, 2021).

David Miller. Photo by Zuzana Parusniková

Welcome dinner. Photo by Zuzana Parusniková

Olival Freire Junior. Photo by Zuzana Parusniková

Denis Noble. Photo by Zuzana Parusniková

Logic, Agency, and Rationality

Organisers: Valentin Goranko (Stockholm University) and Frederik Van De Putte (Erasmus University of Rotterdam)

General Description

The concept of *rational agency* is broadly interdisciplinary, bringing together philosophy, social psychology, sociology, decision and game theory. The scope and impact of the area of rational agency has been steadily expanding in the past decades, also involving technical disciplines such as computer science and AI, where multi-agent systems of different kinds (e.g., robotic teams, computer and social networks, institutions, etc.) have become a focal point for modelling and analysis.

Rational agency relates to a range of key concepts: knowledge, beliefs, knowledge and communication, norms, action and interaction, strategic ability, cooperation and competition, social choice, etc. The use of formal models and logic-based methods for analysing rational agency has become an increasingly popular and successful approach to dealing with this complex diversity.

This symposium brought together different perspectives and approaches to the study of rational agency and interaction in the context of philosophical logic. It was divided into three thematic clusters, each consisting of four to five presentations:

Logic and Game-theoretic Semantics. One general method for interpreting agency, interaction, and related notions in philosophical logic is in terms of game-theoretic semantics. Conversely, formal languages allow us to specify specific properties of and equivalences between games, and hence get better grip on specific solution concepts for such games. The talks that belong to this thematic cluster were:

1. *Antti Kuusisto,* **'Interactive Turing-Complete Logic via Game-Theoretic Semantics'**
2. *Dominik Klein,* **'A Logical Approach to Nash Equilibria'**
3. *Fengkui Ju,* **'Coalitional Logic on Non-Interfering Actions'**
4. *Raine Rönnholm,* **'Rationality Principles in Pure Coordination Games'**
5. *Alexandra Kuncová,* **'Ability and Knowledge'**[25]

---

[25] This talk was canceled last minute, for personal reasons.

Deontic Logic, Agency, and Action. Since Horty's seminal *Agency and Deontic Logic* (Oxford University Press, 2001), the relevance of STIT logic—a logic of agency and strategic interaction—for reasoning about norms is generally acknowledged. Moreover, there is a lasting stream of research in deontic logic on the normative status of actions, as opposed to propositions. This thematic cluster brings together these two traditions, including the following presentations:

6. *Alessandra Marra,* **'From Oughts to Goals'**
7. *Ilaria Canavotto,* **'Introducing Causality in STIT Logic'**
8. *Karl Nygren,* **'Varieties of Permission for Complex Actions'**
9. *Thijs De Coninck,* **'Reciprocal Group Oughts in STIT Logic'**
10. *Grigory Olkhovikov,* **'STIT Heuristics and the Construction of Justification STIT Logic'**

Logic, Social Epistemology, and Collective Decision-making. Rational agency and interaction also presuppose an epistemological dimension. In addition, intentional group agency is inextricably linked to social choice theory. In this thematic cluster, various logical and formal models are discussed that allow us to shed light on these factors and processes. The presentations it includes were:

11. *Rasmus Rendsvig,* **'Dynamic Term-Modal Logic'**
12. *Soroush Rafiee Rad,* **'A Logic for Statistical Learning'**
13. *Olivier Roy & Soroush Rafiee Rad,* **'Deliberation, Single-Peaked Preferences, and Voting Cycles'**
14. *Frederik Van De Putte,* **'Constructive Deliberation: Pooling and Stretching Modalities'**

Brief Descriptions of Talks (in the above order)

Antti Kuusisto, **'Interactive Turing-Complete Logic via Game-Theoretic Semantics'**

We define a simple extension of first-order logic via introducing self-referentiality operators and domain extension quantifiers. We analyse the conceptual properties of this logic, especially the way it links games and computation in a one-to-one fashion.

Dominik Klein, **'A Logical Approach to Nash Equilibria'**

We present first steps towards incorporating MaxEU-like choice rules into logical frameworks for reasoning about games. In particular, we show that enhancing classic strategic logics with modest operators for probabilistic beliefs is sufficient to express when an extended game form has a pure or mixed Nash equilibrium strategy.

Fengkui Ju, **'Coalitional Logic on Non-interfering Actions'**

This work presents a coalitional logic which presupposes that every agent controls a part of the world, and those parts do not overlap.

Raine Rönnholm, **'Rationality Principles in Pure Coordination Games'**

We analyse pure win-lose coordination games in which all players receive the same playoff, either 0 or 1, after every round. Under the assumption of no communication, we study various principles of rationality that can be applied in these games in both one-shot and repeated setting.

Alexandra Kuncová, **'Ability and Knowledge'**

I explore the epistemic qualification of ability and three ways of modelling it. I show that both analyses, of knowing how in epistemic transition systems and of epistemic ability in labelled STIT models, can be simulated using a combination of impersonal possibility and knowingly doing in standard STIT models.

Alessandra Marra, **'From Oughts to Goals'**

This talk focuses on (an interpretation of) the Enkratic principle of rationality, according to which rationality requires that if an agent sincerely and with conviction believes she ought to X, then X-ing is a goal in her plan. We analyse the logical structure of Enkrasia and its implications for deontic logic.

Ilaria Canavotto, **'Introducing Causality in STIT Logic'**

We propose a refinement of STIT semantics in order to represent the causal connection between an agent's actions and their consequences. We do this by supplementing STIT semantics with action types. In this way, we obtain a framework in which we can interpret new STIT operators suitable to represent basic degrees of responsibility of an agent.

Karl Nygren, **'Varieties of Permission for Complex Actions'**

One of the main questions in deontic logic based on propositional dynamic logic is how to decide the normative status of complex actions based on the normative status of atomic actions, transitions and states. Focusing on permission, I will define and discuss a variety of permission concepts for complex actions in propositional dynamic logic.

Thijs De Coninck, **'Reciprocal Group Oughts in STIT Logic'**

We present an alternative semantics for group oughts in a STIT framework, relying on earlier work by Turrini and Grossi. We compare the resulting logic to Horty's utilitarian ought and to the strategic oughts by Kooi and Tamminga.

Grigory Olkhovikov, **'STIT Heuristics and the Construction of Justification STIT Logic'**

We propose a set of heuristic assumptions for the STIT approach, extending the proposal of Belnap et al. in their seminal work. We show how our heuristics is useful in constructing extensions of STIT logic, taking justification STIT logic as an example.

Rasmus Rendsvig, **'Dynamic Term-Modal Logic'**

In term-modal logics, operators double as predicates, making e.g., $\exists x K_x \varphi(x)$ well-formed. We present a well-behaved term-modal semantics with dynamic extension and complete proof system, and illustrate it using epistemic social network dynamics.

Soroush Rafiee Rad, **'A Logic for Statistical Learning'**

We study a dynamic doxastic logic for expressing and analysing the process learning probability distributions by observations as well as learning of higher order information. We study the learning mechanism in terms of its long-term behaviour. We show relevant convergence results and briefly investigate the logical structure and some related validities.

Olivier Roy & Soroush Rafiee Rad, **'Deliberation, Single-peaked Preferences, and Voting Cycles'**

In social choice theory, it is often claimed that preference cycles can be avoided by deliberation, since the latter fosters single-peaked preferences. We will present an agent-based model of deliberation and argue that this model sheds new light on this claim.

Frederik Van De Putte, **'Constructive Deliberation: Pooling and Stretching Modalities'**

In *constructive deliberation*, a group of agents determines which alternatives make up a collective decision problem for that group. After a general discussion of this type of process, we introduce a modal logic that allows us to explicate one specific aspect of it, viz. the way agent-specific, soft constraints are merged among coalitions of agents.

In Conclusion

The symposium went very well, sparked lively and interesting discussions and was quite successful in achieving its goals. The organisers are much indebted to the CLMPST local organizing committee, and to Ondrej Majer in particular, for supporting this symposium and for helping us prepare it in the best possible way. Looking back, and given the current pandemic situation, one cannot help but being nostalgic about the way the entire conference was organised and the many intellectual (in vivo!) contacts it enabled.

Mario Bunge: Appraising his Long-Life's Contribution to Philosophy

Organiser: Michael Matthews (University of New South Wales)

As Mario Bunge celebrates his 100th birthday, this symposium will appraise four different aspects of his life-long contribution to philosophy. The five individual presentations are: Mario Bunge: A Pioneer of the New Philosophy of Science; Mario Bunge's Scientific Approach to Realism; Mach and Bunge on the Principle of Parsimony; Quantifiers and Conceptual Existence; Bunge and the Enlightenment Tradition in Education.

Bunge was born in Argentina on 21st September 1919. He has held chairs in physics and in philosophy at universities in Argentina and the USA, and since 1966 a philosophy chair at McGill University. He has published 70 books (many with revised editions) and 540 articles; with many translated into one or other of twelve languages.

Bunge has made substantial research contributions to an unequalled range of fields: physics, philosophy of physics, metaphysics, methodology and philosophy of science, philosophy of mathematics, logic, philosophy of psychology, philosophy of social science, philosophy of biology, philosophy of technology, moral philosophy, social and political philosophy, management theory, medical philosophy, linguistics, criminology, legal philosophy, and education.

Bunge's remarkable corpus of scientific and philosophical writing is not inert; it has had significant disciplinary, cultural, and social impact. In 1989 the American Journal of Physics asked its readers to vote for their favourite papers from the journal in the sixty years since its founding in 1933. Bunge's 1956 'Survey of the Interpretations of Quantum Mechanics' was among the 20 top voted papers. In 1993, the journal repeated the exercise and this time Bunge's 1966 paper 'Mach's Critique of Newtonian Mechanics' joined his first paper in the journal's top 20 list.

Beyond breadth, Bunge's work is noteworthy for its coherence and systemicity. Through the mid twentieth-century, the most significant Western philosophers were systematic philosophers. But in the past half-century and more, the pursuit of systemic philosophy, 'big pictures', 'grand narratives', or even cross-disciplinary understanding has considerably waned. Bunge has defied this trend. His philosophical system was laid out in detail in his monumental eight-volume Treatise on Basic Philosophy (1974–1989). Individual volumes were devoted to Semantics, Ontology, Epistemology, Systemism, Philosophy of Science, and Ethics. His Political Philosophy: Fact, Fiction and Vision (2009) was originally planned as its ninth volume.

Bunge has applied his systems approach to issues in logic, mathematics, physics, biology, psychology, social science, technology, medicine, legal studies, economics, and science policy.

Bunge's life-long commitment to Enlightenment-informed, socially engaged, systemic philosophy is manifest in his being asked by the Academia Argentina de Ciencias Exactas, Físicas y Naturales to draft its response to the contemporary crisis of anthropogenic global warming. Bunge authored the Manifesto which was signed by numerous international associations. Guided by his own systematism, he wrote that since climate is not regional but global, all the measures envisaged to control it should be systemic rather than sectoral, and they should alter the causes at play—mechanisms and inputs—rather than their effects.

Clearly Bunge is one of the most accomplished, informed, wide-ranging philosophers of the modern age. This symposium, held in the year that he, hopefully, celebrates his 100th birthday on 21st September is an opportunity for the international philosophical community to both celebrate and appraise his contribution to the discipline.

'Mario Bunge: A Pioneer of the New Philosophy of Science'
Rodolfo Gaeta, Philosophy Dept., University of Buenos Aires, Argentina
rodygaeta@gmail.com

Mario Bunge anticipated many of the 'post-positivist' arguments of Hanson, Kuhn, Feyerabend, and the Edinburgh Strong Programme that were used to promote a skeptical view of science, a view that became entrenched in the final decades of the twentieth century, giving rise to the 'New Philosophy of Science' (Brown, 1977).

But Bunge used the arguments to defend the veracity and value of scientific knowledge. Several years before the irruption of the new philosophy of science, Bunge was developing his own view in a place far away from the most important centers of study of philosophy of science. The result of this work was the publication in 1959 of the first edition of Causality (Bunge, 1959). This was a striking event because it was not common for an Argentine philosopher to write a book in English and have it published by a prestigious publisher. But the most important thing is the novelty of the ideas expressed in that book.

At this point it is worthwhile to re-evaluate the image of Bunge. Many believe that Bunge is a physicist who has become a philosopher that defends a positivist doctrine. The reality is quite the opposite. According to his own statements, since a young age he rejected the subjective interpretations of quantum mechanics and devoted himself to the study of physics in order to obtain the necessary elements to support his position.

Bunge defended scientific realism and he argued against naive empiricism as well as against more sophisticated versions that focus knowledge on the activity of the subject. Quantum mechanics also favoured the questioning of the concept of causality and the validity of determinism. Bunge then undertakes a double task: separating science from a narrow empiricism, and reformulating causality and determinism in an adequate way. He proposes to differentiate causation from the causal principle that guides our beliefs and from the formulation of causal laws. It also separates causation from determinism to give rise to non-causal determinisms.

Bunge's position regarding causality explains both his distancing from the interpretation of quantum mechanics provided by some theorists, as well as from empiricist and Kantian conceptions that understood causality as a projection of conditions of the cognizing subject. His criticism of empiricism is based on considerations that advance ideas later exploited by anti-realists like Kuhn. However, Bunge's arguments are aimed at rescuing, along with plausible versions of causality and determinism, a realist view of science.

One of the merits of Bunge's Causality book is the prominence that he early gave to ideas that are usually attributed to the 'new philosophy of science': the thesis of the theory-ladenness of observation; the conviction that no scientific statement has meaning outside a theoretical system; and the belief that scientific development follows a pattern similar to that of biological evolution, so that scientific progress does not represent a progressive reduction but a progressive differentiation. According to Bunge, this differentiation, pace the 'new philosophers of science', means a genuine cognitive improvement rather than a change of beliefs.

'Mario Bunge's Scientific Approach to Realism'
Alberto Cordero, Philosophy Department, Queen's College, USA
acordelec@outlook.com

1. *The Ontological Thesis:* Bunge upholds the existence of a world independent of the mind, external to our thinking and representations (Ontological Thesis). His supporting reasoning on this matter draws from both general considerations as well as some of the special sciences.

2. *The Epistemological Thesis:* Bunge complements the previous proposal with an epistemological thesis made of three major claims:

> 2(a). It is possible to know the external world and describe it at least to a certain extent. Through experience, reason, imagination, and criticism, we can access some truths about the outside world and ourselves.
> 2(b). While the knowledge we thus acquire often goes beyond the reach of the human senses, it is problematic in multiple ways. In particular, the resulting knowledge is indirect, abstract, incomplete, and fallible.
> 2(c). Notwithstanding its imperfections, our knowledge can be improved. Bunge accepts that theories are typically wrong as total, unqualified proposals. In his

opinion, however, history shows with equal force that successful scientific theories are not entirely false, and that they can be improved.

3. *The Semantic Thesis:* This component of Bunge's realism is framed by the previously stated ontological and epistemological theses. It comprises four interrelated ideas:

3(a). Some propositions refer to facts (as opposed to only ideas).
3(b). We can discern the proper ('legitimate') referents of a scientific theory by identifying its fundamental predicates and examining their conceptual connections to determine the role those predicates play in the laws of theory.
3(c). Some factual propositions are approximately true.
3(d). Any advance towards the truth is susceptible to improvement.

4. *Methodological Thesis:* The fourth facet of Bunge's realism I am highlighting focuses on methodology and comprises at least three proposals: (a) Methodological scientism; (b) Bunge's version of the requirement that theories must allow for empirical testing; and (c) a mechanistic agenda for scientific explanation.

4(a). Scientism asserts that the general methods developed by science to acquire knowledge provide the most effective available exploration strategy at our disposal. The methods of science—whose main use is given in the development and evaluation of theories—use reason, experience, and imagination.
4(b). A theoretical proposal should lead to distinctive predictions, and it should be possible to subject at least some of those predictions to demanding empirical corroboration.
4(c). According to Bunge, we cannot be satisfied with merely phenomenological hypotheses of the 'black box' type (i.e., structures that do not go beyond articulating correlations between observable phenomena).

Good methodology, Bunge insists, presses for further exploration, prompting us to search the world for regularities at deeper levels that provide illuminating explanations of the discovered regularities—ideally 'mechanical' ones. The realism project that Bunge articulates seems, therefore, to have some major issues still pending. Meanwhile, however, I think there is no doubt that Mario Bunge will continue to make valuable contributions in this and other areas of the realist project, responding with honesty and clarity to the enigmas posed by the most intellectually challenging fundamental theories of our time.

'Quantifiers and Conceptual Existence'
María Manzano Arjona & Manuel Crescencio Moreno, Philosophy Department, University of Salamanca, Spain
mara@usal.es

The point of departure of our research is María Manzano's paper 'Formalización en Teoría de Tipos del Predicado de Existencia Conceptual de Mario Bunge' (Manzano 1985). We recall the main concepts of this article and propose new perspectives on existence offered by a wide variety of new formal languages.

First, we place Bunge's ideas within the historical debate about existence. It seems to us that Bunge is in favor of combining the traditional view on existence, wherein it was considered a first-order predicate, with the Fregean account, where existence acts as a second-order predicate.

Second, as in Manzano (1985), we make use of the language of Type Theory, TT, to formulate Bunge's distinction between the logical concept of existence and the ontological one. Both the quantifier and the ontological existence are predicates in TT, but to formulate the first one we need only logical constants while for the second one we need non-logical constants. In particular, the existential quantifier could be introduced by definition, using the lambda operator and a logical predicate constant.

Third, we explore another possibility and try to incorporate in the formal system the tools needed to define the ontological existence predicate using only logical constants. In Hybrid Partial Type Theory, HPTT, assuming semantics with various domains, the predicate of existence can be defined by means of the existential quantifier.

Since a modal model contains many possible worlds, the previous formula could be true for a world (for instance, the world of physical objects) but false for another world of the same structure (for instance, the world of conceptual objects). Moreover, thanks to the machinery of hybrid logic we have enhanced our formal system with nominals, such as i, and with satisfaction operators, such as @. Nominals give us the possibility of naming worlds, and satisfaction operators allow us to formalize that a statement is true for a given possible world. In this logic, we have formulae that could be used to express that the individual object named by the term t exists at the world of physical objects named by i.

In HPTT, we could use the existential quantifier, the equality, and the satisfaction operator to express that an object has ontological existence, either physical or conceptual.

We do not need specific non-logical predicate constants given that the satisfaction operator is forcing the formula to be evaluated at i-world.

Lastly, we analyse existence in the language of our Intensional Hybrid Partial Type Theory, IHPTT. This opens a new possibility concerning existence which we have not considered so far. It is related to existence as a predicate of intensions. In our IHPTT, existence can also be predicated of intensions, and we should expand our previous definition to include terms of type (a, s).

Our formal languages have tools for dealing with existence as a predicate and as a quantifier. In fact, it is possible to give a coherent account of both alternatives. Therefore, from the point of view of the logical systems we have presented in this paper, the relevant issue is that we have tools for dealing with Bunge's distinctions in a variety of forms. We have shown that hybridization and intensionality can serve as unifying tools in the areas involved in this research; namely, Logic, Philosophy of Science, and Linguistics.

'Mario Bunge and the Enlightenment Project in Science Education'
Michael R. Matthews, School of Education, UNSW, Australia
m.matthews@unsw.edu.au

The unifying theme of Bunge's life and research is the constant and vigorous advancement of the eighteenth-century Enlightenment project; and energetic criticism of cultural and academic movements that reject the principles of the project or devalue its historical and contemporary value. Bunge is unashamedly a defender of the Enlightenment; while over the past half-century, many intellectuals, academics, educators, and social critics have either rejected it outright or compromised its core to such an extent that it can barely give direction to the kinds of personal, philosophical, political or educational issues that historically it had so clearly and usefully addressed.

In many quarters, including educational ones, the very expression 'the Enlightenment' is derogatory and its advancement is thought misguided and discredited. This paper begins by noting the importance of debates in science education that hinge upon support for or rejection of the Enlightenment project. It then distinguishes the historic eighteenth-century Enlightenment from its articulation and working out in the Enlightenment project; it details Mario Bunge's and others' summations of the core principles of the Enlightenment; and it fleshes out the educational project of the Enlightenment by referring to the works of John Locke, Joseph Priestley, Ernst Mach, Philipp Frank and Herbert Feigl. It indicates commonalities between the Enlightenment education project and that of the liberal education movement, and for both projects it points to the need to appreciate history and philosophy of science.

Modern science is based on Enlightenment-grounded commitments: the importance of evidence; rejection of simple authority, especially non-scientific authority, as the arbiter of knowledge claims; a preparedness to change opinions and theories; a fundamental openness to participation in science regardless of gender, class, race or religion; a recognition of the inter-dependence of disciplines; a pursuit of knowledge for advancement of personal epistemology concerning the objective knowability of the world; questions of ontology concerning the constitution of the world, specifically regarding methodological and ontological naturalism, questions of methodology concerning theory appraisal and evaluation, and the limits, if any, of scientism, questions of ethics concerning the role of values in science all and social welfare.

All this needs to be manifest in science education, along with a willingness to resist the imposition of political, religious and ideological pressures on curriculum development, textbook choice and pedagogy. Defence of the Enlightenment tradition requires serious philosophical work. Questions need to be fleshed out, and Enlightenment answers defended against their many critics. That Enlightenment banner continues to be carried by Mario Bunge. He champions Enlightenment principles, adjusts them, and adds to them. In Latin America of the mid- and late twentieth century, he was one of the outstanding Enlightenment figures, and has been the same in the wider international academic community.

Observation to Causation: The Background Assumptions for Causal Discovery

Organiser: Frederick Eberhardt (California Institute of Technology)

Over the past few years, the Causal Bayes net framework—developed by Spirtes et. al. (2000) and Pearl (2000), and given philosophical expression in Woodward (2004)—has been successfully spun off into the sciences. From medicine to neuro- and climate-science, there is a resurgence of interest in the methods of causal discovery. The framework offers a perspicuous representation of causal relations, and enables development of methods for inferring causal relations from observational data. These methods are reliable so long as one accepts background assumptions about how underlying causal structure is expressed in observational data. The exact nature and justification of these background assumptions has been a matter of debate from the outset. For example, the causal Markov condition is widely seen as more than a convenient assumption, and rather as encapsulating something essential about causation. In contrast, the causal faithfulness assumption is seen as more akin to a simplicity assumption, saying roughly that the causal world is, in a sense, not too complex. There are other assumptions that have been treated as annoying necessities to get methods of causal discovery off the ground, such as the causal sufficiency assumption (which says roughly that every common cause is measured), and the acyclicity (which implies, for example, that there is no case in which X causes Y, Y causes Z, and Z causes X, forming a cycle). Each of these assumptions has been subject to analysis and methods have been developed to enable causal discovery even when these assumptions are not satisfied. But controversies remain, and we are confronted with some longstanding questions: What exactly is the nature of each of those assumptions? Can any of those assumptions be justified? If so, which ones? How do the question of justification and the question of nature relate to each other?

This symposium aimed to address those questions. It brought together a group of researchers all trained in the causal Bayes nets framework, but who have each taken different routes to exploring how we can address the connection between the underlying causal system and the observational data that we use as basis to infer something about that system. In particular, we discussed a variety of different approaches that go beyond the traditional causal Bayes net framework, such as the discovery of dynamical systems, and the connection between causal and constitutive relations. While the approaches are largely driven by methodological considerations, we expect these contributions to have implications for several other philosophical debates in the foundations of epistemology, the metaphysics of causation, and on natural kinds.

Speakers and Talk Titles (in the order they were presented)

1. *Hanti Lin (UC Davis)*, **'Convergence to the Causal Truth and Our Death in the Long Run'**
2. *Jiji Zhang (Lingnan University)*, **'Causal Minimality in the Boolean Approach to Causal Inference'**
3. *Konstantin Genin (University of Toronto)*, **'Progressive Methods for Causal Discovery'**
4. *Frederick Eberhardt (Caltech)*, **'Proportional Causes and Specific Effects'**
5. *Benjamin Jantzen (Virginia Tech)*, **'Finding Causation in Time: Background Assumptions for Dynamical Systems'**

Particles, Fields, or Both?

Organiser: Charles Sebens (California Institute of Technology)

One of the primary tasks of philosophers of physics is to determine what our best physical theories tell us about the nature of reality. Our best theories of particle physics are quantum field theories. Are these theories of particles, fields, or both? In our symposium, we debated this question of ontology both in the context of quantum field theory and in an earlier and closely related context: classical electromagnetism. We began with a historical introduction and then had one defender of each of the three possible answers to the question 'Particles, Fields, or Both?'.

In modern textbooks, classical electromagnetism is normally presented as a theory where charged matter (described either as point charges or continuous distributions of charge) interacts with the electromagnetic field. This interaction is codified in Maxwell's equations and the Lorentz force law. However, it is possible to remove the electromagnetic field and think of electromagnetism as a theory in which charges interact directly with one another across gaps in space and time. Such a theory can take many forms, e.g., a theory of retarded action at a distance (where the forces a charge feels depend on the past locations and motions of charges), a theory of advanced action at a distance (where future locations and motions are what matter), or a theory of half-retarded half-advanced action at a distance (as in the famous Wheeler-Feynman theory).

At the beginning of the 20th century, Albert Einstein and Walther Ritz debated the radiation asymmetry of electromagnetism and the question of whether electromagnetism should be formulated as including an electromagnetic field or instead as a theory of direct interaction between charges. In this debate, they brought up considerations of energy conservation, equality of action and reaction, and self-interaction. Mathias Frisch presented a re-evaluation of this historical episode that kicked off our debate over the ontological status of particles and fields.

Our second speaker, Mario Hubert, then zoomed in on the self-interaction problem, comparing three strategies for avoiding the problem in the context of classical electromagnetism. The basic problem with self-interaction is that the electric field of a point particle is infinite at the location of a particle, and thus it is hard to evaluate the way in which a point particle interacts with its own electromagnetic field. One idea is to remove the electromagnetic field so that there is no danger of a particle interacting with its own field. As mentioned above, such a particle-only theory can take many forms. Hubert discussed the Wheeler-Feynman approach and its precursors in the work of Fokker, Schwarzschild, and Tetrode. A second idea, due to Born and Infeld, keeps fields as part of the ontology and modifies Maxwell's equations to alter the field near a charged particle (replacing Maxwell's equations by non-linear equations). A third idea, due to Bopp and Podolsky, also retains a particle and field ontology but modifies Maxwell's equations in a different way (replacing them by higher-order equations). Hubert explained that in order to make these latter two ideas work, we need to change more than just Maxwell's equations; we also need to change the Lorentz force law (describing how the electromagnetic field acts on particles).

Next, Dustin Lazarovici proposed a pure particle ontology for quantum field theory, representing electrons as particles and using the Wheeler-Feynman manoeuvre to remove the electromagnetic field. Lazarovici sees quantum electrodynamics as an extension of Dirac's single electron relativistic quantum theory to multiple particles. Dirac's theory faces a problem with negative energies which Dirac solved by introducing an infinite sea of negative energy electrons, impressively predicting the existence of positrons by analysing the behavior of holes in this sea. Lazarovici argued that, although it is not currently popular, the Dirac Sea provides an attractive ontology for quantum field theory and ought to be revived. Lazarovici explained the relation of the Dirac Sea to current Fock space formulations of quantum field theory. He used the Dirac Sea to provide an intuitive physical picture of the unitary inequivalence of free and interacting Fock spaces (a puzzle that has been much discussed in the philosophical literature on Haag's theorem).

In our final talk, Charles Sebens defended a pure field ontology. He began by arguing that the point particle approach to quantum field theory (defended by Lazarovici for electrons) cannot be made to work for photons. However, because the equations describing photons and electrons are so similar, Sebens reasoned that we should adopt the same approach for both: treating both as fields. On this view, the electromagnetic field describes photons, and the Dirac field describes electrons and positrons. Sebens defended a field ontology for electrons by analysing the classical theory of the Dirac field and the process of field quantization (showing how negative energies can be avoided without invoking the Dirac Sea). Sebens explained how taking the classical Dirac field to represent charged matter interacting with the electromagnetic field alters the problem of self-interaction (discussed in the first two talks). He also used this classical field description of electrons to argue that electrons should be regarded as actually spinning.

Philosophers of physics have made great strides in understanding non-relativistic quantum mechanics. We have an embarrassment of riches, multiple precise versions of the theory with different ontologies that are all (arguably) capable of reproducing the theory's empirical successes. Quantum field theory is the next frontier. It is not clear how to proceed and there is a wide variety of philosophical projects engaging with the subject.

We sought to demonstrate the fruitfulness of focusing on questions of ontology by presenting the motivations and advantages of our three alternative perspectives on the question 'Particles, Fields, or Both?'. A definitive answer to this question does not immediately settle all the questions one might have about the ontology of quantum field theory, but we believe that it would lay the groundwork for extending existing precise versions of quantum mechanics to quantum field theory.

In our debate, we hoped to spark more philosophical interest in classical electromagnetism by demonstrating the depth of debate one can have about the ontology of the theory and by connecting that debate to questions about quantum field theory.

List of Presentations:

1. *Mathias Frisch (Leibniz University Hannover), mathias.frisch@philos.uni-hannover.de,* **'Particles, Fields, or Both? A Reevaluation of the Ritz-Einstein Debate'**
2. *Mario Hubert (California Institute of Technology), mhubert@caltech.edu,* **'Good Singularities, Bad Singularities'**
3. *Dustin Lazarovici (Université de Lausanne), dustin.lazarovici@unil.ch,* **'Why Field Theories are not Theories of Fields'**
4. *Charles Sebens (California Institute of Technology), csebens@caltech.edu,* **'The Fundamentality of Fields'**

Proof and Translation: Glivenko's Theorem 90 Years After

Organisers: Sara Negri (University of Genoa) and Peter Schuster (University of Verona)

Symposium Abstract:

Glivenko's theorem from 1929 says that if a propositional formula is provable in classical logic, then its double negation is provable within intuitionistic logic. Soon after, Gödel extended this to predicate logic, which requires the double negation shift. As is well-known, with the Gödel-Gentzen negative translation in place of double negation one can even get by with minimal logic. Several related proof translations saw the light of the day, such as Kolmogorov's and Kuroda's.

Glivenko's theorem thus stood right at the beginning of a fundamental change of perspective: that classical logic can be embedded into intuitionistic or minimal logic, rather than the latter being a diluted version of the former. Together with the revision of

Hilbert Programme ascribed to Kreisel and Feferman, this has led to the quest for the computational content of classical proofs, today culminating in agile areas such as proof analysis, dynamical algebra, program extraction from proofs and proof mining. The considerable success of these approaches suggests that classical mathematics will eventually prove much more constructive than is widely thought today.

The symposium includes but is not limited to the following threads of current research: exploring the limits of Barr's theorem about geometric logic; program extraction in abstract structures characterised by axioms; constructive content of classical proofs with Zorn's Lemma; and the algorithmic meaning of programs extracted from proofs.

Symposium Talks:[26]

Itala Maria Loffredo D'Ottaviano and Hércules De Araujo Feitosa, **'On the Historical Relevance of Glivenko's Translation from Classical into Intuitionistic Logic: Is It Conservative and Contextual?'**

Abstract: For several years we have studied interrelations between logics by analysing translations between them. The first known translations concerning classical logic, intuitionistic logic and modal logic were presented by Kolmogorov (1925), Glivenko (1929), Lewis & Langford (1932), Gödel (1933), and Gentzen (1933). In 1999, da Silva, D'Ottaviano and Sette proposed a very general definition for the concept of translation between logics, logics being characterised as pairs constituted by a set and a consequence operator, and translations between logics being defined as maps that preserve consequence relations. In 2001, Feitosa and D'Ottaviano introduced the concept of conservative translation, and Carnielli, Coniglio & D'Ottaviano (2009) proposed the concept of contextual translation. In this paper, providing some brief historical background, we will discuss the historical relevance of the translation from classical logic into intuitionistic logic introduced by Glivenko in 1929, and will show that his interpretation is a conservative and contextual translation.

References

Carnielli, W. A., Coniglio, M. E., & D'Ottaviano, I. M. L. (2009). New dimensions on translations between logics. *Logica Universalis*, *3*, 1–19.

---

[26] In the order in which they were held at the symposium in Prague on August 9, 2019.

da Silva, J. J., D'Ottaviano, I. M. L., & Sette, A. M. (1999). Translations between logics. In Caicedo, X., Montenegro, C.H. (Eds.) *Models, Algebras and Proofs* (pp. 435–448). New York: Marcel Dekker.

Feitosa, H. A., & D'Ottaviano, I. M. L. (2001). Conservative translations. *Annals of Pure and Applied Logic, 108*, 205–227.

Gentzen, G. (1936). Die Widerspruchsfreiheit der reinem Zahlentheorie. *Mathematische Annalen, 112*, 493–565. Translation into English in Gentzen (1969, Szabo, M. E. (Eds.)).

Gentzen, G. (1969). On the relation between intuitionist and classical arithmetic (1933). In M. E. Szabo (Ed.), *The Collected Papers of Gerhard Gentzen* (pp. 53–67). Amsterdam: North-Holland.

Glivenko, V. (1929). Sur quelques points de la logique de M. Brouwer. *Académie Royale de Belgique, Bulletins de la Classe de Sciences, 5*(15), 183–188.

Gödel, K. (1986) On intuitionistic arithmetic and number theory (1933). In S. Feferman et al. (Eds.), *Collected Works* (pp. 287–295). Oxford: Oxford University Press.

Gödel, K. (1986). An interpretation of the intuitionistic propositional calculus (1933). In S. Feferman et al. (Eds.), *Collected Works* (pp. 301–303). Oxford: Oxford University Press.

Kolmogorov, A. N. (1977). On the principle of excluded middle (1925). In J. Hejenoort (Ed.), *From Frege to Gödel: A Source Book in Mathematical Logic 1879–1931* (pp. 414–437). Cambridge: Harvard University Press.

Lewis, C. I., & Langford, C. H. (1932). *Symbolic Logic*. New York. Reprinted in 1959.

Sara Negri, **'A Simple Proof of Barr's Theorem for Infinitary Geometric Logic'**

Abstract: Geometric logic has gained considerable interest in recent years: contributions and applications areas include structural proof theory, category theory, constructive mathematics, modal and non-classical logics, automated deduction. Geometric logic is readily defined by stating the structure of its axioms. A coherent implication (also known in the literature as a geometric axiom, a geometric sentence, a coherent axiom, a basic geometric sequent, or a coherent formula), is a first-order sentence that is the universal closure of an implication of formulas built up from atoms using conjunction, disjunction and existential quantification. The proper geometric theories are expressed in the language of infinitary logic and are defined in the same way as coherent theories, except for allowing infinitary disjunctions in the antecedent and the consequent. Gentzen's systems of deduction, sequent calculus and natural deduction, have been considered an answer to Hilbert's 24th problem in providing the basis for a general theory of proof methods in mathematics that overcomes the limitations of axiomatic systems. They provide a transparent analysis of the structure of proofs that works to perfection for pure logic. When such systems of deduction are augmented with axioms for mathematical theories, much of the strong properties are lost. However, these properties can be regained through a transformation of axioms into rules of inference of a suitable form. Coherent theories are very well placed into this program, in fact, they can be translated as inference rules in a natural fashion: in the context of a sequent calculus such as G3c (Negri & von Plato, 2001; Troelstra & Schwichtenberg, 2001), special coherent implications as axioms can be converted directly (Negri, 2003) to inference rules without affecting the admissibility of the structural rules; this is essential in the quest of applying the methods of structural proof theory to geometric logic. Coherent implications form sequents that give a Glivenko class (Orevkov, 1968; Negri, 2016). In this case, the result Negri (2003), known as the first-order Barr's Theorem (the general form of Barr's theorem (Barr, 1974; Wraith, 1978; Rathjen, 2016) is higher-order and includes the axiom of choice) states that if each I_i, $0 \leq i \leq n$ is a coherent implication and the sequent $I_1, \ldots, I_n \Rightarrow I_0$ is classically provable then it is intuitionistically provable. By these results, the proof-theoretic study of coherent theories gives a general twist to the problem of extracting the constructive content of mathematical proofs. In this talk, proof analysis is extended to all such theories by augmenting an infinitary classical sequent calculus with a rule scheme for infinitary geometric implications. The calculus is designed in such a way as to have all the rules invertible and all the structural rules admissible. An intuitionistic infinitary multisuccedent sequent calculus is also introduced and it is shown to enjoy the same structural properties as the classical calculus. Finally, it is shown that by bringing the classical and intuitionistic calculi close together, the infinitary Barr theorem becomes an immediate result.

References

Barr, M. (1974). Toposes without points. *Journal of Pure and Applied Algebra*, 5, 265–280.
Negri, S. (2003). Contraction-free sequent calculi for geometric theories, with an application to Barr's theorem. *Archive for Mathematical Logic*, 42, 389–401.
Negri, S. (2016). Glivenko sequent classes in the light of structural proof theory. *Archive for Mathematical Logic*, 55, 461–473.
Negri, S. & von Plato, J. (2001). *Structural Proof Theory*. Cambridge: Cambridge University Press.
Orevkov, V. P. (1968). Glivenko's sequence classes, logical and logico-mathematical calculi 1. *Proc. Steklov Inst. of Mathematics*, 98, 147–173 (131–154 in Russian original).

Rathjen, M. (2016). Remarks on Barr's Theorem: Proofs in geometric theories. In P. Schuster, & Probst, D. (Eds.), *Concepts of Proof in Mathematics, Philosophy, and Computer Science* (pp. 347–374). De Gruyter.

Skolem, T. (1970). *Selected Works in Logic*. Edited by J. E. Fenstad. Oslo: Universitetsforlaget.

Troelstra, A. S., & Schwichtenberg, H. (2001). *Basic Proof Theory* (2nd eds.). Cambridge: Cambridge University Press.

Wraith, G. (1978). Intuitionistic algebra: Some recent developments in topos theory. *Proceedings of International Congress of Mathematics* (pp. 331–337). Helsinki.

Michael Rathjen, **'Proof Theory of Infinite Geometric Theories'**

Abstract: A famous theorem of Barr's yields that geometric implications deduced in classical (infinitary) geometric theories also have intuitionistic proofs. Barr's theorem is of a category-theoretic (or topos-theoretic) nature. In the literature one finds mysterious comments about the involvement of the axiom of choice. In the talk I would like to speak about the proof-theoretic side of Barr's theorem and aim to shed some light on the AC part.

Luiz Carlos Pereira, Elaine Pimentel and Valeria de Paiva, **'Ecumenism: A New Perspective on the Relation between Logics'**

Abstract: A traditional way to compare and relate logics (and mathematical theories) is through the definition of translations/interpretations/embeddings. In the late twenties and early thirties of last century, several such results were obtained concerning some relations between classical logic (CL), intuitionistic logic (IL) and minimal logic (ML), and between classical arithmetic (PA) and intuitionistic arithmetic (HA). In 1925 Kolmogorov proved that classical propositional logic (CPL) could be translated into intuitionistic propositional logic (IPL). In 1927 Glivenko proved two important results relating (CPL) to (IPL). Glivenko's first result shows that A is a theorem of CPL if A is a theorem of IPL. His second result establishes that we cannot distinguish CPL from IPL with respect to theorems of the form A. In 1933 Gödel defined an interpretation of PA into HA, and in the same year Gentzen defined a new interpretation of PA into HA. These interpretations/translations/embeddings were defined as functions from the language of PA into some fragment of the language of the HA that preserve some important properties, like theoremhood. In 2015 Dag Prawitz (see Prawitz, 2015) proposed an ecumenical system, a codification where classical logic and the intuitionistic logic could coexist in peace. The main idea behind this codification is that the classical logic and the intuitionistic logic share the constants for conjunction, negation, and the universal quantifier, but each has its own disjunction, implication and existential quantifier. Similar ideas are present in Dowek (2015) and Krauss (1992), but without Prawitz philosophical motivations. The aims of the present paper are: (1) to investigate the proof theory and the semantics for Prawitz Ecumenical system (with a particular emphasis on the role of negation), (2) to compare Prawitz system with other ecumenical approaches, and (3) to propose new ecumenical systems.

References

Dowek, G. (2015). On the definitions of the classical connective and quantifiers. In E. Haeusler, W. Sanz, & B. Lopes Bruno (Eds.), *Why is this a Proof* (pp. 228–238). College Books.
Krauss, P., H. (1992). A constructive interpretation of classical mathematics. *Mathematische Schriften Kassel*, preprint 5/92.
Prawitz, D. (2015). Classical versus intuitionistic logic. In E. Haeusler, W. Sanz, & B. Lopes Bruno (Eds.), *Why is this a Proof* (pp. 15–32). College Books.

Tadeusz Litak, **'Modal Negative Translations as a Case Study in The Big Programme'**

Abstract: This talk is about negative translations (Kolmogorov, Gödel-Gentzen, Kuroda, Glivenko) and their variants in propositional logics with a unary normal modality. More specifically, it addresses the question whether negative translations as a rule embed faithfully a classical modal logic into its intuitionistic counterpart. As it turns out, even the Kolmogorov translation can go wrong with rather natural modal principles. Nevertheless, one can isolate sufficient syntactic criteria for axioms (enveloped implications) ensuring adequacy of well-behaved (or, in our terminology, regular) translations. Furthermore, a large class of computationally relevant modal logics, namely, logics of type inhabitation for applicative functors (a.k.a. idioms) turns out to validate the modal counterpart of the Double Negation Shift, thus ensuring adequacy of even the Glivenko translation. All the positive results mentioned above can be proved purely syntactically, using the minimal natural deduction system of Bellin, de Paiva and Ritter extended with Sobociski-style additional axioms/combinators. Hence, "mildly proof-theoretic methods can be surprisingly successfully used in the Big Programme" (to borrow F. Wolter and M. Zakharyaschev's phrase from the *Handbook of Modal Logic)*. Most of this presentation is based on results published with my former students, who provided formalization in the Coq proof assistant. In the final part, however, I will discuss variants of a semantic approach based either on a suitable notion of subframe preservation or on a generalization of Wolter's describable operations. An account of this semantic approach and comparison with the scope of the syntactic one remains unpublished.

Ulrich Berger, **'On the Constructive Content of Proofs in Abstract Analysis'**

Abstract: Can a proof in analysis that does not refer to a particular constructive model of the real numbers have computational content? We show that this is the case by considering a formulation of the Archimedean property as an induction principle: For any property P of real numbers, if for all x, $(x > 0 \to P(x-1)) \to P(x)$, then for all x, $P(x)$. This principle is constructively valid and has as computational content the least fixed point combinator, even though real numbers are considered abstract, that is, only specified by the axioms of a real closed field. We give several applications of this principle connected with concurrent computation.

Monika Seisenberger, **'Program Optimisation through Proof Transformation'**

Abstract: In earlier work (Berger, Lawrence, Nordvall Forsberg & Seisenberger, 2015) we have shown that the well-known DPLL SAT solving algorithm can be extracted from a soundness and completeness proof of the corresponding proof system. We carry this work further by showing that also program optimisation techniques such as clauselearning can be obtained by a transformation on the proof level.

References

Berger, A., Lawrence, F., Nordvall Forsberg, M., & Seisenberger, M. (2015). Extracting verified decision procedures: DPLL and resolution. *Logical Methods in Computer Science, 11*(1).

Daniel Wessel, **'Ideals, Idealization, and a Hybrid Concept of Entailment Relation'**

Abstract: The inescapable necessity of higher-type ideal objects, which more often than not are brought into being by one of the infamously elegant combinations of classical logic and maximality (granted by principles as the ones going back to Kuratowski and Zorn), is, it may justly be argued, a self- fulfilling prophecy. Present-day classical mathematics thus finds itself at times clouded by strong ontological commitments. But what is at stake here is mere pretence, and techniques as multifarious as the ideal objects they are meant to eliminate have long borne witness to the fact that unveiling computational content is all but a futile endeavour. Abstract entailment relations have come to play an important role, most notably the ones introduced by Scott (1974) which subsequently have been brought into action in commutative algebra and lattice theory by Cederquist & Coquand (2000). The utter versatility of entailment relations notwithstanding, some potential applications, e.g., with regard to injectivity criteria like Baer's, seem to call for yet another concept that allows for arbitrary sets of succedents (rather than the usual finitely enumerable ones), but maintains the conventional concept's simplicity. In this talk, we discuss a possible development according to which an entailment relation is to be understood (within Aczel's constructive set theory) as a class relation between finitely enumerable and arbitrary subsets of the underlying set, the governing rules for which, e.g., transitivity, to be suitably adjusted. At the heart of our approach, we find van den Berg's finitary non-deterministic inductive definitions (van den Berg, 2013), on top of which we consider inference steps so as to give account of the inductive generation procedure and cut elimination (Rinaldi & Wessel, forthcoming). Carrying over the strategy of Coquand and Zhang (2000) to our setting, we associate set-generated frames (Azcel, 2006) to inductively generated entailment relations, and relate completeness of the latter with the former's having enough points. Once the foundational issues have been cleared, it remains to give evidence why all this might be a road worth taking in the first place, and we will do so by sketching several case studies, thereby revisiting the extension-conservation paradigm, which in the past successfully guided the quest for constructivisation in order theory, point-free topology, and algebra. The intended practical purpose will at least be twofold: infinitary entailment relations might complement the approach taken in dynamical algebra, and, sharing aims, may ultimately contribute to the revised Hilbert programme in abstract algebra.

References

Aczel, P. (2006). Aspects of general topology in constructive set theory. *Ann. Pure Appl. Logic*, *137*(13), 3–29.

Benno van den, B. (2013). Non-Deterministic inductive definitions. *Arch. Math. Logic*, *52*(12), 113–135.

Cederquist, J., & Coquand, T. (2000). Entailment relations and distributive lattices. In S. R. Buss, P. Hájek, & P. Pudlák (Eds.), *Logic Colloquium 98. Proceedings of the Annual European Summer Meeting of the Association for Symbolic Logic, Prague, Czech Republic, August 915, 1998, volume 13 of Lecture Notes Logic* (pp. 127–139). Natick, MA: A. K. Peters.

Coquand, T., & Zhang G.-Q. (2000). Sequents, frames, and completeness. In P. G. Clote & H. Schwichtenberg, (Eds.), *Computer Science Logic (Fischbachau, 2000), volume 1862 of Lecture Notes in Comput. Science* (pp. 277–291). Berlin: Springer.

Rinaldi D., & Wessel, D. (forthcoming). Cut elimination for entailment relations. *Arch. Math. Logic*, in press.

Scott Dana (1974). Completeness and axiomatizability in many-valued logic. In L. Henkin, J. Addison, C.C. Chang, W. Craig, D. Scott, & R. Vaught (Eds.), *Proceedings of the Tarski Symposium (Proc. Sympos. Pure Math., Vol. XXV, Univ. California, Berkeley, Calif., 1971)* (pp. 411–435). Providence, RI: Amer. Math. Soc.

Peter Schuster, Giulio Fellin and Daniel Wessel, **'The Jacobson Radical and Glivenko's Theorem'**

Abstract: Alongside the analogy between maximal ideals and complete theories, the Jacobson radical carries over from ideals of commutative rings to theories of propositional calculi. This prompts a variant of Lindenbaum's Lemma that relates classical validity and intuitionistic provability, the syntactical counterpart of which is Glivenko's Theorem. Apart from shedding fresh light on intermediate logics, this eventually prompts a non-trivial interpretation in logic of Rinaldi, Schuster and Wessel's conservation criterion for Scott-style entailment relations (BSL 2017 & Indag. Math., 2018).

Also, Olivia Caramello and Hajime Ishihara intended to attend the symposium but unfortunately could not; they had planned to give talks entitled 'Grothendieck Topologies and Deductive Systems' and 'On the Gödel-Gentzen Translation', respectively.

Some Recent Directions in Model Theory

Organiser: John Baldwin (University of Illinois at Chicago)

Speakers were invited to discuss recent trends in model theory that forge connections between model theory and algebra, probability, computer science, and category theory. Here are descriptions of the seven talks by invitees who were able to attend.

James Freitag (University of Illinois at Chicago, United States), **'Some Recent Applications of Model Theory'**

After some general remarks we will explain recent applications of model theory which use, in an essential way, structural results coming from stability theory. The first application centers around automorphic functions on the upper half plane, for instance, the j-function mapping the generator of a lattice to the j-invariant of the associated elliptic curve. The central problem of interest involves understanding which algebraic varieties V have the property that $j(V)$ is an algebraic variety. We call such varieties bialgebraic. The philosophy is that the bi-algebraic varieties should be rare and reveal geometric information about the analytic function. At least two general sort of approaches using model theory have emerged in the last decade. The first involves o-minimality and the second involves the model theory of differential fields, applied to the algebraic differential equations satisfied by the analytic functions. We concentrate on the second approach in this talk. The second application is related to machine learning. In the last several years, the dividing line between learnability/nonlearnability in various settings of machine learning (online learning, query learning, private PAC learning) has proved to be related to dividing lines in classification theory. By using structural results and inspiration from model theory, various new results in machine learning have been established. We will survey some of the recent results and raise a number of open questions.

Rehana Patel (Harvard University, Boston, United States), **'Towards a Model Theory of Symmetric Probabilistic Structures'**

Logic and probability bear a formal resemblance, and there is a long history of mathematical approaches to unifying them. One such approach is to assign probabilities to statements from some classical logic in a manner that respects logical structure. Early twentieth century efforts in this direction include, as a partial list, work of Lukasiewicz, Keynes, Masukiewicz, Hosiasson and Los, all essentially attaching measures to certain algebras. Carnap goes somewhat further in his influential 1950 treatise *Logical Foundations of Probability*, where he considers a limited monadic predicate logic and finite domains. The key model-theoretic formalisation is due to Gaifman, in work that was presented at the 1960 Congress of Logic, Methodology and Philosophy of Science held at Stanford—the first in the present conference series—and that appeared in his 1964 paper *Concerning Measures in First Order Calculi*. This work stipulates coherence conditions for assigning probabilities to formulas from a first order language that are instantiated from some fixed domain and shows the existence of an assignment fulfilling these conditions for any first order language and any domain. Shortly thereafter, Scott and Krauss extended these results to an infinitary setting that provides a natural parallel to countable additivity. In his 1964 paper Gaifman also introduced the notion of a symmetric probability assignment, where the measure given to a formula is invariant under finite permutations of the instantiating domain. When the domain is countable, such an assignment is an exchangeable structure, in the language of probability theory, and may be viewed as a symmetric probabilistic analogue of a countable model-theoretic structure. There is a rich body of work within probability theory on exchangeability beginning with de Finetti in the 1930s and culminating in the representation theorems of Aldous, Hoover and Kallenberg and this can be brought to bear on the study of such symmetric probabilistic structures. A joint project of Nathanael Ackerman, Cameron Freer and myself, undertaken over the past ten years, investigates the model theory of these exchangeable structures. In this talk I will discuss the historical context for this project, and its current status.

Jiří Rosický (Masaryk University, Czechia), **'Accessible Categories and Model Theory'**

Accessible categories were introduced by M. Makkai and R. Paré as a framework for infinitary model theory. They have turned out to be important in algebra, homotopy theory, higher category theory and theoretical computer science. I will survey their connections with abstract elementary classes and discuss how model theory of abstract elementary classes can be extended to that of accessible categories. In particular, I will present a hierarchy beginning with finitely accessible categories and ending with accessible category having directed colimits.

Michael Lieberman (Masaryk University, Czechia), **'Tameness, Compactness, and Cocompleteness'**

We discuss the emerging characterization of large cardinals in terms of the closure of images of accessible functors under particular kinds of colimits. This effects a unification, in particular, of large-cardinal compactness and colimit cocompleteness, bringing the former somewhat closer to the structuralist heart of modern mathematical practice. Mediating these equivalences is the phenomenon of tameness in abstract elementary classes, which, not least for historical reasons, has provided an indispensable bridge between the set-theoretic and category-theoretic notions, beginning with work of myself and Rosický, Brooke-Taylor and Rosický, and Boney and Unger. We summarise the current state of knowledge, with a particular focus on my paper 'A Category-Theoretic Characterization of almost Measurable Cardinals' and forthcoming joint work with Boney.

Sebastien Vasey (Harvard University, United States), **'Forking and Categoricity in Non-elementary Model Theory'**

The classification theory of elementary classes was started by Michael Morley in the early sixties, when he proved that a countable first-order theory with a single model in some uncountable cardinal has a single model in all uncountable cardinals. The proof of this result, now called Morley's categoricity theorem, led to the development of forking, a notion of independence jointly generalizing linear independence in vector spaces and algebraic independence in fields, and is now a central pillar of modern model theory.

In recent years, it has become apparent that the theory of forking can also be developed in several non-elementary contexts. Prime among those are the axiomatic frameworks of accessible categories and abstract elementary classes (AECs), encompassing classes of models of any reasonable infinitary logics. A test question to judge progress in this direction is the forty-year-old eventual categoricity conjecture of Shelah, which says that a version of Morley's categoricity theorem should hold of any AEC. I will survey recent developments, including the connections with category theory and large cardinals, a theory of forking in accessible categories (joint with M. Lieberman and J. Rosický), as well as the resolution of the eventual categoricity conjecture from large cardinals (joint with S. Shelah).

Tibor Beke (University of Massachusetts Lowell, United States), **'Feasible Syntax, Feasible Proofs, and Feasible Interpretations'**

Recursion theory in the guise of the Entscheidungsproblem, or the arithmetic coding of the syntax of first-order theories has been a part of symbolic logic from its very beginning. The spectacular solution of Post's problem by Friedberg and Muchnik, as well as the many examples of decidable and essentially undecidable theories found by Tarski, focused logicians' attention on the poset of Turing degrees, among which recursive sets appear as the minimal element. Starting with the work of Cook and others on computational complexity in the 1970s, computer scientists' attention shifted to resource-bounded notions of effective computation, under which primitive recursive in fact, elementary recursive algorithms may be deemed unfeasible. The threshold of feasible computability is reduced to polynomial-time and/ or polynomial-space computations, or possibly their analogues in singly or doubly exponential times. Under this more stringent standard, for example, Tarski's decision algorithm for the first order theory of the reals is not feasible, and it took considerable effort to discover a feasible alternative. This talk examines what happens to the classical notion of bi-interpretability when the translation between formulas, and between proofs, is required to be feasibly computable. The case of propositional logic is classical, and the extension to classical first order logic is not hard. Interesting and, I believe, open problems arise when one compares two theories with different underlying logics. The most intriguing case is when the theories do not share a common syntax, such as when one compares first order logic with the lambda calculus, or ZFC with Voevodsky's Univalent Foundations. The case of category theory is yet more interesting, since the syntax of category theory is not clearly defined. The language of category theory, as understood by the working category theorist, certainly includes diagrams of objects and arrows. We will also outline some theorems on the computational complexity of verifying the commutativity of diagrams. Bibliography [partial]: Boolos: 'Don't Eliminate Cut'; Mathias: 'A Term of Length 4,523,659,424,929'; Cook, Reckhow: 'The Relative Efficiency of Propositional Proof Systems'; Cavines, Johnson: 'Quantifier Elimination and Cylindrical Algebraic Decompositions'.

Cameron Hill (Wesleyan University, United States), **'Towards a Characterization of Pseudo-Finiteness'**

Methods with ultraproducts of finite structures have been used extensively by model theorists to prove theorems in extremal graph theory and additive combinatorics. In those arguments, they exploit ultralimits of the counting measures of finite structures, turning asymptotic analyses into questions about dimension and measure in an infinite structure. Looked at in reverse, pseudo-finite structures always have meaningful notions of dimension and measure associated with them, so it seems valuable to characterise pseudo-finiteness itself. The best-known existing theorem of this kind is Ax's characterisation of pseudo-finite fields. I will discuss an ongoing project to find a characterisation of pseudo-finiteness for countably categorical theories in which algebraic closure is trivial. Our approach to proving such a characterisation is, in a sense, the standard one for model theorists, but the details are novel. First, we would like to identify certain primitive building blocks out of which models of pseudo-finite theories are made. Second, we will need to understand the program for actually putting those building blocks together. Our working hypothesis is that pseudo-finite theories are those that are approximable in a certain sense by almost-sure theories (those arising from 0,1-laws for classes of finite structures), which we also speculate are precisely the rank-1-super-simple theories. In a loose sense, randomness seems to take the place of combinatorial geometry in the primitive building blocks of this discussion, and the process of assembling those building blocks into a model has a more analytic flavor than one usually seen in model theory. I will discuss the current state of this work and try to point out some of the interesting contrasts between this program and other classification programs we have seen.

Symposium of the Spanish Society of Logic, Methodology and Philosophy of Science

Organiser: Cristina Corredor (UNED Madrid)

The *Spanish Society of Logic, Methodology and Philosophy of Science* (SLMFCE in its Spanish acronym) is a scientific association formed by scholars working in the common field of Logic and Philosophy of Science, understood in a broad sense, and including the domains of logic, history and philosophy of logic; philosophy and methodology of science; history of science; science, technology and society; philosophy of language; philosophy of mind and epistemology; and argumentation theory. Its members are university teachers, researchers, and other specialists. They are affiliated with almost all universities and research centres in Spain, with some also in Latin America. Presently, there are about 240 members, a significant proportion of them being junior researchers. The board of directors is elected every three years.

The SLMFCE was founded in 1994 by a group of scholars that aimed to establish the Spanish branch of the IUHPST/DLMPST. Its objectives are the following,

- to encourage, sustain and disseminate research and study in logic, methodology and philosophy of science, and other close disciplines;
- to organise, sponsor and promote conferences and meetings, both national and international, within its fields of expertise;
- to provide a meeting point for specialists doing teaching and research in logic, methodology and philosophy of science, etc., and scientists interested in the foundations of their disciplines, as well as for other institutions and associations with closely related aims; and
- to issue a journal (*Revista de la SLMFCE*) and other publications related to its activities (conference proceedings, awarded works by junior members, etc).

The SLMFCE is eager to promote and support junior researchers and scholars. To this aim, the Society has developed a policy of grants and awards for its younger members. Moreover, every three years, the SLMFCE organises a conference which has become an indispensable meeting for its associated members and other scholars, not only from Spain but also from Latin America and other European countries. The last such conference took place in Salamanca in November 2021.

Following the invitation by the Program Committee of CLMPST 2019, the SLMFCE symposium was intended to showcase the work carried out by some senior researchers and research groups associated with the Society. It featured four contributions in different subfields of specialization, allowing the audience at the CLMPST 2019 to form an idea of the plural research interests and relevant outcomes of our members. Before these works were presented, José Martínez Fernández, Vice Chair of the SLMFCE board of directors, briefly introduced the Society and its aims and scope.

The participants and topics addressed in the symposium were the following:

1. *José Martínez Fernández,* **'On Revision-theoretic Semantics for Special Classes of Circular Definitions'**
   Short abstract: The aim of the talk is to define and analyse some special classes of circular definitions that have simple revision-theoretic semantics. The classes generalise the finite definitions studied by Gupta, Chapuis and others.
2. *Sergi Oms,* **'Common Solutions to Several Paradoxes. What Are They? When Should They Be Expected?'**
   Short abstract: In this paper we examine what a common solution to more than one paradox is, and why, in general, such a solution should be expected. In particular, it is explored why a common solution to the Liar and the Sorites should be expected.
3. *Lilian Bermejo-Luque,* **'What Should a Normative Theory of Argumentation Look Like?'**
   Short abstract: In this paper, an analysis is carried out on the rewards and shortcomings of two different epistemological conceptions of Argumentation Theory, taking into account their corresponding criteriological and transcendental accounts of the sort of objectivity that good argumentation is able to provide.
4. *María Cerezo,* **'Issues at the Intersection between Metaphysics and Biology'**
   Short abstract: In this contribution, some examples are presented in which the interaction between Metaphysics and Biology takes place, and the different ways in which such interaction takes place are explored. The examples include interactions between Evolutionary Biology, Genetics or Developmental Biology and metaphysical notions such as dispositions, identity and persistence or teleology.

The four contributions represented the fields of philosophical logic and semantics, argumentation theory, and philosophy of the life sciences. The contributors are recognised scholars with a relevant track record of published research. Among SLMFCE affiliated members are very active researchers working in these and other closely related subfields who have frequent participation in international meetings, including the activities organised by the IUHPST/DLMPST.

On behalf of the SLMFCE, I should like to thank the Program Committee of the CLMPST 2019 for the opportunity given to us to present both our Society and the quality research work done by its members.

Symposium on John Baldwin's Model Theory and the Philosophy of Mathematical Practice

Organiser: John Baldwin (University of Illinois at Chicago)

1) The Symposium

The symposium provided an opportunity for Juliette Kennedy (University of Helsinki), Maryanthe Malliaris (University of Chicago), and Andrew Arana (Université de Lorraine, France) to engage John Baldwin's recently published monograph, Model Theory and the Philosophy of Mathematical Practice (Cambridge, 2017), a ground-breaking contribution to the philosophy of mathematical practice. We summarise below the contributions of the three commenters and Baldwin's responses.

2) Kennedy

Kennedy asked the following questions:

What should be the central philosophical concerns of philosophy of model theory? What central questions it can shed light on? What insight can it give us about mathematical practice that is not available in other approaches?

The general philosophical point of view on which the book relies is built on the idea of seeking local foundations for mathematics, as opposed to the idea of a global foundation. What is the nature of this seemingly anti-foundationalist view, and how deep does this anti-foundationalist stance reach? Does localizing in this case mean rejecting any kind of global framework?

Shelah's dividing lines, exemplified by the Main Gap Theorem, play a central role in the book. How do we know that this or indeed any classificatory scheme in mathematics tracks the actual contours of the subject?

Theories on the structure side of Shelah's Main Gap Theorem admit dimension-like geometric invariants. Do these theories track our geometric or spatial intuition more closely than theories on the nonstructure side, if they can be said to track these intuitions at all? Is the Main Gap Theorem a foundational theorem in the sense that Hilbert imagined, demarcating the tractable vs the untractable in mathematics? Or are the theories on the non-structure side tractable from some other point of view?

3) Response to Kennedy

The book addressed two issues. On the one hand, the classification identifies similar (often unrecognised) themes in distinct areas of mathematics and so illuminates mathematical practice. On the other hand, it is a pattern case study of a detailed account of one area of mathematics.

The book presumes a ZFC background but is deliberately agnostic about the philosophical import of this presumption.

The book does not attempt to track the current sociological contours of modern mathematics; it describes a framework that is justified by its applicability in large areas of mathematics.

On the structure side, classification theory generalises the study of modern real and complex geometry. Clearly any complete theory with the independence property cannot track our geometric and spatial intuitions. As o-minimality demonstrates, the main gap does not demarcate the 'tractable' from the 'untractable'. It explicitly distinguishes models using 'geometric' dimension.

4) Malliaris

Malliaris, a model theorist, addressed the question: Should a mathematician read this book, a book which, at first glance, appears to be a book explaining model theory to philosophers? Answers which were not defended included: that the book is of general intellectual interest and we should read it for culture, that it is interesting to see how our field interacts with another field, that it is a good opportunity for mathematicians to get a feel for how philosophers think because the examples analysed are ones we know well, or that it can call our attention to philosophical aspects of our own work. These are all reasonable answers but when we weigh them against the work we have in front of us on any given day, they may have little urgency.

Malliaris suggested that the answer which does have urgency is the most interesting answer: for its mathematics. This is a book written by a mathematician who has been doing core work in the subject for almost fifty years and who, under the umbrella of discussing various philosophical ideas, gives detailed mathematical information to illustrate his impressions of how various advances arose. Malliaris then responded to some mathematical aspects of the book. These included some long-term influences of Hilbert's work in the field, some reasons the mathematics around first-order logic is so developed, and some key moves in work of Robinson and Shelah. She emphasized the creativity of the logical point of view and the flexibility of model theoretic classification, and she considered the role of probability and randomness to complement the book's discussions of geometry.

5) Response to Malliaris

Different classifications may be appropriate for different problems. Remarkably the stability classification solves problems far beyond the one it was designed for. Malliaris and Shelah greatly refined the connection of Keisler order with the Shelah classification while, not entirely coincidentally making a major advance in set theory. The ordering, $T_1 \leq_{univ} T_2$, of theories by the set of cardinals in which they have universal models not only addresses a natural model theoretic question but shows that such questions are not decidable in ZFC; this has motivated new techniques, e.g., club guessing in set theory.

Bourbaki offers a rough framework for mathematics via the three great mother structures: group, order, topology. The classification of theories incorporated a crucial fourth, geometrical, dimension. Malliaris adds combinatorics, probability, and randomness. Malliaris emphasized the flexibility of the classification notion. The 21st century explosion of 'neo-stability' theory reflects this adaptability. Such properties of theories as dp-minimality, distality, mutual algebraicity, and monadic stability/nip are applied to such topics as combinatorial geometry, enumerative combinatorics, and field theory.

6) Arana

Andrew Arana focused on Baldwin's treatment of the relationship between a data set, the received propositions in an area of mathematics such as Diophantine geometry or motivic integration, and axiomatisations of that data set. Baldwin observed that logic, and in particular model theory, helps identify axiomatisations of data sets with virtuous properties, such as categoricity in power or stability, that are especially helpful in mathematical practice. For instance, the recognition following Morley by Baldwin-Lachlan and Zilber that strong minimality and geometricity are unexpectedly related has advanced work in both model theory and in algebraic geometry. Strong minimality is thus such a virtuous property of an axiomatic theory. In observing this contribution of logic to mathematical practice Baldwin aimed to counter the view that logical analysis does not contribute to our mathematical understanding, a view that one can find in Poincaré and Bourbaki, for instance.

Arana then focused on Hilbert and Bernays' description of axiomatization as finding the extract of a subject matter. In the case of geometry this involves the stripping away of all visual content, following Pasch. Arana asked what the relation between an axiomatization (as extract) and its data set should be. He observed that chemists do not expect that the original thing can be restored from the extract, but logicians are expected to be able to do this: this is why completeness is judged virtuous. He suggested that it could be helpful to think of mathematical extraction like chemists do, in order to understand how mathematicians use axiomatic theories without stressing the relationship to the original data set that these theories axiomatize. In this latter practice the abstraction created by extraction becomes an object of interest in its own right. This way of doing mathematics should not be seen as the only way, but one that has become important since the end of the nineteenth century and one for which logic has played an important role.

7) Response to Arana

I focus here on the notion of extraction in respect to the study of geometry. Mangling the metaphor a bit, one might say, the goal is to extract the essence. And the essence can be taken as the collection of theorems. Thus, the first order axiomatization of, say, Tarski, succeeds as an axiomatic program. As argued in the book, it grounds a complete first order theory containing all the theorems of Euclid. By avoiding Archimedes, it also avoids Dedekind's postulation of limits to all convergent sequences that strays from the Greek proof of existence of well-described numbers such as π. In contrast, the immodest axiomatization given by Hilbert implies both the axiom of Archimedes (roughly known to and used by Euclid), and Dedekind (foreign to Euclid). One might well argue that Euclid was only interested in universal sentences (Avigad, Dead & Mumma, 2009) and so Tarski is immodest as well.

References

Avigad, J., Dean, E., & Mumma, J. (2009). A formal system for Euclid's Elements. *Review of Symbolic Logic, 2*, 700–768.

Text-Driven Approaches to the Philosophy of Mathematics

Organisers: Carolin Antos (University of Konstanz), Deborah Kant (University of Konstanz) and Deniz Sarikaya (University of Hamburg)

General information

Conference Homepage: https://tdphima.weebly.com/.

Support: The workshop was possible due to the generous support of the DVMLG: Deutsche Vereinigung für Mathematische Logik und für Grundlagenforschung der Exakten Wissenschaften and the GWP: Gesellschaft für Wissenschaftsphilosophie [The German Society for Philosophy of Science].

Topic of the Workshop

Text is a crucial medium to transfer mathematical ideas, agendas and results among the scientific community and in educational context. This makes the focus on mathematical texts a natural and important part of the philosophical study of mathematics. Moreover, it opens up the possibility to apply a huge corpus of knowledge available from the study of texts in other disciplines to problems in the philosophy of mathematics.

This symposium aimed to bring together and build bridges between researchers from different methodological backgrounds to tackle questions concerning the philosophy of mathematics. This included approaches from philosophical analysis, linguistics (e.g., corpus studies), and literature studies; but also methods from computer science (e.g., Big Data approaches and natural language processing), artificial intelligence, cognitive sciences and mathematics education (cf. Fisseni et al. to appear; Giaquinto, 2007; Mancosu et al., 2005; Schlimm, 2008; Pease et al., 2013).

The right understanding of mathematical texts might also become crucial due to the fast successes in natural language processing on one side, and automated theorem proving on the other side. Mathematics as a technical jargon or as natural language, with quite rich structure, and semantic labelling (via LaTeX) is from the other perspective an important test-case for practical and theoretical study of language.

Herein we understand text in a broad sense, including informal communication, textbooks and research articles.

References

Fisseni, B., Schröder, B., Sarikaya, D., & Schmitt, M. (2019). How to frame a mathematician. Modelling the cognitive background of proofs. In S. Centrone, D. Kant, & D. Sarikaya (Eds.), *Reflections on the Foundations of Mathematics: Univalent Foundations, Set Theory and General Thoughts* (pp. 417–436). Berlin: Springer.

Giaquinto, M. (2007). *Visual Thinking in Mathematics. An Epistemological Study.* Oxford: Oxford University Press.

Mancosu, P., Jørgensen, K. F., Pedersen, S. A. (Eds.) (2005). *Visualization, Explanation and Reasoning Styles in Mathematics.* Dordrecht: Springer.

Pease, A., Guhe, M., Smaill, A. (2013). Developments in research on mathematical practice and cognition. *Topics in Cognitive Science*, 5(2), 224–230.

Schlimm, D. (2008). Two ways of analogy. Extending the study of analogies to mathematical domains, *Philosophy of Science*, 75(2), 178–200.

List of Invited Speakers and the Topics of the Talks

We heard the following talks during the conference:

1. *Anna Steensen (ETH Zurich),* **'Semiotic Analysis of Dedekind's Arithmetical Strategies'**
2. *Bernhard Fisseni (IDS),* **'Perspectives on Proofs'**
3. *Marcos Cramer (TU Dresden),* **'Bridging the Gap Between Proof Texts and Formal Proofs Using Frames and PRSs'**
4. *Karl Heuer (TU Berlin),* **'Text-driven Variation as a Vehicle for Generalisation, Abstraction, Proofs and Refutations: An Example about Tilings and Escher within Mathematical Education'**
5. *Mikkel Willum Johansen (Univ. Copenhagen),* **'Entering the Valley of Formalism: Results from a Large-scale Quantitative Investigation of Mathematical Publications'**
6. *Juan Luis Gastaldi (Univ. Paris-Diderot) and Luc Pellissier (IRIF),* **'A Structuralist Framework for the Automatic Analysis of Mathematical Texts'**
7. *Fanner Stanley Tanswell (Univ. Loughborough) and Matthew Inglis (Univ. Loughborough),* **'Studying Actions and Imperatives in Mathematical Texts'**
8. *Juan Pablo Mejía-Ramos (Rutgers Univ.) and Matthew Inglis (Univ. Loughborough),* **'Using Linguistic Corpora to Understand Mathematical Explanation'**

Short Summary of the Talks

Marcos Cramer presented in his talk 'Bridging the Gap Between Proof Texts and Formal Proofs Using Frames and PRSs' (based on joint work with Bernhard Fisseni, Deniz Sarikaya and Bernhard Schröder) linguistic frames for induction proofs and how they could help to fill gaps in mathematical texts or even to recognise induction proofs in texts computationally.

The second talk 'Perspectives on Proofs' by Bernhard Fisseni started with the question whether textual and formal proofs are different perspectives on the same thing and proposed to look for the common core of both.

After a short break, Juan Luis Gastaldi told us in 'A Structuralist Framework for the Automatic Analysis of Mathematical Texts' about joint work with Luc Pellissier. For their automatic analysis of texts, one determines first the smallest unit (e.g., letter, syllable, word), and then obtains types by orthogonality: two units are of the same type if they are both orthogonal to the same thing.

Mikkel Willum Johansen continued the symposium with his talk 'Entering the Valley of Formalism: Results from a large-scale Quantitative Investigation of Mathematical Publications', in which he presented a categorisation of mathematical diagrams used in research articles from 1940 until now in major maths journals and showed that the use of diagrams decreased significantly from 1950 to 1980.

In the fifth talk 'Using Linguistic Corpora to Understand Mathematical Explanation', Juan Pablo Mejía-Ramos illustrated how he and Matthew Inglis identify explanatory talk in maths and physics texts. Their results include that 'explain why' talk appears more often in physics, whereas 'explain how' talk is more often used by mathematicians.

A similar method was applied by Fenner Tanswell and Matthew Inglis to investigate the use of imperatives in mathematical texts. Fenner Tanswell explained in 'Studying Actions and Imperatives in Mathematical Texts' that they look for capitalised imperatives since lower case imperatives are not distinguishable from the non-imperative forms of the verb.

Anna Steensen presented in 'Semiotic Analysis of Dedekind's Arithmetical Strategies' her investigation how the text works as a medium between author and reader, and that the reader has an active part as well. She gave an example of a sentence by Dedekind which included algebraic expressions but also had set content which has to be added by the reader.

The last talk of the Symposium by Karl Heuer, 'Text-driven Variation as a Vehicle for Generalisation, Abstraction, Proofs and Refutations: an Example about Tilings and Escher within Mathematical Education' (based on joint work with Deniz Sarikaya), was about the design of exercises for pupils; which, for example, required them to replace words by other suitable words and to learn in that way about generalisation.

Toward the Reconstruction of Linkage between Bayesian Philosophy and Statistics

Organiser: Masahiro Matsuo (Hokkaido University)

In philosophy of science, Bayesianism has long been tied to subjective interpretation of probability, or probability as a degree of belief. Although several attempts have been made to construct an objective kind of Bayesianism, most of the issues and controversies concerning Bayesianism have been focused on this subjectivism, particularly on subjective priors. Because of this tradition, philosophers of science are likely to assume that Bayesian statistics, which is now popular in many fields of science, can be considered as part of subjective Bayesianism.

True, not only Bayesian philosophy but Bayesian statistics as well can be seen as rooted in subjectivism like Savage's 'Foundation of Statistics', but how subjectivity is involved in current usage of Bayesian statistics is not so obvious. Apparently, scientists who use Bayesian statistics as their basic tool of data analyses would now take it as just one of the mathematical techniques, which is based on a simple updating rule of parameter distribution, without knowing historical arguments of Bayesian philosophy. They seldom refer to subjectivity or even to a degree of belief, and besides, the lack of this seems to be no obstacle to their analyses. Therefore, we should say there is a considerable gap between Bayesianism typically discussed in philosophy of science, and Bayesian statistics used in science.

How can we treat this gap? We cannot miss this since Bayesian philosophy would be practically meaningless if it merely sticks to traditional subjectivism neglecting recent development of Bayesian statistical methods. On the other hand, to rely on an updating rule superficially without considering philosophical bases would be epistemically unsound, even if it seems to provide useful methods to science. So, it is inevitable for any Bayesians to fill this gap, and particularly Bayesian philosophers should take the responsibility of restoring the linkage between Bayesian philosophy and statistics (if not, who else?).

In our symposium, we presented perspectives which we think will provide some of the necessary steps for this restoration. To achieve this goal, trying to fully examine the history of Bayesianism is one promising approach we can take. But there is also a risk of losing our way if we focus too much attention on the history, because it is tremendously complicated to unravel, particularly when Bayesian statistics emerged as a standard of statistics in the late 20th century. It seems more promising a way to start from the current situation to recognise what kind of gaps there actually are, and to find a toehold for filling each gap from the available philosophical and statistical perspectives, and from a historical perspective when necessary. Our focus is not upon tracing back to the divergent point between Bayesian philosophy and statistics to find out the reason why, but mainly upon reconstructing the two Bayesian camps. The perspectives we presented in the symposium are: a parallelism between Bayesianism as a whole and inductive logic; a complementary relation between Bayesian philosophy and statistics; a resolution of a mismatch between Bayesian philosophy and statistics by contrasting them with frequentism; and a linkage between Bayesian philosophy and statistics through statistical theories based on both Bayesianism and frequentism. Abstracts of each speaker are as follows (in order of presentation).

Kazutaka Takahashi, in his talk 'Examination of the Linkage between Bayesian Philosophy and Statistics from a Logical Point of View', presented an approach for the linkage by way of Carnap's inductive logic. Carnap's system of inductive logic involves not only the logical or empirical assumptions in our inductive inferences but also a close relation between inductive logic and Bayesianism through his λ-continuum. Festa (1993) tried to show a parallelism between Bayesian statistics and Carnap's system, which holds in the case of multinomial distribution. Takahashi showed two possible ways to extend Festa's argument. One is to extend it to cases of likelihood other than multinomial distribution. The other is to extend it to the more general Bayesian philosophy. He focused on the latter and talked about the possibility of the reduction of Bayesianism as a whole to a system of inductive logic.

Masahiro Matsuo, in his talk 'Constructing a Complementary Relation between Bayesian Philosophy and Statistics', separated the conceptual linkage from the practical one, and examined each. The key to the conceptual linkage between the two is, according to Matsuo, the Likelihood Principle. Despite the apparent difference between Bayesian philosophy and statistics, literally, belief updating or updating of parameter distribution almost independent of belief, it can be shown they are both in accord with the Likelihood Principle, though in different ways. Thus, one way to reconstruct their linkage is to recapture the Likelihood Principle as the basis for their methodologies. Practical linkage, on the other hand, could be reconstructed through a complemental way. Bayesian statistics is dedicated to the analysis of parameter, while Bayesian philosophy is usually committed to the analysis of a more general hypothesis. This mixture, which is desirable in science, is hindered by difficulty of general application of Bayesian philosophy (Bayesian inference). Matsuo suggested some approaches to avoid this.

Yusaku Ohkubo tried to articulate a common feature of Bayesian philosophy and statistics by way of contrasting it with frequentists' basic feature in his talk 'Revisiting the Two major Statistical Problems, Stopping-rule and the Catch-all Hypothesis, from the Viewpoint of neo-Bayesian Statistics'. Looking back at the history of the statistics and of the philosophy of statistics, we find that a fundamental conflict between Frequentist (or error statistician) and Bayesian theory has often resulted from different interpretation of probability. But recently we find frequentists who do not cling to frequentist interpretation of probability, and Bayesian statisticians who adopt frequentist properties to justify their procedures. This does not mean, however, that they are merging with each other. Rather, statisticians (and scientists) are beginning to use two distinct statistics in a complementary way in scientific analyses. The target of statistical assessment for frequentists is data, while that for Bayesians is hypothesis. From this simple distinction we can sometimes create a practical combination of the two statistics, and through this we can clarify a basic role Bayesianism plays, which could unite Bayesian philosophy and statistics.

Finally, Kenichiro Shimatani, in his talk 'The Linkage between Bayesian and Frequentism Statistics is easier than between Bayesian Statistics and Philosophy', also refers to a combination of frequentist and Bayesian statistics, but in contrast with Ohkubo and also with Matsuo, he, as a statistician, pointed out the difficulty of application of Bayesian philosophy (Bayesian inference) in science. He showed in detail how a combinatory usage of frequentist statistics and Bayesian statistics (including BIC) is an example of ecology's analysis. Depending on this example, he also showed a diagram in which scientific inferences are composed as a whole, and what is the part of this whole that statistics (Bayesian statistics and frequentist statistics as well) generally plays. In his view, scientific inferences consist of logic, statistical inferences, and 'others' which may be usually called expert judgment. He suggested Bayesian philosophy can be linked to these 'others', but the Bayesian philosophy available now is far from practical usage. At the end of his talk, he showed a list of problems which Bayesian philosophy has to overcome to become a practical method of science.

What Method for Conceptual Engineering?
Organiser: Manuel Gustavo Isaac (University of Zurich)

The Symposium

'Conceptual engineering' is chief among the most popular labels at the cutting edge of philosophical research. The phrase was independently coined in Carnap scholarship (Creath, 1991; Carus, 2007; Wagner, 2012) and in metaphilosophy (Blackburn, 1999; Brandom, 2001; Floridi, 2011). Since then, these two trends have connected (Brun, 2016), and further expanded on the side of social philosophy (Haslanger, 2012). In less than a decade, the movement has spanned philosophy to become a proper field of its own, whose attraction is still growing with an explosive intensity.[27] The basic idea behind it is that sometimes our conceptual apparatuses need to be ameliorated in the attainment of some beneficial consequences. Accordingly, conceptual engineers are guided by a normative agenda: they aim to prescribe the concepts we *ought to* have and use, rather than merely describing those we *do* have and use. To this end, one of their main purposes is to develop a methodological framework for assessing and improving our conceptual devices—that is, in particular, for identifying deficiencies in our conceptual apparatuses, and for fixing them (Cappelen, 2018).

Despite its centrality to research in conceptual engineering, little had been said about how we could develop its methodological framework. The rationale for the 'What Method for Conceptual Engineering?' (MET4CE) symposium was to initiate the first forays into this topic. Against this background, the symposium intended to focus on two core issues: first, can we devise the method of conceptual engineering as a staged and parametrised process; that is, as a set of step-by-step guidelines for ameliorating our conceptual devices supplemented by a set of adjustable parameters for measuring their functional efficacy? The common framework to tackle this issue is Carnapian explication, procedurally reconstructed, and complemented by other compatible frameworks and methods (e.g., reflective equilibrium, levels of abstraction, metalinguistic negotiations). Second, how could the process of conceptual engineering be assisted by other compatible methods at its different stages? For instance, we discussed how the tools and techniques of experimental philosophy could be used in the assessment and improvement stages of the conceptual engineering process. Additionally, the symposium also addressed a variety of foundational issues in the vicinity of conceptual engineering's methodological framework, its development, and its implementation.

---

[27] See the PhilPapers entry 'Conceptual Engineering' edited by Steffen Koch: https://philpapers.org/browse/conceptual-engineering.

The MET4CE symposium was comprised of ten 30-minute talks by established and up-and-coming scholars from Austria, Belgium, Germany, the Netherlands, Norway, Spain, Switzerland, and the United Kingdom. The talks presented at the symposium have produced 6 articles in peer-reviewed journals of philosophy (see Section 3 for the full reference list). In addition, the symposium was the starting point for a number of collaborative projects between several of its panel speakers including the Conceptual Engineering Network (https://www.conceptualengineering.xyz). The symposium was organised by Manuel Gustavo Isaac.

The Talks

In his talk titled *Broad-Spectrum Conceptual Engineering*, Manuel Gustavo Isaac introduced a variant of conceptual engineering that is expected to be appropriately applicable to any of our representation-involving cognitive activities, with major consequences for our whole cognitive life. Isaac focused his talk on the theoretical foundations of conceptual engineering thus characterised. With a view to ensuring the actionability of conceptual engineering as a broad-spectrum method, he addressed the issue of how best to construe the subject matter of conceptual engineering, and argued that conceptual engineering should be: (i) about concepts, (ii) psychologically theorised, and (iii) as multiply realised functional kinds. Thereby, Isaac claimed, we would theoretically secure and justify the maximum scope, flexibility, and impact for the method of conceptual engineering on our representational devices in our whole cognitive life—in other words, a broad-spectrum version of conceptual engineering.

In the same vein, Steffen Koch asked what are concepts, and how does one engineer them? Answering these questions, Koch observed, is of central importance for implementing and theorising about conceptual engineering. In his talk titled *On Two Kinds of Conceptual Engineering and their Methodological Counterparts*, he discussed and criticised two influential views of this issue: semanticism, according to which conceptual engineers aim to change linguistic meanings; and psychologism, according to which conceptual engineers aim to change psychological structures. Koch argued that neither of these accounts can give us the full story. Instead, he proposed and defended the Dual Content View of Conceptual Engineering. On this view, conceptual engineering targets concepts, where concepts are understood as having two (interrelated) kinds of contents: referential content and cognitive content. Koch showed that this view is independently plausible and that it gives us a comprehensive account of conceptual engineering that helps to make progress on some of the most difficult problems surrounding conceptual engineering.

In her talk titled *Conceptual Engineering and Semantic Control*, Joey Pollock defended an internalist approach to conceptual engineering in response to an argument from Cappelen (2018). Cappelen proposes a radically externalist framework for conceptual engineering, which embraces the following two theses. First, the mechanisms that underlie conceptual engineering are inscrutable: they are too complex, unstable, and non-systematic for us to grasp. Second, the process of conceptual engineering is largely beyond our control. Cappelen argues that these two commitments—'Inscrutability' and 'Lack of Control'—must be accepted by both externalist and internalist views of meaning and concepts. For the internalist to avoid commitment to these theses, she must provide arguments for three claims: (a) there are inner states that are scrutable and within our control; (b) concepts supervene on these inner states; and (c) the determination relation from supervenience base to content is itself scrutable and within our control. Pollock responded to Cappelen by demonstrating how some kinds of internalism can meet these challenges. She argued that (a) it is plausible that we have a weak sort of control over some of our inner states, some of the time; (b) it is reasonable to treat concepts as supervening on these states, as the resultant view is largely in keeping with widely accepted desiderata on a theory of concepts; and (c) we should appeal, not to mere supervenience, but to alternative relations such as identity or realisation in order to secure the result that the relation from determination base to content is both scrutable and within our control.

Delia Belleri's talk, titled *Downplaying the Topic-Change Objection to Conceptual Engineering*, touched on yet another foundational issue in conceptual engineering. Projects of conceptual engineering may face the following Strawsonian objection: once a concept, 'C', has been revised, one cannot have continuity in inquiry with the newly engineered concept, 'C'. The conceptual engineer 'has changed the subject'. Cappelen's (2018) answer to this objection invokes topics, which are representations of what a concept 'is about', that are coarser grained than intensions and extensions. Cappelen argues that we can have continuity of topic even if a concept's intension or extension undergoes a change. After pointing out some difficulties for Cappelen's approach, Belleri argued that inquirers can ask their questions while operating in different contexts. In contexts of Type 1, the questions they ask are to be interpreted as object-level and descriptive; in these contexts, change of subject is indeed a problem. In contexts of Type 2, however, the questions they ask are to be interpreted as meta-level and normative. Belleri argued that subject-change need not be a problem in Type-2 contexts. Indeed, it can be expected or even welcomed. This leads to conceding the Strawsonian objection in contexts of Type 1, but also to a downplaying, or dismissal, of the same objection in contexts of Type 2. In closing, Belleri suggested that conceptual engineers explicitly acknowledge that their inquiry is of Type 2, to neutralise the dialectical threat posed by the Strawsonian objection.

With Georg Brun and Kevin Reuter's talk titled *The Common-Sense Notion of 'Truth' as a Challenge for Conceptual Re-Engineering*, the symposium moved on to applied case studies in conceptual engineering. Tarski claims, the speakers recalled, that his theory of truth provides an explication of 'true' that is sufficiently similar to the ordinary notion of truth, which he interpreted in the sense of correspondence with reality. In the first part of their talk, Brun and Reuter presented results of experimental studies which challenge the idea that—within the empirical domain—the common-sense notion of truth is rooted in correspondence. When participants were presented with situations in which correspondence and coherence come apart, a substantial number (in some experiments up to 60%) responded in line with the predictions of the coherence account. These results challenge monistic accounts of truth as well as their most popular alternative: scope pluralism. In the second part of their talk, Brun and Reuter explored the consequences of these results for the project of re-engineering *truth*. Three proposals were discussed. (i) Defending a unique explication of truth might seem attractive for theoretical reasons, but would, given the results of the presented studies, amount to dismissing a great deal of applications of the truth-predicate. (ii) The idea of re-engineering truth as a non-classical concept (e.g., as a family resemblance concept) raises the challenge of finding such a concept which does not only explain the data of the presented studies but also has a convincing and theoretically fruitful structure. (iii) Giving more than one explicatum for *true* is promising in light of the data and substantiates the claim that 'truth' is ambiguous, but we need to know more about the mechanisms that play a role in ordinary discourses on truth.

Finally, Lieven Decock presented in his talk another insightful case study for future applications of the method of conceptual engineering. Decock analysed conceptual change and conceptual engineering in the case of color concepts. This special case raises the prospects of conceptual engineering because a precise standard for measuring the amelioration of the structure of concepts is available. On the other hand, the study highlights the problems with controlling conceptual engineering pointed out by Cappelen. Decock argued that in the case of conceptual change of color concepts varying degrees of optimisation, design and control are possible. This observation can be generalised to other classes of concepts. As a result, the scope of conceptual engineering is reduced considerably; conceptual engineering appears as a limit case of conceptual change, Decock concluded.

Besides the above contributions, the MET4CE symposium also included the four following talks: *Conceptual Engineering in the Philosophy of Information* by Patrick Allo (Free University of Brussels), *The Methodological Tradition of Explication* by Moritz Cordes (University of Greifswald), *Concepts and Replacement: What Should the Carnapian Model of Conceptual Re-Engineering Be?* by Mark Pinder (Open University), and *The Semantic Account of Slurs, Appropriation, and Meta-linguistic Negotiations* by Esa Díaz-León (University of Barcelona).

Symposium outputs

The MET4CE symposium has resulted in the following peer-reviewed publications:

Belleri, D. (2021). Downplaying the change of subject objection to conceptual engineering. *Inquiry*. Online first. DOI: 10.1080/0020174X.2021.1908161.
Decock, L. (2021). Conceptual change and conceptual engineering: The case of colour concepts. *Inquiry*, *64*(1–2), 168–185. DOI: 10.1080/0020174X.2020.1784783.
Isaac, M. G. (2021). Broad-spectrum conceptual engineering. *Ratio, 34*(4), 286–302. DOI: 10.1111/rati.12311.
Koch, S. (2021). Engineering what? On concepts in conceptual engineering. *Synthese*, *199*(1–2), 1955–1975. DOI: 10.1007/s11229-020-02868-w.
Pollock, J. (2021). Content internalism and conceptual engineering. *Synthese, 198*(12), 11587–11605. DOI: 10.1007/s11229-020-02815-9.
Reuter, K., & Brun, G. (2021). Empirical studies on truth and the project of re-engineering truth. *Pacific Philosophical Quarterly*, *103*(3), 493–517. DOI: 10.1111/papq.12370.

References

Blackburn, S. (1999). *Think: A Compelling Introduction to Philosophy*. Oxford: Oxford University Press.
Brandom, R. (2001). Modality, normativity, and intentionality. *Philosophy and Phenomenological Research, 63*(3), 587–609.
Brun, G. (2016). Explication as a method of conceptual re-engineering. *Erkenntnis, 81*(6), 1211–1241.
Cappelen, H. (2018). *Fixing Language: An Essay on Conceptual Engineering*. Oxford: Oxford University Press.
Carus, A. (2007). *Carnap and the Twentieth-Century Thought: Explication as Enlightenment*. Cambridge: Cambridge University Press.

Creath, R. (Ed.) (1991). *Dear Carnap, Dear Van: The Quine- Carnap Correspondence and Related Work*. Berkley: University of California Press.

Floridi, L. (2011). A defence of constructionism: Philosophy as conceptual engineering. *Metaphilosophy*, *42*(3), 282–304.

Haslanger, S. (2012). *Resisting Reality: Social Construction and Social Critique*. Oxford: Oxford University Press.

Wagner, P. (Ed.) (2012). *Carnap's Ideal of Explication and Naturalism*. Basingstoke: Palgrave MacMillan.

Authors of the symposium summary: Manuel Gustavo Isaac (University of Zurich), Delia Belleri (University of Lisbon), Georg Brun (University of Bern), Lieven Decock (Vrije Universiteit Amsterdam), Steffen Koch (Bielefeld University), Joey Pollock (University of Oslo), Kevin Reuter (University of Zurich).

(b) Other Contributed Symposia

(i) DLMPST/IUHPST Symposia

Identity in Computational Formal and Applied Systems
Symposium of the Commission for the History and Philosophy of Computing of the DLMPST (HaPoC)
Organisers: Giuseppe Primiero (University of Milan) and Nicola Angius (University of Messina)

Communication and Exchanges among Scientific Cultures
Symposium of the IUHPST Commission International Association for Science and Cultural Diversity
Organisers: Nina Atanasova (The University of Toledo), Karine Chemla (CNRS), Vitaly Pronskikh (Fermi National Accelerator Laboratory) and Peeter Müürsepp (Tallinn University of Technology)

Climate Change: History and Philosophy of Science and Nature of Science Challenges
Symposium of the Inter-Divisional Teaching Commission of the IUHPST
Organiser: Paulo Maurício (Lisbon School of Education)

Academic Means-End Knowledge in Engineering, Medicine and other Practical Sciences
Symposium of the DLMPST Commission on the Philosophy of Technology and Engineering Sciences
Organiser: Sjoerd Zwart (Technical University Delft)

Adolf Grünbaum Memorial Symposium
Organiser: Sandra Mitchell (University of Pittsburgh)

(ii) Contributed Symposia

Commitments of Foundational Theories
Organiser: Mateusz Łełyk (University of Warsaw)

New Directions in Connexive Logic
Organisers: Hitoshi Omori (Ruhr University Bochum) and Heinrich Wansing (Ruhr University Bochum)

Science as a Profession and Vocation. On STS's Interdisciplinary Crossroads
Organisers: Ilya Kasavin (Russian Academy of Sciences), Alexandre Antonovskiy (Russian Academy of Sciences), Liana Tukhvatulina (Russian Academy of Sciences), Anton Dolmatov (Russian Academy of Sciences), Eugenia Samostienko (Nizhny Novgorod State University), Svetlana Shibarshina (Lobachevsky State University), Elena Chebotareva (Saint Petersburg State University) and Lada Shipovalova (Saint Petersburg State University)

Styles in Mathematics
Organisers: Erich Reck (University of California, Riverside) and Georg Schiemer (University of Vienna)

Substructural Epistemology
Organisers: Dominik Klein (University of Bamberg), Soroush Rafiee Rad (University of Amsterdam) and Ondrej Majer (Czech Academy of Sciences)

Symposium on the Philosophy of the Historical Sciences
Organiser: Aviezer Tucker (Harvard University)

Symposium on Higher Baire Spaces
Organisers: Lorenzo Galeotti (University of Amsterdam) and Philipp Lücke (University of Bonn)

Theories and Formalization. Symposium of the Italian Society for Logic and Philosophy of Science
Organisers: Giovanni Valente (Polytechnic University of Milan) and Roberto Giuntini (University of Cagliari)

Appendices

Appendix A: Congress Sections

A. Logic

A.1 Mathematical Logic
A.2 Philosophical Logic
A.3 Computational Logic and Applications of Logic
A.4 Historical Aspects of Logic

B. General Philosophy of Science

B.1 Methodology
B.2 Formal Philosophy of Science and Formal Epistemology
B.3 Empirical and Experimental Philosophy of Science
B.4 Metaphysical Issues in the Philosophy of Science
B.5 Ethical and Political Issues in the Philosophy of Science
B.6 Historical Aspects of the Philosophy of Science
B.7 Educational Aspects of the Philosophy of Science

C. Philosophical Issues of Particular Disciplines

C.1 Philosophy of the Formal Sciences (including Logic, Mathematics, Statistics)
C.2 Philosophy of the Physical Sciences (including Physics, Chemistry, Earth Science, Climate Science)
C.3 Philosophy of the Life Sciences
C.4 Philosophy of the Biomedical and Health Sciences
C.5 Philosophy of the Cognitive and Behavioral Sciences
C.6 Philosophy of Computing and Computation
C.7 Philosophy of the Humanities and the Social Sciences
C.8 Philosophy of the Applied Sciences and Technology
C.9 Philosophy of Emerging Sciences

Appendix B: List of Plenary, Invited and Contributed Talks

Plenary Lectures

Heather Douglas (Michigan State University)
Scientific Freedom and Scientific Responsibility

Joel D. Hamkins (University of Oxford)
Can Set-Theoretic Mereology Serve as a Foundation of Mathematics?

Sandra D. Mitchell (University of Pittsburgh)
Integrating Perspectives: Learning from Model Divergence

Hannes Leitgeb (Ludwig Maximilian University, Munich)
What If Meaning Is Indeterminate? Ramsification and Semantic Indeterminacy
(joint event with Logic Colloquium 2019)

Invited Lectures

Christina Brech (University of São Paulo)
Indiscernibility and Rigidity in Banach Spaces
A.1

Maryanthe Malliaris (University of Chicago)
Complexity and Model Theory
A.1

Valentin Goranko (Stockholm University)
Logic-Based Strategic Reasoning in Social Context
A.2

Heinrich Wansing (Ruhr University Bochum) & Sergey Drobyshevich (Sobolev Institute of Mathematics, Novosibirsk)
Proof Systems for Various FDE-Based Modal Logics
A.2

Jan Krajíček (Charles University, Prague)
What Is Proof Complexity?
A.3

Anna Brożek (University of Warsaw)
Formal and Informal Logic in the Lvov-Warsaw School
A.4

Atocha Aliseda Llera (UNAM, Mexico City)
A Plurality of Methods in the Philosophy of Science: How Is That Possible?
B.1

Franz Dietrich (Paris School of Economics, CNRS)
Beyond Belief: Logic in Multiple Attitudes
(joint work with A. Staras & R. Sugden)
B.2

Dunja Šešelja (Munich Centre for Mathematical Philosophy)
Understanding Scientific Inquiry via Agent-Based Modeling
B.3

Tarja Knuuttila (University of Vienna)
Modeling Biological Possibilities in Multiple Modalities
B.4

Jonathan Okeke Chimakonam (University of Calabar)
Decolonising Scientific Knowledge: Morality, Politics and a New Logic
B.5

Gürol Irzık & Sibel Irzık (Sabanci University)
Kuhn's Wide-Ranging Influence on the Social Sciences, Literary Theory, and the Politics of Interpretation
B.6

Michael Matthews (University of New South Wales)
Philosophy in Science Teacher Education
B.7

Gerhard Heinzmann (Université de Lorraine)
Mathematical Understanding by Thought Experiments
C.1

Hans Halvorson (Princeton University)
How to Describe Reality Objectively: Lessons from Einstein
C.2

Sabina Leonelli (University of Exeter)
The Shifting Semantics of Plant (Data) Science
C.3

Alex Broadbent (University of Johannesburg)
The Inquiry Model of Medicine
C.4

Jacqueline Anne Sullivan (University of Western Ontario)
Creating Epistemically Successful Interdisciplinary Research Infrastructures: Translational Cognitive Neuroscience as a Case Study
C.5

Ray Turner (University of Essex)
Computational Abstraction
C.6

Anna Alexandrova (University of Cambridge)
On the Definitions of Social Science and Why They Matter
C.7

Julia Bursten (University of Kentucky)
Scale Separation, Scale Dependence, and Multiscale Modeling in the Physical Sciences
C.8

Contributed Talks[1]

A.1 Mathematical Logic

Badia, Guillermo & Carles Noguera – A Generalized Omitting Type Theorem in Mathematical Fuzzy Logic
Baldwin, John – Mathematical and Philosophical Problems Arising in the Context of the Book
Bazhenov, Nikolay, Manat Mustafa & Mars Yamaleev – Semilattices of Numberings
Berger, Ulrich – On the Constructive Content of Proofs in Abstract Analysis
Cheng, Yong – Some Formal and Informal Misunderstandings of Gödel's Incompleteness Theorems
Cieslinski, Cezary – Commitments of Foundational Theories: Introduction
Enayat, Ali – Feasible Reducibility and Interpretability of Truth Theories
Engler, Mirko – Generalized Interpretability and Conceptual Reduction of Theories
Fiori Carones, Marta – A Theorem of Ordinary Mathematics Equivalent to Ads
Freitag, James – Some Recent Applications of Model Theory
Galeotti, Lorenzo – Higher Metrisability in Higher Descriptive Set Theory
Godziszewski, Michał Tomasz – Some Semantic Properties of Typed Axiomatic Truth Theories Built over Theory of Sets
Herrera González, José Rafael – Combining Temporal and Epistemic Logic: A Matter of Points of View
Hill, Cameron – Towards a Characterization of Pseudo-Finiteness
Honzík, Radek – The Indestructibility of the Tree Property
Kiouvrekis, Yiannis – Remarks on Abstract Logical Topologies: An Institutional Approach
Łełyk, Mateusz – The Contour of the Tarski Boundary
Lieberman, Michael – Tameness, Compactness, and Cocompleteness
Litak, Tadeusz – Modal Negative Translations as a Case Study in the Big Programme
Loffredo D'Ottaviano, Itala Maria & Hércules de Araujo Feitosa – On the Historical Relevance of Glivenko's Translation from Classical into Intuitionistic Logic: Is It Conservative and Contextual?
Lücke, Philipp – Definable Bistationary Sets
Malliaris, Maryanthe – Should a Mathematician Read This Book?
Markhabatov, Nurlan & Sergey Sudoplatov – On Calculi and Ranks for Definable Families of Theories

---

[1] Talks in multiple categories are listed according to their first category.

Martínez Herrera, Francisco – Dialogical Justication Logic. A Basic Approach
Martino, Giovanni Marco – An Algebraic Model for Frege's Basic Law V
Negri, Sara – A Simple Proof of Barr's Theorem for Infinitary Geometric Logic
Nicolás-Francisco, Ricardo Arturo – A Non-trivial Extension for MS
Patel, Rehana – Towards a Model Theory of Symmetric Probabilistic Structures
Pavlyuk, Inessa & Sergey Sudoplatov – On Ranks for Families of Theories of Abelian Groups
Pellissier, Luc & Juan-Luis Gastaldi – Duality and Interaction: A Common Dynamics behind Logic and Natural Language
Pereira, Luiz Carlos, Elaine Pimentel & Valeria de Paiva – Ecumenism: A New Perspective on the Relation between Logics
Piccolomini D'Aragona, Antonio – A Class of Languages for Prawitz's Epistemic Grounding
Radzki, Mateusz – The Tarski Equipollence of Axiom Systems
Rathjen, Michael – Proof Theory of Infinite Geometric Theories
Rosický, Jiří – Accessible Categories and Model Theory
San Mauro, Luca – Inductive Inference and Structures: How to Learn Equality in the Limit
Schuster, Peter, Giulio Fellin & Daniel Wessel – The Jacobson Radical and Glivenko's Theorem
Seisenberger, Monika – Program Optimisation through Proof Transformation
Shami, Ziv – On the Forking Topology of a Reduct of a Simple Theory
Sorina, Galina & Irina Griftsova – The Alienated/Subjective Character of Scientific Communication
Stejskalová, Šárka – Easton's Function and the Tree Property below $\aleph_\alpha$
Sziráki, Dorottya – The Open Dihypergraph Dichotomy for Definable Subsets of Generalized Baire Spaces
Tarafder, Sourav, Benedikt Löwe & Robert Passmann – Constructing Illoyal Algebra-Valued Models of Set Theory
Vasey, Sebastien – Forking and Categoricity in Non-elementary Model Theory
Wcisło, Bartosz – Models of Truth Theories
Wessel, Daniel – Ideals, Idealization, and a Hybrid Concept of Entailment Relation
Wohofsky, Wolfgang – Can We Add Kappa-Dominating Reals without Adding Kappa-Cohen Reals?

A.2 Philosophical Logic

Angelova, Doroteya – Logical Approaches to Vagueness and Sorites Paradoxes

Angius, Nicola & Giuseppe Primiero – Second Order Properties of Copied Computational Artefacts

Ansari, Mahfuz Rahman & Avr Sarma – Counterfactuals and Reasoning about Action

Arazim, Pavel – Are Logical Expressions Ambiguous and Why?

Baltag, Alexandru, Soroush Rafiee Rad & Sonja Smets – Learning Probabilities: A Logic of Statistical Learning

Běhounek, Libor – A Formalism for Resource-Sensitive Epistemic Logic

Bílková, Marta – Common Belief Logics Based on Information

Bozdag, Ayse Sena – Modeling Belief Base Dynamics Using Hype Semantics

Brîncuș, Constantin C. – Open-Ended Quantification and Categoricity

Canavotto, Ilaria, Alexandru Baltag & Sonja Smets – Introducing Causality in Stit Logic

Christoff, Zoé, Olivier Roy & Norbert Gratzl – Priority Merge and Intersection Modalities

Cintula, Petr, Carles Noguera & Nicholas Smith – Formalizing the Sorites Paradox in Mathematical Fuzzy Logic

De Coninck, Thijs & Frederik Van de Putte – Reciprocal Group Oughts

Diaz Montilla, Francisco – Fuzzy Logic and Quasi-Legality

Domínguez, Daniel Álvarez – Splicing Logics: How to Combine Hybrid and Epistemic Logic to Formalize Human Reasoning

Dumitru, Mircea – New Thoughts on Compositionality. Contrastive Approaches to Meaning: Fine's Semantic Relationism vs. Tarski-Style Semantics

Edwards, Adam – Seeing and Doing, Or, Why We Should All Be Only Half-Bayesian

Elgin, Samuel – The Semantic Foundations of Philosophical Analysis

Estrada-González, Luis & Claudia Tanús – Variable Sharing Principles in Connexive Logic

Faust, Don – Predication Elaboration: Providing Further Explication of the Concept of Negation

Ferrario, Roberta – Organisations and Variable Embodiments

Frijters, Stef & Thijs de Coninck – If Killing Is Forbidden, Do I Have to Ensure That No One Is Killed?

Fu, Haocheng – Iterated Belief Revision and Dp Postulates

Galliani, Pietro – What Is It like to Be First Order? Lessons from Compositionality, Teams and Games

García, María D. – Justification of Basic Inferences and Normative Freedom

Godziszewski, Michał Tomasz & Rafal Urbaniak – Modal Quantifiers, Potential Infinity, and Yablo Sequences

Grimau, Berta – Fuzzy Semantics for Graded Adjectives

Günther, Mario – Learning Subjunctive Conditional Information

Hîncu, Mihai – Intensionality, Reference, and Strategic Inference

Ipakchi, Sara – Even Logical Truths Are Falsifiable

Ju, Fengkui – Coalitional Logic on Non-interfering Actions
Kapsner, Andreas – Connexivity and Conditional Logic
Karpinskaia, Olga – Abstract and Concrete Concepts: An Approach to Classification
Klein, Dominik – A Logical Approach to Nash Equilibria
Klev, Ansten – Definitional Identity in Arithmetic
Kozachenko, Nadiia – Critical Thinking and Doxastic Commitments
Kuncová, Alexandra – Ability and Knowledge
Kuusisto, Antti – Interactive Turing-Complete Logic via Game-Theoretic Semantics
Lee, Jui-Lin – Model Existence in Modal Logics
Liggins, David – Semantic Paradoxes of Underdetermination
Majer, Ondrej, Dominik Klein & Soroush Rafiee Rad – Non-classical Probabilities over Dunn-Belnap Logic
Manzano, Maria & Manuel Crescencio Moreno Gomez – Quantifiers and Conceptual Existence
Marra, Alessandra & Dominik Klein – From Oughts to Goals
Méndez, José M., Gemma Robles & Francisco Salto – Expansions of Relevant Logics with a Dual Intuitionistic Type Negation
Mishra, Meha & A. V. Ravishankar Sarma – An Attempt to Highlight Ambiguities in Approaches to Resolve Chisholm Paradox
Moreno, Luis Fernández – Rigidity and Necessity: The Case of Theoretical Identifications
Narita, Ionel – Logic of Scales
Negri, Sara & Edi Pavlovic – Dstit Modalities through Labelled Sequent Calculus
Nygren, Karl – Varieties of Permission for Complex Actions
Olkhovikov, Grigory – Stit Heuristics and the Construction of Justification Stit Logic
Omori, Hitoshi – Towards a Bridge over Two Approaches in Connexive Logic
Parker, Matthew – Comparative Infinite Lottery Logic
Pavlenko, Andrey – Frege Semantics or Why Can We Talk about Deflation of False?
Pavlov, Sergey – On Conditions of Inference in Many-Valued Logic Semantics of CL2
Pawlowski, Pawel & Rafal Urbaniak – Combining Truth Values with Provability Values: A Non-deterministic Logic of Informal Provability
Pezlar, Ivo – Analysis of Incorrect Proofs
Pfeifer, Niki – Are Connexive Principles Coherent?
Piecha, Thomas & Peter Schroeder-Heister – Abstract Semantic Conditions and the Incompleteness of Intuitionistic Propositional Logic with Respect to Proof-Theoretic Semantics
Pizzi, Claudio E. A. – Tableaux Procedures for Logics of Consequential Implication
Porwolik, Marek – The Axiomatic Approach to Genidentity According to Z. Augustynek

Pribram-Day, Ivory – The Problem of the Variable in Quine's Lingua Franca of the Sciences
Punčochář, Vít – Algebraic Semantics for Inquisitive Logics
Raclavský, Jiří – Type Theory, Reducibility and Epistemic Paradoxes
Rafiee Rad, Soroush & Olivier Roy – Deliberation, Single-Peakedness and Voting Cycles
Redmond, Juan & Rodrigo Lopez-Orellana – Clasical Logic and Schizophrenia: For a Neutral Game Semantics
Rendsvig, Rasmus K. – Dynamic Term-Modal Logic
Rey, David & Pablo Cubides – Expressive Power and Intensional Operators
Rivello, Edoardo – Definite Truth
Roberts, Alice – A Bridge for Reasoning: Logical Consequence as Normative
Robles, Gemma – Basic Quasi-Boolean Expansions of Relevant Logics with a Negation of Intuitionistic Kind
Rönnholm, Raine, Valentin Goranko & Antti Kuusisto – Rationality Principles in Pure Coordination Games
Rudnicki, Konrad & Piotr Łukowski – Empirical Investigation of the Liar Paradox. Human Brain Perceives the Liar Sentence to be False.
Rybaříková, Zuzana – Łukasiewicz's Concept of Anti-psychologism
Sanyal, Manidipa & Debirupa Basu – Attack at Dawn If the Weather Is Fine
Sedlar, Igor – Substructural Propositional Dynamic Logic
Shramko, Yaroslav – First-Degree Entailment and Structural Reasoning
Simons, Peter – Leśniewski, Lambda, and the Problem of Defining Operators
Speitel, Sebastian G. W. – A Notion of Semantic Uniqueness for Logical Constants
Svoboda, Vladimír – Language Games and Paradoxes of Deontic Logic(s)
Świętorzecka, Kordula & Marcin Łyczak – A Bimodal Logic of Change with Leibnizian Hypothetical Necessity
Tabakov, Martin – Reflections on the Term 'Philosophical Logic'
Tanús, Claudia – The Irrelevance of the Axiom of Permutation
Taşdelen, İskender – Free Logic and Unique Existence Proofs
Tedder, Andrew & Igor Sedlár – Residuals and Conjugates in Positive Substructural Logic
Tranchini, Luca & Alberto Naibo – Harmony, Stability, and the Intensional Account of Proof-Theoretic Semantics
Tsai, Hsing-Chien – Classifying First-Order Mereological Structures
Valor Abad, Jordi – Is the Liar Sentence Meaningless?
Van de Putte, Frederik – Constructive Deliberation: Pooling and Stretching Modalities
Vázquez, Margarita & Manuel Liz – Reasoning about Perspectives. New Advances
Wang, Ren-June – Knowledge, Reasoning Time, and Moore's Paradox
Wang, Wen-Fang – A Three-Valued Pluralist Solution to the Sorites Paradox

Wood, Nathan & Thijs de Coninck – Probabilistic Agent-Dependent Oughts
Yang, Syraya Chin-Mu – Higher-Order Identity in the Necessitism-Contingentism Debate in Higher-Order Modal Logic
Zhou, Xunwei – Mutually Inverse Implication Inherits from and Improves on Material Implication

A.3 Computational Logic and Applications of Logic

Beke, Tibor – Feasible Syntax, Feasible Proofs, and Feasible Interpretations
Duží, Marie – Hyperintensions as Abstract Procedures
Fuenmayor, David & Christoph Benzmüller – Automated Reasoning with Complex Ethical Theories – A Case Study Towards Responsible AI
Garbayo, Luciana, Wlodek Zadrozny & Hossein Hematialam – Measurable Epistemological Computational Distances in Medical Guidelines Peer Disagreement
Groza, Adrian – Differences of Discourse Understanding between Human and Software Agents
Hertel, Joachim – Hypercomputing Minds: New Numerical Evidence
Kutz, Oliver & Nicolas Troquard – A Logic for an Agentive Naïve Proto-Physics
Lampert, Timm – Theory of Formalization: The Tractarian View
Ospichev, Sergey & Denis Ponomaryov – On the Complexity of Formulas in Semantic Programming
Ozaki, Yuki – Sensory Perception Constructed in Terms of Carnap's Inductive Logic: Developing Philosophy of Computational Modeling of Perception
Trela, Grzegorz – Logic as Metaphilosophy? Remarks on the Mutual Relations of Logic and Philosophy
Yamasaki, Susumu – Multi-Modal Mu-Calculus with Postfix Modal Operator Abstracting Actions

A.4 Historical Aspects of Logic

Aray, Başak – Sources of Peano's Linguistics
Bar-On, Kati Kish – Towards a New Philosophical Perspective on Hermann Weyl's Turn to Intuitionism
Bertran-San Millán, Joan – Frege and Peano on Axiomatisation and Formalisation
Besler, Gabriela – Transcriptions of Gottlob Frege's Logical Formulas into Boole's Algebra and Language of Modern Logic. Similarities and Differences
Beziau, Jean-Yves – Tarski's Two Notions of Consequence
Chatti, Saloua – An Introduction to Arabic Hypothetical Logic

Dasdemir, Yusuf – Reception of Absolute Propositons in the Avicennian Tradition: Ibn SahlāN Al-SāWī on the Discussions of the Contradiction and Conversion of Absolute Propositions
Finley, James – Medieval Debates over the Infinite as Motivation for Pluralism
Haniková, Zuzana – On Vopěnka's Ultrafinitism
Hodges, Wilfrid – Abū Al-BarakāT and His 12th Century Logic Diagrams
Jetli, Priyedarshi – Hilbert and the Quantum Leap from Modern Logic to Mathematical Logic
Khudoydodov, Farrukh – Similarities and Differences in the Logic of Aristotle and Avicenna
Kurokawa, Hidenori – On Takeuti's View of the Concept of Set
Levina, Tatiana – In Defense of Abstractions: Sofia Yanovskaya between Ideology and Cybernetics
Neumann, Jared – Deductive Savages: The Oxford Noetics on Logic and Scientific Method
Palomäki, Jari – The Intensional and Conceptual Content of Concepts
Ren, Xiaoming & Xianhua Liang – Three Ways to Understand the Inductive Thoughts of Whewell
Schlimm, Dirk – On the 'Mechanical' Style in 19th-Century Logic
Šebela, Karel – Sortal Interpretation of Aristotelian Logic
Utrero, Víctor Aranda – The Universalism of Logic and the Theory of Types
Vandoulakis, Ioannis – Pythagorean Arithmetic as a Model for Parmenidean Semantics

B.1 Methodology

Ahlskog, Jonas & Giuseppina D'Oro – Collingwood, the Narrative Turn, and the Cookie Cutter Conception of Historical Knowledge
Allo, Patrick – Conceptual Engineering in the Philosophy of Information
Antonovskiy, Alexander – Max Weber's Distinction Truth/Value and 'Old-European' Semantics
Antos, Carolin & Daniel Kuby – Mutually Inconsistent Set Theoretic-Universes: An Analysis of Universist and Multiversist Strategies
Apolega, Dennis – Does Scientific Literacy Require a Theory of Truth?
Arenhart, Jonas Becker & Décio Krause – Quasi-Truth and Defective Situations in Science
Artamonov, Denis – Media Memory as the Object of Historical Epistemology
Badiei, Sina – Karl Popper's Three Interpretations of the Epistemological Peculiarities of the Social Sciences

Banchetti-Robino, Marina Paola – Early Modern Chemical Ontologies and the Shift from Vitalism to Mechanicism

Barker, Matt – Using Norms to Justify Theories within Definitions of Scientific Concepts

Barseghyan, Hakob & Jamie Shaw – Integrating Hps: What's in It for a Philosopher of Science?

Batens, Diderik – Handling of Defectiveness in a Content-Guided Manner

Belleri, Delia – In Defense of a Contrastivist Approach to Evidence Statements

Bobadilla, Hernán – Two Types of Unrealistic Models: Programatic and Prospective

Boon, Mieke – How Scientists Are Brought Back into Science – The Error of Empiricism

Boyd, Nora – Constraining the Unknown

Brown, Martin – Chunk and Permeate: Reasoning Faute de Mieux

Brun, Georg & Kevin Reuter – The Common-Sense Notion of Truth as a Challenge for Conceptual Re-Engineering

Bustos, Moises Macias – Lewis, Stalnaker and the Problem of Assertion & Defective Information in the Sciences

Černín, David – Experiments in History and Archaeology: Building a Bridge to the Natural Sciences?

Chall, Cristin – Abandoning Models: When Non-empirical Theory Assessment Ends

Chebotareva, Elena – An Engineer: Bridging the Gap between Mechanisms and Values

Chen, Rueylin – Natural Analogy: A Hessean Approach to Analogical Reasoning in Theorizing

Coko, Klodian – Robustness, Invariance, and Multiple Determination

Cordero, Alberto – Four Realist Theses of Mario Bunge

Cordes, Moritz – The Methodological Tradition of Explication

D'Oro, Giuseppina – Why Epistemic Pluralism Does Not Entail Relativism

Daniell, Paul – Equilibrium Theory and Scientific Explanation

David-Rus, Richard – Considering the Factivity of Non-explanatory Understanding

Davis, Cruz & Benjamin Jantzen – Do Heuristics Exhaust the Methods of Discovery?

Dietz, Bettina – Tinkering with Nomenclature. Textual Engineering, Co-authorship, and Collaborative Publishing in Eighteenth-Century Botany

Dolmatov, Anton – Moral Achievement of a Scientist

Donato-Rodriguez, Xavier De – Inconsistency and Belief Revision in Cases of Approximative Reduction and Idealization

Dong, Zili – Discovering Unfaithful Causal Structures from Observations and Interventions

Durlacher, Thomas – Idealizations and the Decomposability of Models in Science

Eberhardt, Frederick – Proportional Causes and Specific Effects

Erasmus, Adrian – Expected Utility, Inductive Risk, and the Consequences of P-Hacking

Fahrbach, Ludwig – Is the No-Miracles Argument an Inference to the Best Explanation?

Feldbacher-Escamilla, Christian J. – Simplicity in Abductive Inference
Fischer, Mark – Pluralism and Relativism from the Perspective of Significance in Epistemic Practice
Friend, Michèle – Disturbing Truth
Frolov, Igor & Olga Koshovets – Rethinking the Transformation of Classical Science in Technoscience: Ontological, Epistemological and Institutional Shifts
Gebharter, Alexander & Christian J. Feldbacher-Escamilla – Modeling Creative Abduction Bayes Net Style
Genin, Konstantin & Kevin Kelly – Progressive Methods for Causal Discovery
Ginammi, Michele – Applicability Problems Generalized
Grobler, Adam – How Science Is Knowledge
Hansson, Sven Ove – The Philosophical Roots of Science Denialism
Hasan, Yousuf – Carnap on the Reality of Atoms
Hattiangadi, Jagdish – Inductive Method, or the Experimental Philosophy of the Royal Society
Hendry, Robin – The History of Science and the Metaphysics of Chemistry
Hijmans, Sarah – The Building Blocks of Matter: The Chemical Element in 18th and 19th-Century Views of Composition
Hirvonen, Ilmari, Rami Koskinen & Ilkka Pättiniemi – Epistemology of Modality without Metaphysics
Hladky, Michal – Mapping vs. Representational Accounts of Models and Simulations
Huss, John – Tool-Driven Science
Isaac, Manuel Gustavo – Broad-Spectrum Conceptual Engineering
Jackson, Rebecca – Sending Knowns into the Unknown: Towards an Account of Positive Controls in Experimentation
Jantzen, Benjamin – Finding Causation in Time: Background Assumptions for Dynamical Systems
Jaster, Romy & David Lanius – Truth and Truthfulness, Part I: What Fake News Is and What It's Not
Jaster, Romy & David Lanius – Truth and Truthfulness, Part II: Why They Matter
Jordan, Roman Otto – The Evolutionary Epistemology of Rupert Riedl – A Consequent Realization of the Program of Naturalizing Epistemology?
Kasavin, Ilya – The Scientist's Dilemma: After Weber
Kashyap, Abhishek – Underdetermination of Theories, Theories of Gravity, and the Gravity of Underdetermination
Kelly, Kevin, Hanti Lin, Konstantin Genin & Jack Parker – A Learning Theoretic Argument for Scientific Realism
Kendig, Catherine – Messy Metaphysics: The Individuation of Parts in Lichenology

Koch, Steffen – On Two Kinds of Conceptual Engineering and Their Methodological Counterparts

Kornmesser, Stephan – Frames – A New Model for Analyzing Theories

Kozlova, Natalia – The Problem of Figurativeness in Science: From Communication to the Articulation of Scientific Knowledge

Křepelová, Tereza – Positivisation of Political Philosophy and Its Impact on the Whole Discipline

Kuukkanen, Jouni-Matti – Truth, Incoherence and the Evolution of Science

Kuznetsov, Vladimir & Alexander Gabovich – Commutative Transformations of Theory Structures

Lê, François – Characterizing as a Cultural System of the Organization of Mathematical Knowledge: A Case Study from the History of Mathematics

Lehtinen, Aki & Jani Raerinne – Simulated Data

León, Esa Díaz – The Semantic Account of Slurs, Appropriation, and Metalinguistic Negotiations

Lin, Chia-Hua – The Increasing Power of Chomsky Hierarchy: A Case Study of Formal Language Theory Used in Cognitive Biology

Lin, Hanti – Convergence to the Causal Truth and Our Death in the Long Run

Linsbichler, Alexander – In Defense of a Thought-Stopper: Relativizing the Fact/Value Dichotomy

Lyons, Tim – The Reach of Socratic Scientific Realism: From Axiology of Science to Axiology of Exemplary Inquiry

Mäki, Uskali – Asymmetries in Interdisciplinarity

Marasoiu, Andrei – The Truth in Understanding

Meheus, Joke – Logic-Based Ontologies in the Biomedical Domain: From Defects to Explicit Contradictions

Mets, Ave – The Pluralist Chemistry and the Constructionist Philosophy of Science

Miranda, Rogelio – Three Problems with the Identification of Philosophy with Conceptual Analysis

Mutanen, Arto – On Explanation and Unification

Northcott, Robert – Prediction Markets and Extrapolation

Ondráček, Tomáš – Science as Critical Discussion and Problem of Immunizations

Onishi, Yukinori – Does Research with Deep Neural Networks Provide a New Insight to the Aim of Science Debate?

Ordaz, Maria Del Rosario Martinez & Otávio Bueno – Making Sense of Defective Information: Partiality and Big Data in Astrophysics

Parkkinen Veli-Pekka – Robustness in Configurational Causal Modelling

Pinder, Mark – Concepts and Replacement: What Should the Carnapian Model of Conceptual Re-Engineering Be?

Pollock, Joey – Conceptual Engineering and Semantic Control
Portides, Demetris & Athanasios Raftopoulos – Abstraction in Scientific Modeling
Pulkkinen, Karoliina – Some Sixty or More Primordial Matters: Chemical Ontology and the Periodicity of the Chemical Elements
Ramírez-Cámara, Elisángela – Is Biased Information Ever Useful (In the Philosophy of Science)?
Reddy, Daya – Fake News, Pseudoscience, and Public Engagement
Reinhard, Franziska – Realism and Representation in Model-Based Explanation
Ropolyi, László – Technoscience and Philoscience
Ruttkamp-Bloem, Emma – A Dynamic Neo-Realism as an Active Epistemology for Science
Schickore, Jutta – Blur Science through Blurred Images. What the Diversity of Fuzzy Pictures Can Do for Epistemic, Methodological and Clinical Goals
Schickore, Jutta – Scientists' Reflections on Messy Science
Scholl, Raphael – Scenes from a Marriage: On the Confrontation Model of History and Philosophy of Science
Schwed, Menashe – Truth Lies: Taking Yet Another Look at the Theory-Laden Problem
Shaw, Jamie – Feyerabend's Well-Ordered Science: How an Anarchist Distributes Funds
Shibarshina, Svetlana – Scientists' Social Responsibilities in the Context of Science Communication
Shinod, N. K. – Evidential Relations in a Trading Zone
Shipovalova, Lada – M. Weber's 'Inconvenient Facts' and Contemporary Studies of Science-Society Communication
Sidiropoulos, Michael – Philosophical and Demarcation Aspects of Global Warming Theory
Syrjänen, Pekka – Some Problems in the Prediction vs Accommodation Debate
Tambolo, Luca – The Problem of Rule-Choice Redux
Tikhonova, Sophia – Knowledge Production in Social Networks as the Problem of Communicative Epistemology
Tukhvatulina, Liana – Scientist as an Expert: Breaking the Ivory Tower
Tvrdý, Filip – Mysterianism and the Division of Cognitive Labour
Veigl, Sophie Juliane – An Empirical Challenge for Scientific Pluralism – Alternatives or Integration?
Veit, Walter – Who Is Afraid of Model Pluralism?
Virmajoki, Veli – The Science We Never Had
Vos, Bobby – Integrated Hps? Formal versus Historical Approaches to Philosophy of Science
Wang, Linton, Ming-Yuan Hsiao & Jhih-Hao Jhang – Unfalsifiability and Defeasibility
Wray, K. Brad – Setting Limits to Chang's Pluralism

Zhang, Jiji & Kun Zhang – Causal Minimality in the Boolean Approach to Causal Inference

B.2 Formal Philosophy of Science and Formal Epistemology

Almpani, Sofia, Petros Stefaneas & Ioannis Vandoulakis – On the Significance of Argumentation in Discovery Proof-Events
Andreas, Holger – Explanatory Conditionals
Arpaia, Salvatore Roberto – Incompleteness-Based Formal Models for the Epistemology of Complex Systems
Benedetto, Matteo De – Explicating 'Explication' via Conceptual Spaces
Bielik, Lukáš – Abductive Inference and Selection Principles
Cevolani, Gustavo & Roberto Festa – Approaching Deterministic and Probabilistic Truth: A Unified Account
Chajda, Ivan, Davide Fazio & Antonio Ledda – The Generalized Orthomodularity Property: Configurations, Pastings and Completions
Cocco, Lorenzo & Joshua Babic – Theoretical Equivalence and Special Relativity
Cresto, Eleonora – A Constructivist Application of the Condorcet Jury Theorem
Dimarogkona, Maria & Petros Stefaneas – A Meta-Logical Framework for Philosophy of Science
Dotan, Ravit – Machine Learning, Theory Choice, and Non-epistemic Values
Douven, Igor – Optimizing Group Learning of Probabilistic Truths
Fischer, Florian & Alexander Gebharter – Dispositions and Causal Bayes Nets
Fletcher, Samuel – The Topology of Intertheoretic Reduction
Kilinc, Berna – Deterministic and Indeterministic Situations
King, Martin – Towards the Reconciliation of Confirmation Assessments
Koscholke, Jakob – Siebel's Argument against Fitelson's Measure of Coherence Reconsidered
Kuipers, Theo – Inductively Approaching a Probabilistic Truth and a Deterministic Truth, the Latter in Comparison with Approaching It in a Qualitative Sense.
Lapeña, Alfonso García – Scientific Laws and Closeness to the Truth
Lobovikov, Vladimir – A Formal Axiomatic Epistemology Theory and the Controversy between Otto Neurath and Karl Popper about Philosophy of Science
Makhova, Maria & Anastasia Arinushkina – Scientific Communication in the Problematic Field of Epistemology: Inside And/or outside (In French)
Marfori, Marianna Antonutti – Formalisation and Proof-Theoretic Reductions
Nicholson, Daniel – Schrödinger's 'What Is Life?' 75 Years On
Niiniluoto, Ilkka – Approaching Probabilistic Laws
Oddie, Graham – Credal Accuracy in an Indeterministic Universe

Pinto, Victor Hugo – The Problem of Scientific-Epistemological Racism and the Contributions of Southern Global Epistemologies in the Construction of Paradigmatic Transformations of the Philosophy of Science

Proszewska, Agnieszka – Is Semantic Structuralism Necessarily 'Set-Theoretical' Structuralism? A Case of Ontic Structural Realism.

Rhee, Young E. – On Howson's Bayesian Approach to the Old Evidence Problem

Rodriguez, Jorge Luis Garcia – A Naturalized Globally Convergent Solution to the Problem of Induction

Sarma, A. V. Ravishankar – On a Structuralist View of Theory Change: Study of Some Semantic Properties in Formal Model of Belief Revision

Schurz, Gerhard – Approaching Objective Probabilities by Meta-Inductive Probability Aggregation

Sikimić, Vlasta – Optimal Team Structures in Science

Stefaneas, Petros & Ioannis Vandoulakis – Mathematical Proving as Spatio-Temporal Activity of Multi-Agent Systems

Tagliaferri, Mirko – How to Build a Computational Notion of Trust

Thorn, Paul & Gerhard Schurz – Meta-Inductive Prediction Based on Attractivity Weighting: Mathematical and Empirical Performance Evaluation

Vignero, Leander – A Computational Pragmatics for Weaseling

Vojtas, Peter & Michal Vojtas – Problem Reduction as a General Epistemic Reasoning Method

Weingartner, Paul – A Defence of Pluralism of Causality

Zhou, Liqian – Mutual Misunderstanding in Signalling Games

B.3 Empirical and Experimental Philosophy of Science

Cho, In-Rae – Toward a Coevolutionary Model of Scientific Change

Cordero, Alberto – Functional Ontologies and Realism: The Case of Nuclear Physics

Holman, Bennett – Dr. Watson: The Impending Automation of Diagnosis and Treatment

Johansen, Mikkel Willum – Entering the Valley of Formalism: Results from a Large-Scale Quantitative Investigation of Mathematical Publications

Ma, Lei – Empirical Identity as an Indicator of Theory Choice

Michel, Nicolas – Avatars of Generality: On the Circulation and Transformation of List-Making Practices in the Context of Enumerative Geometry

Muntersbjorn, Madeline – Notations & Translations as Catalysts of Conceptual Change

Paitlová, Jitka & Petr Jedlička – Objectivity of Science from the Perspective of X-Phi

Petrovich, Eugenio – Bridging across Philosophy of Science and Scientometrics: Towards an Epistemological Theory of Citations

Petrukhina, Polina & Vitaly Pronskikh – High-Energy Physics Cultures during the Cold War: Between Exchanges and Translations

Pronskikh, Vitaly, Kaja Damnjanović, Polina Petruhina, Arpita Roy & Vlasta Sikimić – Are In-Depth Interviews a Must for Ethnography of Hep Labs?

Reva, Nataliia – Does Analogical Reasoning Imply Anthropomorphism?

Roure, Pascale – Logical Empiricism in Exile. Hans Reichenbach's Research and Teaching Activities at Istanbul University (1933–1938)

Sauzet, Romain – Cognitive and Epistemic Features: A Tool to Identify Technological Knowledge

Starikova, Irina – 'Thought Experiments' in Mathematics?

Tovar-Sánchez, Guillermo Samuel & Luis Mauricio Rodríguez-Salazar – From Subject Natural Logic to Scientist Logic in Natural Science: An Epistemological Reflexion

Vinokurov, Vladimir & Marina Vorontsova – Geometry, Psychology, Myth as Aspects of the Astrological Paradigm

Wang, Xiaofei – The Epistemic Values Pursued by J.-L. Lagrange in His Teaching of Analysis at the Ecole Polytechnique

Zheng, Fanglei – When Arabic Algebraic Problems Met Euclidean Norms in the 13th Century – A Case Study on the Scientific Innovation by the Transformation in Cross-Cultural Transmission

Zhou, Xiaohan – Elements of Continuity in the Circulation of Mathematical Knowledge and Practices in Chapter 'Measures in Square' in Mathematical Writings in China

B.4 Metaphysical Issues in the Philosophy of Science

Akcin, Haktan – Structural Modality as the Criterion for Naturalistic Involvement in Scientific Metaphysics

Anta, Javier – Sympletic Battlefronts. Phase Space Arguments for (And Against) The Physics of Computation

Benda, Thomas – Change, Temporal Anticipation, and Relativistic Spacetime

Boulter, Stephen – On the Possibility and Meaning of Truth in the Historical Sciences

Brugnami, Enrico – Scientific Ways of Worldmaking. Considerations on Philosophy of Biology from Goodman's Theory of Worlds

Castro, Eduardo – Laws of Nature and Explanatory Circularity

Chabout-Combaz, Babette – A Philosophy of Historiography of the Earth. Metaphor and Analogies of 'Natural Body'

Chen, Ruey-Lin & Jonathon Hricko – Pluralism about Criteria of Reality

Coffey, Kevin – The Ontology of Mass and Energy in Special Relativistic Particle Dynamics

Córdoba, Mariana, Cristian López & Hernán Accorinti – Metaphysical Pluralism: Between Skepticism and Trivialization?
Denisova, Tatiana – Metaphysical Issues in Modern Philosophy of Time: V. I. Vernadsky's Idea of 'Cause' of Time ('Source' of Time)
Egg, Matthias – Scientific Metaphysics and the Manifest Image
Gan, Nathaniel – Explanation and Ontology
Gentile, Nelida & Susana Lucero – On the Unifying Character of Dispositional Realism
Ghins, Michel – Scientific Realism and the Reality of Properties
Golovko, Nikita – Second Pattern Existence and Truth-Making
Honma, Souichiro – Free Will and the Ability to Change Laws of Nature
Iakovlev, Vladimir – Metaphysics and Physics of Consciousness as a Problem of Modern Science
Karmaly, Sajjad – Why Cognitive Kinds Can't Be the Kind of Kinds That Are Natural Kinds? A New Hypothesis for Natural Kinds
Leshkevich, Tatiana – Digital Determination and the Search for Common Ground
Livanios, Vassilis – Can Categorical Properties Confer Dispositions?
Liz, Manuel & Margarita Vázquez – Scientific Perspectivism. Metaphysical Aspects
Luty, Damian – Regarding Minimal Structural Essentialism in Philosophy of Spacetime
Matarese, Vera – Super-Humeanism: A Naturalized Metaphysical Theory?
Maziarz, Mariusz – Econometric Modeling Falsifies Structural Realism
McCoy, C. D. – Counterfeit Chance
Milkov, Nikolay – Towards an Analytic Scientific Metaphysics
Müürsepp, Peeter – Practical Realism and Metaphysics in Science
Nevvazhay, Igor – Transcendental in Physical Theory
Oleksowicz, Michal – Some Philosophical Remarks on the Concept of Structure. Case of Ladyman's and Heller's View
Onnis, Erica – Discontinuity and Robustness as Hallmarks of Emergence –
Penner, Myron & Amanda Nichols – Realism about Molecular Structures
Piccinini, Gualtiero – Levels of Being: An Egalitarian Ontology
Reydon, Thomas – How Do Evolutionary Explanations Explain?
Rusu, Mihai & Mihaela Mihai – Modal Notions and the Counterfactual Epistemology of Modality
Schäfer, Leon-Philip – Objectivity as Mind-Independence – Integrating Scientific and Moral Realism
Slipkauskaitė, Julita – Theory of Impetus and Its Significance to the Development of Late Medieval Notions of Place
Sterpetti, Fabio – Non-causal Explanations of Natural Phenomena and Naturalism
Szabo, Laszlo E. – Intrinsic, Extrinsic, and the Constitutive a Priori
Teller, Paul – Processes and Mechanisms

Vega, Pablo Vera – Rethinking the Given. Sellars on the First Principles.
Vivanco, Melisa – Numbers as Properties. Dissolving Benacerraf's Tension
Wachter, Tina – Can Conventionalism Safe the Identity of Indiscernibles?
Yan, Chunling – Underdetermination and Empirical Equivalence: The Standard Interpretation and Bohmian Mechanics
Zorzato, Lisa & Antonella Foligno – An Attempt to Defend Scientific Realism

B.5 Ethical and Political Issues in the Philosophy of Science

Bouwel, Jeroen van – Are Transparency and Representativeness of Values Hampering Scientific Pluralism?
Bschir, Karim – Corporate Funding of Public Research: A Feyerabendian Perspective
Cuevas-Badallo, Ana & Daniel Labrador-Montero – How Pragmatism Can Prevent from the Abuses of Post-truth Champions
García, Laura & Abraham Hernández – The Role of TV Series in the Democratization of Science
Kano, Hiroyuki – Updating Scientific Adviser Models from Policy-Maker Perspectives: A Limit of Debate of 'Science and Value' and Norms of Public Policy
Małecka, Magdalena – Why the Behavioural Turn in Policy Takes Behavioural Science Wrong and What It Means for Its Policy Relevance
Neemre, Eveli – Values in Science and Value Conflicts
Pégny, Maël & Mohamed Issam Ibnouhsein – Can Machine Learning Extend Bureaucratic Decisions?
Puyol, Angel – Solidarity and Regulatory Frameworks in (Medical) Big Data
Schroeder, Andrew – Values in Science: Ethical vs. Political Approaches
Shang, Zhicong – The Competition of Interests in Public Scientific Knowledge Production – An Analysis of Chinese Case
Viatkina, Nataliia – Deference as Analytic Technique and Pragmatic Process

B.6 Historical Aspects of the Philosophy of Science

Barany, Michael – Experts and Expertise in North-South Circulation in Mid-Twentieth Century Mathematics
Bellomo, Anna – Bolzano's Real Numbers: Sets or Sums?
Betti, Arianna & Annapaola Ginammi – Bolzano's Theory of Ground and Consequence and the Traditional Theory of Concepts
Blok, Johan – Did Bolzano Solve the Eighteenth-Century Problem of Problematic Mathematical Entities?
Brîncuș, Constantin C. – Comment on 'Karl R. Popper: Logical Writings'

Cantu, Paola – Bolzano's Requirement of a Correct Ordering of Concepts and Its Inheritance in Modern Axiomatics
Childers, Timothy – Comment on 'de Finetti Meets Popper'
Clatterbuck, Hayley – Darwin's Causal Argument against Intelligent Design
Collodel, Matteo – Feyerabend and the Reception and Development of Logical Empiricism
Crippa, Davide, Elias Fuentes Guillén & Jan Makovský – Bernard Bolzano's 1804 Examination: Mathematics and Mathematical Teaching in Early 19th Century Bohemia
Cunha, Ivan F. da – Utopias in the Context of Social Technological Inquiry
Donohue, Christopher – The Monogenesis Controversy: A Historical and Philosophical Investigation
Freire Junior, Olival – Popper and the Quantum Controversy
Fuentes-Guillén, Elías – On Bolzano's Early Rejection of Infinitesimals
Gaeta, Rodolfo – Mario Bunge: A Pioneer of the New Philosophy of Science
Heller, Henning – Structuralist Abstraction and Group-Theoretic Practice
Koterski, Artur – The Nascency of Ludwik Fleck's Polemics with Tadeusz Bilikiewicz
Krauss, Alexander – How Early Humans Made the Sciences Possible
Kutrovatz, Gabor – What Mature Lakatos Learnt from Young Lakatos
Lazutkina, Anastasiia – Comment on 'Popper and Modern Cosmology'
Madgwick, Philip – Agency in Evolutionary Biology
Miller, David – Comment on 'Popper on the Mind-Brain Problem'
Murr, Caroline E. – Defamiliarization in Science Fiction: New Perspectives on Scientific Concepts
Noble, Denis – The Rehabilitation of Karl Popper's Views of Evolutionary Biology and the Agency of Organisms
Oseroff, Nathan – Don't Be a Demarc-Hater: Correcting Popular Misconceptions Surrounding Popper's Solution to the Demarcation Problem
Otte, Michael – Bolzano, Kant and the Evolution of the Concept of Concept
Peterman, Alison – Margaret Cavendish on Corporeal Qualities
Piecha, Thomas – Karl R. Popper: Logical Writings
Potschka, Martin – What Is an Hypothesis?
Rendl, Lois – Peirce on the Logic of Science – Induction and Hypothesis
Santo, Flavio Del – Comment on 'Popper and the Quantum Controversy'
Sebestik, Jan – Looking at Bolzano's Mathematical Manuscripts
Senn, Stephen – De Finetti Meets Popper or Should Bayesians Care about Falsificationism?
Simbotin, Dan Gabriel – The Unity of Science: From Epistemic Inertia to Internal Need
Simons, Peter – On the Several Kinds of Number in Bolzano

Sinelnikova, Elena & Vladimir Sobolev – V. N. Ivanovsky's Conception of Science
Sokolova, Tatiana – What's in a Name? To the History of a 'Scientist'
Stump, David – Poincaré Read as a Pragmatist
Trlifajová, Kateřina – Bernard Bolzano and the Part-Whole Principle for Infinite Collections
Zemplen, Gabor – Evolving Theories and Scientific Controversies: A Carrier-Trait Approach

B.7 Educational Aspects of the Philosophy of Science

Bozin, Dragana – Teaching Conceptual Change: Can Building Models Explain Conceptual Change in Science?
Jiao, Zhengshan – The History of Science-Related Museums: A Comparative and Cultural Study
Matthews, Michael – Mario Bunge and the Enlightenment Project in Science Education
Puncochar, Judith – Reducing Vagueness in Linguistic Expression
Štěpánek, Jan, Tomáš Ondráček, Iva Svačinová, Michal Stránský and Paweł Łupkowski – Impact of Teaching on Acceptance of Pseudo-Scientific Claims

C.1 Philosophy of the Formal Sciences (including Logic, Mathematics, Statistics)

Aberdein, Andrew – Virtues, Arguments, and Mathematical Practice
Benis-Sinaceur, Hourya – Granger's Philosophy of Style
Błaszczyk, Piotr – On How Descartes Changed the Meaning of the Pythagorean Theorem
Buldt, Bernd – Abstraction by Parametrization and Emdedding. A Contribution to Concept Formation in Modern and Contemporary Mathematics
Buzzoni, Marco – Mathematical vs. Empirical Thought Experiments: Between Informal Mathematics and Formalization
Chartier, Cian Guilfoyle – A Practice-Oriented Logical Pluralism
Chemla, Karine – Comparing the Geometric Style and Algebraic Style of Establishing Equations in China, 11th–13th Centuries
Conti, Ludovica – Extensionalist Explanation and Solution of Russell's Paradox
Cramer, Marcos, Bernhard Fisseni, Deniz Sarikaya & Bernhard Schröder – Bridging the Gap between Proof Texts and Formal Proofs Using Frames and Prss
Drekalović, Vladimir – New Versions of the Mathematical Explanation of the Cicada Case – Ad Hoc Improvements with Uncertain Outcomes or the Way to a Full Explanation?
Eastaugh, Benedict & Marianna Antonutti Marfori – Epistemic Aspects of Reverse Mathematics

Ehrlich, Philip – Are Points (Necessarily) Unextended?

Fernandes, Diego – On the Elucidation of the Concept of Relative Expressive Power among Logics

Fila, Marlena – On Continuity in Bolzano's 1817 Rein Analytischer Beweis

Fisseni, Bernhard – Perspectives on Proofs

Fogliani, Maria Paola Sforza – Revising Logic: Anti-exceptionalism and Circularity

Folina, Janet – The Philosophy and Mathematical Practice of Colin Maclaurin

Friedman, Michael – Heterogeneous Mathematical Practices: Complementing or Translation?

Gastaldi, Juan Luis & Luc Pellissier – A Structuralist Framework for the Automatic Analysis of Mathematical Texts

Giardino, Valeria – The Practice of Proving a Theorem: From Conversations to Demonstrations

Hasselkuß, Paul – Computers and the King's New Clothes. Remarks on Two Objections against Computer Assisted Proofs in Mathematics

Heuer, Karl & Deniz Sarikaya – Text-Driven Variation as a Vehicle for Generalisation, Abstraction, Proofs and Refutations: An Example about Tilings and Escher within Mathematical Education.

Heuer, Karl & Deniz Sarikaya – The Development of Epistemic Objects in Mathematical Practice: Shaping the Infinite Realm Driven by Analogies from Finite Mathematics in the Area of Combinatorics.

Islami, Arezoo – Who Discovered Imaginaries? On the Historical Nature of Mathematical Discovery

Kvasz, Ladislav – On the Relations between Visual Thinking and Instrumental Practice in Mathematics

Larvor, Brendan & Gila Hanna – As Thurston Says

Lorenat, Jemma – 'Cultured People Who Have Not a Technical Mathematical Training': Audience, Style, and Mathematics in the Monist (1890–1917)

Marquis, Jean-Pierre – Designing the Structuralist Style: Bourbaki, from Chevalley to Grothendieck

Martínez-Vidal, Concha & Ismael Ordóñez – Thin Objects and Dynamic Abstraction versus Possible Structures

Máté, András – Lakatos' Philosophy of Mathematics and 'Political Ideologies'

Matsuo, Masahiro – Constructing a Complimentary Relation between Bayesian Philosophy and Statistics

Mensik, Josef – How Are Mathematical Structures Determined

Mikami, Onyu – An Attempt at Extending the Scope of Meaningfulness in Dummett's Theory of Meaning

Moshier, Michael Andrew – The Independence of Excluded Middle from Double Negation via Topological Duality

Mukhopadhyay, Ranjan – Natural Deduction Rules as Means of Knowing

Mumma, John – The Computational Effectiveness of Geometric Diagrams

Naibo, Alberto – A Formalization of Logic and Proofs in Euclid's Geometry

Okamoto, Kengo – How Should We Make Intelligible the Coexistence of the Different Logics? — an Attempt Based on a Modal Semantic Point of View

Osimani, Barbara – A Game-Theoretic Approach to Evidence Standards in Medicine

Pantsar, Markus – Complexity of Mathematical Cognitive Tasks

Panza, Marco – Formalisation and Understanding in Mathematics

Perez, Oscar – Paths of Abstraction: Between Ontology and Epistemology in Mathematical Practice. The Zilber's Trichotomy through the Lens of Lautman and Cavaillès

Ramos, Juan Pablo Mejía & Matthew Inglis – Using Linguistic Corpora to Understand Mathematical Explanation

Reck, Erich – Dedekind, Number Theory, and Methodological Structuralism: A Matter of Style?

Reznikov, Vladimir – Frequency Interpretation of Conditions for the Application of Probability Theory According to Kolmogorov

Rodin, Andrei – Formal Proof-Verification and Mathematical Intuition: The Case of Univalent Foundations

Schiemer, Georg – Structuralism as a Mathematical Style: Klein, Hilbert, and 19th-Century Geometry

Secco, Gisele – The Interaction between Diagrams and Computers in the First Proof of the Four-Color Theorem

Sereni, Andrea & Luca Zanetti – Modelling Minimalism and Trivialism in the Philosophy of Mathematics through a Notion of Conceptual Grounding

Sereni, Andrea, Maria Paola Sforza Fogliani & Luca Zanetti – A Roundabout Ticket to Pluralism

Shimatani, Kenichiro – The Linkage between Bayesian and Frequentism Statistics Is Easier than between Bayesian Statistics and Philosophy

Sørensen, Henrik Kragh & Mikkel Willum Johansen – Employing Computers in Posing and Attacking Mathematical Problems: Human Mathematical Practice, Experimental Mathematics, and Proof Assistants

Soysal, Zeynep – Independence and Metasemantics

Steensen, Anna Kiel – Semiotic Analysis of Dedekind's Arithmetical Strategies

Stemeroff, Noah – Symmetry, General Relativity, and the Laws of Nature

Szabó, Máté & Patrick Walsh – Gödel's and Post's Proofs of Incompleteness

Takahashi, Kazutaka – Examination of the Linkage between Bayesian Philosophy and Statistics from a Logical Point of View
Tanswell, Fenner & Matthew Inglis – Studying Actions and Imperatives in Mathematical Texts
Toffoli, Silvia De – The Epistemic Basing Relation in Mathematics
Vineberg, Susan – Mathematical Depth and Explanation
Wagner, Pierre – Gödel and Carnap on the Impact of Incompleteness on Formalization and Understanding
Yamasaki, Sakiko – What Is the Common Conceptual Basis of Gödel Embedding and Girard Embedding?
Yusaku, Ohkubo – Revisiting the Two Major Statistical Problems, Stopping-Rule and the Catch-All Hypothesis, from the Viewpoint of Neo-Bayesian Statistics.

C.2 Philosophy of the Physical Sciences (including Physics, Chemistry, Earth Science, Climate Science)

Acuña, Pablo – Dynamics and Chronogeometry in Spacetime Theories
Bozin, Dragana & Anna Smajdor – Bridging the Gap between Science and Public through Engineering Environmental Concepts
Corgini, Marco – Anomalous Averages, Bose-Einstein Condensation and Spontaneous Symmetry Breaking of Continuous Symmetries, Revisited
Corral-Villate, Amaia – On the Infinite Gods Paradox via Representation in Classical Mechanics
Dizadji-Bahmani, Foad – In Defence of Branch Counting In Everettian Quantum Mechanics
Fortin, Sebastian, Jesús A. Jaimes Arriaga & Hernán Accorinti – About the World Described by Quantum Chemistry
Frisch, Mathias – Particles, Fields, or Both? A Reevaluation of the Ritz-Einstein Debate
Fujita, Sho – Spacetime and Fundamental Parts
Gömöri, Márton – Why Do Outcomes in a Long Series of Rolling a Fair Dice Approximately Follow the Uniform Distribution?
Hopster, Jeroen – Real Climate Possibilities: Proximate vs. Remote
Hubert, Mario – Good Singularities, Bad Singularities
Johansson, Lars-Göran – The Direction of Time
Kountaki, Dimitra – Anthropocentrism in Science
Lazarovici, Dustin – Why Field Theories Are Not Theories of Fields
López, Cristian – What Time Symmetry Can (And Cannot) Tell Us about Time's Structure
Panagiotatou, Maria – The Quantum Measurement as a Physical Interaction

Sebens, Charles – The Fundamentality of Fields
Stefanov, Anguel – Spacetime: Substantive or Relational?
Stergiou, Chrysovalantis – Empirical Underdermination for Physical Theories in C* Algebraic Setting: Comments to an Arageorgis's Argument
Terekhovich, Vladislav – Does the Reality of the Wave Function Follow from the Possibility of Its Manipulation?
Touze-Peiffer, Ludovic – History and Epistemology of Climate Model Intercomparison Projects
Yao, Yuting – Fragmented Authoritarian Environmentalism, Nationalism, Ecological Civilisation and Climate Change in China
Zambon, Alfio & Mariana Córdoba – Chemical Reactivity: Causality or Reciprocal Action?

C.3 Philosophy of the Life Sciences

Ankeny, Rachel & Sabina Leonelli – Organisms as Situated Models
Berg, Hein Van Den – Theoretical Virtues in Eighteenth-Century Debates on Animal Cognition
Fabris, Flavia – Rethinking Cybernetics in Philosophy of Biology
Gim, Jinyeong – Category Theory as a Formal Language of the Mechanistic Philosophy
Gontier, Nathalie – Time, Causality and the Transition from Tree to Network Diagrams in the Life Sciences
Greslehner, Gregor – What Is the Explanatory Role of the Structure-Function Relationship in Immunology?
Kohár, Matej – Top-down Inhibitory Experiments: The Neglected Option
Lopez-Orellana, Rodrigo & David Cortés-García – A Scientific-Understanding Approach to Evo-Devo Models
Lorenzano, Pablo – Laws, Causation and Explanations in Classical Genetics: A Model-Theoretic Account
Prado, Javier González de & Cristian Saborido – Organizational Etiological Teleology: A Selected-Effect Approach to Biological Self-Regulation
Regt, Henk De, Linda Holland and Benjamin Drukarch – Modeling in Neuroscience: Can Complete and Accurate Understanding of Nerve Impulse Propagation Be Achieved?
Ruiz, María Ferreira – Parity Claims in Biology and a Dilemma for Informational Parity
Saldaña, David Villena – Theoretical and Methodological Differences in the Evolutionary Analysis of Human Behavior
Stencel, Adrian – Fitness Incommensurability and Evolutionary Transitions in Individuality

Suárez, Javier – Stability of Traits as the Kind of Stability That Matters among Holobionts
Vesterinen, Tuomas – An Explanatory View of Individuating Natural Kinds
Wasmer, Martin – Bridging between Biology and Law: European GMO Law as a Case for Applied Philosophy of Science
Witteveen, Joeri – 'Taxonomic Freedom' and Referential Practice in Biological Taxonomy
Yan, Karen – Understanding Causal Reasoning in Neurophysiology
Yavuz, Mustafa – Definition and Faculties of Life in Medieval Islamic Philosophy
Zhao, Shimin – Process, Not Just Product: The Case of Network Motifs Analysis

C.4 Philosophy of the Biomedical and Health Sciences

Arruda, Renata – Multicausality and Manipulation in Medicine
Auker-Howlett, Daniel – In Defence of the Evidential Role of Mechanistic Reasoning
Battilotti, Giulia, Milos Borozan & Rosapia Lauro Grotto – A Discussion of Bi-logic and Freud's Representation Theory in Fromal Logic
Casacuberta, David – Innovative Tools for Reaching Agreements in Ethical and Epistemic Problems in Biosciences
Chen, Chi-Hsiang – Using Repertoire to Understand Psychotherapy
Dong, Xiaoju – How Should We Treat Human Enhancement Technology: Acceptance or Rejection?
Estany, Anna – Design Epistemology as Innovation in Biomedical Research
Qiu, Renzong & Ruipeng Lei – Intergenerational Justice Issues in Germline Genome Editing
Rodríguez, Natividad Garrido – Intimate Diary of an Aids Patient. An Approximation to the 'Medical Gaze' with Foucault and Guibert
Sans, Alger – The Incompleteness of Explanatory Models of Abduction in Diagnosis: The Case of Mental Disorders
Sikka, Tina – Food, Identity and End of Life
Vreese, Leen De – Risk Factors, Explanation and Scientific Understanding

C.5 Philosophy of the Cognitive and Behavioral Sciences

Artiga, Marc – Doing without Structural Representations
Atanasova, Nina – Eliminating Pain
Bazhanov, Valentin & Tatyana Shevchenko – Numerical Cognition in the Perspective of the Kantian Program in Modern Neuroscience
Chis-Ciure, Robert & Francesco Ellia – Is There a Hard Problem for the Integrated Information Theory of Consciousness?

Decock, Lieven – Varieties of Conceptual Change: The Evolution of Color Concepts
Fabry, Regina – Turing Redux: An Enculturation Account of Calculation and Computation
Guardo, Andrea – The Privilege Problem for Semantic Dispositionalism
Gundersen, Ståle – Can Neuropsychoanalysis Save the Life of Psychoanalysis?
Hardalupas, Mahi – What Is 'Biological' about Biologically-Inspired Computational Models in Cognitive Science? Implications for the Multiple Realisation Debate
Hvorecký, Juraj – Disputing Unconscious Phenomenality
Ko, Insok – Prerequisite for Employing Intelligent Machines as Human Surrogate
Marin, Lavinia – Online Misinformation as a Problem of Embodied Cognition
Matsumoto, Shunkichi – How Can We Make Sense of the Relationship between Adaptive Thinking and Heuristic in Evolutionary Psychology?
Matyja, Jakub – Music Cognition and Transposition Heuristics: A Peculiar Case of Mirror Neurons
Nowakowski, Przemysław – Problematic Interdisciplinarity of the Cognitive Science
Quinon, Paula & Peter Gardenfors – Situated Counting
Santos-Sousa, Mario – Grounding Numerals
Schubart, Tomasz – Neuroscience: Science without Disguise. A Critique of Manzotti's and Moderato's Dualistic Account of Neuroscience
Skrzypulec, Błażej – What Is Constitutive for Flavour Experiences?
Songhorian, Sarah – The Role of Cognitive and Behavioral Research on Implicit Attitudes in Ethics

C.6 Philosophy of Computing and Computation

Archer, Ken – The Historical Basis for Algorithmic Transparency as Central to AI Ethics
Hansen, Jens Ulrik – Philosophizing on Big Data, Data Science, and AI
Kipper, Jens – Intuition, Intelligence, Data Compression
Kossowska, Helena – Big Data in Life Sciences
Leonelli, Sabina – Semantic Interoperability: The Oldest Challenge and Newest Frontier of Big Data
Napoletani, Domenico, Marco Panza & Daniele C. Struppa – Finding a Way Back: Philosophy of Data Science on Its Practice
Pietsch, Wolfgang – On the Epistemology of Data Science – The Rise of a New Inductivism

C.7 Philosophy of the Humanities and the Social Sciences

Alexandrova, Anna – On the Definitions of Social Science and Why They Matter

Barbashina, Evelina – Schematism of Historical Reality
Caamaño, María – Pseudoscience within Science? The Case of Economics
Chakrabarty, Manjari – Karl Popper, Prehistoric Technology and Cognitive Evolution
Crespo, Ricardo – Economic Sciences and Their Disciplinary Links
Dementavičienė, Augustė – Challenges of New Technologies: The Case of Digital Vigilantism
Hoyningen-Huene, Paul – Do Abstract Economic Models Explain?
Jorge, Hugo Tannous – The Problem of Causal Inference in Clinical Psychoanalysis: A Response to the Charges of Adolf Grünbaum Based on the Inductive Principles of the Historical Sciences.
Martínez, Germán Hevia – Can We Apply the Science/Technology Distinction to the Social Sciences? A Brief Analysis of the Question
Medvedev, Vladimir – Explanation in Humanities
Muntanyola-Saura, Dafne – Against Reductionism: Naturalistic Methods in Pragmatic Cognitive Sociology
Murphy, Dominic – Psychology, Anthropology and Delusions
Orekhov, Andrey – Cmw-Revolution in Social Sciences as a Type of 'Scientific Revolution'
Ouzilou, Olivier – Social Sciences and Moral Biases
Pérez-González, Saúl – Mechanistic Explanations and Components of Social Mechanisms
Sequeiros, Sofia Blanco – External Validity and Field Experiments in Economics
Skripnik, Konstantin & Ekaterina Shashlova – Philosophy (And Methodology) Of the Humanities: Towards Constructing a Glossary
Špecián, Petr – Thou Shalt Not Nudge: Towards an Anti-psychological State
Tucker, Aviezer – Origins
Wang, Nan & Bo Cong Li – On Social Reality: Taking the Enterprise as an Example
Wang, Wei – Methodological Individualism and Holism in the Social Sciences
Weaver, Max – Pegging Levels
Yamada, Tomoyuki – Count-as Conditionals, Background Conditions and Hierarchy of Constitutive Rules
Zamečník, Lukáš – Non-causal Explanations in Quantitative Linguistics

C.8 Philosophy of the Applied Sciences and Technology

Alijauskaitė, Agnė – Liability without Consciousness? The Case of a Robot
Danielyan, Naira – Prospect of Nbics Development and Application
Durán, Juan M. & Nico Formanek – Computational Reliabilism: Building Trust in Medical Simulations

Franssen, Maarten – Truth-Values for Technical Norms and Evaluative Judgements: A Comparative Analysis

Fursov, Aleksandr – The Anthropic Technological Principle

Kopecký, Robin & Michaela Košová – How Virtue Signalling Makes Us Better: Moral Preference of Selection of Types of Autonomous Vehicles.

Salha, Henri – Declarative and Procedural Knowledge: A Recent Mutation of the Theory/Practice Duality and Its Significance in the Era of Computational Science

Seger, Elizabeth – Taking a Machine at Its Word: Applying Epistemology of Testimony to the Evaluation of Claims by Artificial Speakers

Thuermel, Sabine – Smart Systems: The Power of Technology

Wang, Dazhou – A Phenomenological Analysis of Technological Innovations

Wiejak, Paulina – On Engineering Design. A Philosophical Inquiry

Zwart, Sjoerd – Interlocking Models Validating Engineering Means-End Knowledge

C.9 Philosophy of Emerging Sciences

Kantor, Jean-Michel – Machine Learning: A New Technoscience

Appendix C: Mario Bunge (1919–2020)

Mario Bunge. Photo by Marta Bunge

On 24 February 2020, the physicist and philosopher Mario Bunge passed away aged 100 in Montréal, Canada. Bunge was born in Buenos Aires on 21 September 1919 and obtained his degrees in physics from the *Universidad Nacional de La Plata* in 1942. After holding chairs in physics and philosophy at the *Universidad Nacional de La Plata* and the Universidad de Buenos Aires, he was the *Frothingham Professor of Logic and Metaphysics* at McGill University in Montréal from 1966 to his retirement.

Bunge was a prolific writer, publishing 70 books and 540 articles and making substantial contributions to physics, philosophy of science, metaphysics, moral philosophy, and political philosophy. He was best known for his realist interpretation of quantum mechanics and his strong conviction that science and philosophy should be brought together for the advancement of human welfare. His philosophical core principles were the search for non-subjective truth, the universality of science, the value of rationality, and the respect for individuals, and he vigorously fought for them. Particularly important were his battles against pseudo-science in all forms: this part of his work is particularly urgent and relevant right now and we consider the fight against pseudo-science and science denialism to be one of the current scientific priorities of our Division.

In his autobiography, Quine recalled encountering Bunge for the first time at the 1956 Inter-American Philosophical Congress in Santiago de Chile: "The star of the philosophical congress was Mario Bunge, an energetic and articulate young Argentinian of broad background and broad, if headstrong, intellectual concerns. He seemed to feel that the burden of bringing South America up to a northern scientific and intellectual level rested on his shoulders. He intervened eloquently in the discussion of almost every paper."

In 2016, Bunge published his own autobiography, entitled *Between Two Worlds*, recalling events and conversations from his prodigious memory and laying out in fascinating detail his personal, family, cultural and scholarly life. In particular, he devoted two pages to his attendance at our Division's first Congress in Stanford in 1960, summing up in his characteristic pithy style: "The Stanford Congress was of a high level, taught me a lot, and gave me the chance of making some interesting acquaintances. Nobody seemed to regret the absence of Thomists, Kantians, Hegelians, Marxists, phenomenologists, or existentialists. But I regretted the pre-eminence of logical positivists, who attempted to fill their ontological vacuum with logic, or to find use for non-standard logics."

Last year, just a month before his centenary, our Division celebrated his work at the sixteenth Congress in Prague with a symposium in his honour. The General Assembly of DLMPST in Prague recognised his centenary acknowledging his immense, diversified yet systematic, science-informed, life-time contribution to philosophy of science with admiration. His passing is a loss for the scholarly world and our field of logic, philosophy and methodology of science and technology.

Cambridge & Sydney, 2 March 2020
Benedikt Löwe & Michael Matthews

Appendix D: Division of Logic, Methodology and Philosophy of Science and Technology of the International Union of History and Philosophy of Science and Technology Bulletin No. 23[1]

1 Executive Committee and Assessors of the Division, 2016-2019

Executive Committee:
- *President*. Menachem Magidor, Jerusalem, Israel.
- *First Vice-President*. Helen Longino, Stanford, Calif., United States of America.
- *Second Vice-President*. Amita Chatterjee, Kolkata, India.
- *Secretary General*. Benedikt Löwe, Amsterdam, The Netherlands, Hamburg, Germany, & Cambridge, United Kingdom.
- *Treasurer*. Peter Schroeder-Heister, Tübingen, Germany.
- *Past President*. Elliott Sober, Madison, Wisc., United States of America.

Assessors. Samson Abramsky (United Kingdom), Rachel Ankeny (Australia), Verónica Becher (Argentina), Heather Douglas (United States of America), Hannes Leitgeb (Germany), Mitsuhiro Okada (Japan), Katarzyna Paprzycka-Hausman (Poland), Charlotte Werndl (Austria).

Former Presidents († = deceased). Elliott Sober (United States of America), Wilfrid Hodges (United Kingdom), Adolf Grünbaum† (United States of America), Michael Rabin (United States of America), Wesley Salmon† (United States of America), Jens-Erik Fenstad (Norway), Lawrence J. Cohen† (United Kingdom), Dana S. Scott (United States of America), Jerzy Łoś† (Poland). Patrick Suppes† (United States of America), Jaakko Hintikka† (Finland & United States of America), Andrzej Mostowski† (Poland), Stephan Körner† (United Kingdom), Yehoshua Bar-Hillel† (Israel), Georg Henrik von Wright† (Finland), Stephen C. Kleene† (United States of America).

---

[1] The DLMPST Bulletin No. 23 was published in Volume 65, Issue 4 (December 2019, pages 394–406) of the journal *Mathematical Logic Quarterly* and is re-printed in this volume with kind permission of the publisher and copyright holder, *Wiley-VCH Verlag GmbH & Co. KGaA*. The editors would like to thank the publisher for this permission.

The *Executive Committee* of the Division is composed of the President, the Vice-Presidents, the Secretary General, the Treasurer, and the immediate Past President. The *Council* consists of the Executive Committee plus the Assessors.

Adolf Grünbaum passed away on 15 November 2018; his life and work were remembered during the *Adolf Grünbaum Memorial Symposium* organised by Sandra Mitchell during CLMPST XVI in Prague.

2 16th International Congress (Prague, Czech Republic, 5–10 August 2019)

2.1 Committees

Programme Committee.

- *Chair*: Hanne Andersen (Copenhagen, Denmark).
- *Representing the Executive Committee*: Benedikt Löwe (Amsterdam, The Netherlands, Hamburg, Germany & Cambridge, United Kingdom).
- *Representing the Local Organising Committee*: Tomáš Marvan (Prague, Czech Republic).
- *Representing the Joint Commission*: Hasok Chang (Cambridge, United Kingdom).
- *Members*: Rachel Ankeny (Adelaide, Australia), Theodore Arabatzis (Athens, Greece), Verónica Becher (Buenos Aires, Argentina), Craig Callender (La Jolla, Calif., United States of America), Hasok Chang (Cambridge, United Kingdom), Xiang Chen (Thousand Oaks, Calif., United States of America), Eleonora Cresto (Buenos Aires, Argentina), Zoubeida Dagher (Newark, Del., United States of America), Liesbeth De Mol (Lille, France), Valeria Giardino (Nancy, France), Zuzana Haniková (Prague, Czech Republic), Paul Humphreys (Charlottesville, Va., United States of America), Maria Kronfeldner (Budapest, Hungary), Sabina Leonelli (Exeter, United Kingdom), Fenrong Liu (Beijing, China & Amsterdam, The Netherlands), Endla Lõhkivi (Tartu, Estonia), Michiru Nagatsu (Helsinki, Finland), Dhruv Raina (Delhi, India), R. Ramanujam (Chennai, India), Adriane Rini (Palmerston North, New Zealand), Federica Russo (Amsterdam, The Netherlands), Dirk Schlimm (Montréal, Canada), Yaroslav Shramko (Kryvyi Rih, Ukraine), Smita Sirker (Delhi, India), Andrés Villaveces (Bogotá, Colombia).

Appendices

The Chair of the programme committee together with the representatives of the Executive Committee, the Local Organising Committee, and the Joint Commission formed the *Core Team* of the programme committee.

Local Organising Committee.

- *Chair.* Tomáš Marvan.
- *Members.* Joan Bertran-San Millán, Markéta Báčová, Marta Bílková, Vít Gvoždiak, Martin Haloun, Vladimír Havlík, Petr Koťátko, Ondrej Majer, Vera Matarese, Ondřej Ševeček, Kateřina Trlifajová, Denisa Valentová, Martin Zach (Congress Secretary).

2.2 Programme structure

The special theme of the programme was "*Bridging across academic cultures*".

- Logic
    - A.1 Mathematical Logic
    - A.2 Philosophical Logic
    - A.3 Computational Logic and Applications of Logic
    - A.4 Historical Aspects of Logic
- General Philosophy of Science
    - B.1 Methodology
    - B.2 Formal Philosophy of Science and Formal Epistemology
    - B.3 Empirical and Experimental Philosophy of Science
    - B.4 Metaphysical Issues in the Philosophy of Science
    - B.5 Ethical and Political Issues in the Philosophy of Science
    - B.6 Historical Aspects in the Philosophy of Science
    - B.7 Educational Aspects of the Philosophy of Science
- Philosophical Issues of Particular Disciplines
    - C.1 Philosophy of the Formal Sciences (including Logic, Mathematics, Statistics)
    - C.2 Philosophy of the Physical Sciences (including Physics, Chemistry, Earth Science, Climate Science)
    - C.3 Philosophy of the Life Sciences
    - C.4 Philosophy of the Biomedical and Health Sciences
    - C.5 Philosophy of the Cognitive and Behavioural Sciences
    - C.6 Philosophy of Computing and Computation
    - C.7 Philosophy of the Humanities and the Social Sciences

- C.8 Philosophy of the Applied Sciences and Technology
- C.9 Philosophy of Emerging Sciences

In addition, the Congress hosted four symposia of the IUHPST Joint Commission chaired by Hasok Chang (Cambridge, United Kingdom): *What is the value of history of science for philosophy of science?* (JC1), *Can the history of science be used to test philosophy?* (JC2), *Messy science* (JC3), and *The history and ontology of chemistry* (JC4).

3 Unconfirmed Minutes of the General Assembly of IUHPST/DLMPST in Prague on 8 August 2019

Ordinary Members Present (number of votes in parentheses; total votes: 65 before agenda item 4; 75 after agenda item 4). Argentina (2), Australia (3), Austria (1), Belgium (1), Brazil (2), Canada (3), P. R. China (3), Croatia (1), Czech Republic (1; 2 after agenda item 4), Denmark (2), Eire (1), Estonia (1), Finland (2), France (4), Germany (4), Hungary (1), India (1), Israel (1), Italy (1; 4 after agenda item 4), Japan (4), Republic of Korea (2), The Netherlands (2), Norway (1), Poland (2), Serbia (1), South Africa (1), Spain (2), Sweden (3), Switzerland (1), Taiwan (2), United Kingdom (4), and United States of America (5). After agenda item 4, also the new Ordinary Members Mexico (2), Moldova (1), and Russia (3).

Ordinary Members Absent. Iran (1), New Zealand (2), Romania (1).

International Members Present (total votes: 18 before agenda item 4; 24 after agenda item 4). Association Computability in Europe (1), Association for Symbolic Logic (6), European Philosophy of Science Association (2), Association for Logic, Language and Information (2), Institut Wiener Kreis (2), *Polskie Towarzystwo Logiki i Filozofii Nauki* (1), *Société de Philosophie des Sciences* (4). After agenda item 4, also the new International Members *Académie Internationale de Philosophie des Sciences* (2), Association for the Philosophy of Mathematical Practice (1), *Gesellschaft für Wissenschaftsphilosophie* (2), and Scandinavian Logic Society (1).

International Members Absent. Charles S. Peirce Society (1).

Commissions Present (total votes: 5). Commission on Arabic Logic (1), Commission on the Philosophy of Technology and Engineering Sciences (1), History and Philosophy of Computation (1), International Association for Science and Cultural Diversity (1), Joint Commission (1).

Commissions Absent. Inter-Division Teaching Commission (1).

Membership Candidates Present. Académie Internationale de Philosophie des Sciences (2), Association for the Philosophy of Mathematical Practice (1), *Gesellschaft für Wissenschaftsphilosophie* (2), Mexico (2), Moldova (1), Russian Federation (3), and Scandinavian Logic Society (1).

Membership Candidates Absent. None.

Others. According to tradition, the General Assembly was open to all Congress Participants. The *Division for History of Science and Technology* (DHST/IUHPST) was represented by DHST President Elect Marcos Cueto (Brazil).

Total number of votes present. 88 before agenda item 4; 104 after item 4 (6 votes absent). Therefore, the number of votes present was at least half of of the valid voting power and thus the General Assembly was validly constituted according to Article 15 of the Division's statutes.

Agenda item 1. Opening and confirmation of the agenda

The General Assembly took place in the Kotěra Lecture Hall of the Faculty of Architecture of Czech Technical University in Prague. After verification of the delegates and distribution of the paper ballots, the President opened the Assembly at 6:33pm. The agenda was presented and unanimously confirmed by the Assembly:

1. Opening and confirmation of the agenda.
2. Confirmation of the minutes of the 2015 General Assembly.
3. Reports by the President, the Secretary General, and the Treasurer.
4. Membership Issues.
5. Commission Issues.
6. Election of the next Council.
7. Hosting the next Congress in 2023.
8. Any other business.

Nina Atanasova and Dirk Schlimm were unanimously approved by the Assembly as tellers for all votes with paper ballots cast. The Secretary General announced that Giuseppe Primiero was taking notes during the Assembly and gave an overview of the voting rules for the Assembly and the use of the distributed paper ballots.

Agenda item 2. Confirmation of the minutes of the 2015 General Assembly

The minutes of the previous General Assembly (Helsinki 2015) had been made available on the Division's website and published in print as part of the DLMPST Bulletin No. 22 in the journal *Mathematical Logic Quarterly* (Vol. 61, Issue 6, 2015, pp. 383–398) in unconfirmed form. The Assembly unanimously approved the confirmation of the minutes.

Agenda item 3. Reports by the President, the Secretary General, and the Treasurer

3a. President's Report

The President gave a brief report about the activities of the Division. He remembered four past officers of the Division and four past Council members of the Division who passed away between the 2015 congress in Helsinki and the 2019 congress in Prague: Jaakko Hintikka (12 January 1929–12 August 2015; Assessor 1969–1971; First Vice President 1971–1975; President 1975), András Hajnal (13 May 1931–30 July 2016; Second Vice President 1983–1987), Petr Hájek (6 February 1940–26 December 2016; First Vice President 1995–1999), and Adolf Grünbaum (15 May 1923–15 November 2018; Assessor 1971–1975; President 2004–2007), as well as Mary Hesse (15 October 1924–2 October 2016; Assessor 1971–1975), Marcel Guillaume (1928–25 Oct 2016; Assessor 1975–1979), Boris G. Yudin (14 August 1943–6 August 2017; Assessor 1995–1999), and Roberto Cignoli (1937–2018; Assessor 2000–2003). The Assembly stood in silence in memory of the deceased. [After the General Assembly, it was brought to the Secretary General's attention that the past Council member Myroslaw Popowych (12 April 1930–10 February 2018; Assessor 1971–1975) also passed away between the last two congresses.]

The President informed the Assembly that the year 2019 marked the centenary of Mario Bunge (born 21 September 1919, Buenos Aires). The Assembly acknowledged Bunge's immense, diversified yet systematic, science-informed, life-time contribution to philosophy of science with admiration.

The President reported that the past term saw an increase in the participation of the Division in the activities of other international bodies of science governance as well as an increase in the membership engagement in the activities of the Division. He encouraged the Division and its members to continue on this path of increased participation and engagement. The President was particularly pleased with the increased cooperation of the

Division with its sister division DHST and encouraged the Division and its members to intensify this connection in future years.

3b. Secretary General's Report

The Secretary General gave a detailed report on the activities of the Division. A written version of the report had been made available to the delegates in advance. The following includes the complete text of the report:

Internal matters

The *Executive Committee* during the term 2016–2019 consisting of the President, the two Vice Presidents, the Past President, the Secretary General, and the Treasurer had regular e-mail discussions. There were no separate physical meetings of the Executive Committee.

The *Council*, consisting of the members of the Executive Committee and the eight Assessors met twice during the term: in the context of a workshop "Global Perspectives on Reasoning and Scientific Method" in Salzburg (Austria) on 30 November and 1 December 2017, and during CLMPST XVI in Prague on 7 August 2019.

Relationship with members

In the last term, the Executive Committee has increased its efforts to connect to the members. Already the last Executive Committee had decided to set up a scheme of "Small conference grants" where member institutions can support research workshops and conferences to receive funding from the Division. This programme was continued in 2017 and 2018 (cf. below); the Executive Committee decided not to issue a call for funding of events in 2019 in order to avoid competition with the Congress.

During the last General Assembly, the Council had announced that the election procedure would "strengthen the involvement of the Division's members in the nomination of candidates in the future". This was implemented by the Nominations Committee in 2018: members were given the opportunity to recommend candidates for the new Council in December 2018 with a deadline of 31 January 2019. All candidates recommended by members were added to the proposed slate by the Nominations Committee (cf. agenda item 6).

The IUHPST

The Division is one of the two Divisions of the *International Union of History and Philosophy of Science and Technology* (IUHPST). The governance of the Union is determined by a *Memorandum of Cooperation* between its two Divisions, the DLMPST and the *Division for History of Science and Technology* (DHST). The Memorandum is available on the DLMPST webpage.

The Memorandum determines that the Presidency of the Union rotates between the DLMPST President and the DHST President. The following were the officers of the Union in the past four years:

	2016 & 2017	2018 & 2019
IUHPST President	Efythymios Nicolaidis DHST President	Menachem Magidor DLMPST President
IUHPST Vice President	Menachem Magidor DLMPST President	Michael Osborne DHST President
IUHPST Secretary General	Catherine Jami DHST Secretary General	Benedikt Löwe DLMPST Secretary General
IUHPST Treasurer	Jeffrey Hughes DHST Treasurer	Peter Schroeder-Heister DLMPST Treasurer

The Divisions send representatives to each other's Council meetings and Congresses. DLMPST was represented by Menachem Magidor at the DHST Council meeting in Beijing (2015), by Benedikt Löwe at the DHST Council meetings in Rio de Janeiro (2016) and Princeton (2017) and at the DHST Congress in Rio de Janeiro (2017), and by Mitsuhiro Okada at the DHST Council meeting in Tokyo (2018). DHST was represented by Catherine Jami at the DLMPST Council meeting in Salzburg (2017) and by Marcos Cueto at the DLMPST Congress in Prague (2019).

The Secretary General used this opportunity to invite Marcos Cueto, President Elect of DHST to deliver words of greeting to the Assembly from the sister division DHST.

The DLMPST and the DHST share four *inter-division commissions*: History and Philosophy of Computing (HaPoC), the International Association for Science and Cultural Diversity (IASCUD), the Inter-Division Teaching Commission (IDTC), and the Joint Commission (JC). The Joint Commission has a special status and is governed by the Memorandum. According to a decision of the Councils of both divisions (DLMPST in Helsinki, August 2015 and DHST in Beijing, December 2015), a joint committee of the two divisions re-evaluated the Joint Commission and proposed a re-organisation of the Joint Commission to the General Assembly of the DHST in Rio de Janeiro in 2017. The members of this committee were Hasok Chang, Jean Gayon, Catherine Jami, Benedikt Löwe, Menachem Magidor, and Efthymios Nicolaidis. The committee proposed opening JC membership to additional people in order to give the JC the opportunity to develop more activities: this proposal was implemented by the two Councils in 2018. The JC also instituted the *IUHPST Essay Prize in History and Philosophy of Science*; the prize was won by Theodore Arabatzis (Athens) for his essay "What's in it for the historian of science? Reflections on the value of philosophy of science for history of science" in 2017 and by Agnes Bolinska and Joseph D. Martin (Cambridge) for their essay "Negotiating History: Contingency, Canonicity, and Case Studies" in 2019.

Relationship with ICSU/ISC

The IUHPST has been a member of the *International Council for Science* (ICSU) since 1955. At the last General Assembly, the Executive Committee reported that ICSU was planning to merge with the *International Social Science Council* (ISSC). In October 2016, an extraordinary General Assembly of ICSU was held (where Catherine Jami and Benedikt Löwe represented IUHPST) during which the executive of ICSU and ISSC were given the mandate to plan the merger.

The merger was implemented at great speed: ICSU and ISSC formed joint working groups, a *Strategy Working Group* (SWG) and a *Transition Task Force* (TTF). IUHPST nominated members for both of these groups; Benedikt Löwe was appointed as a member of the SWG. The SWG jointly wrote the strategy document of the new council, entitled *Advancing Science as a Global Public Good*.

At the 32nd ICSU General Assembly in Taipei in October 2017 where IUHPST was represented by Benedikt Löwe, the members of ICSU and ISSC decided to merge the two councils into the new *International Science Council* (ISC).

The founding General Assembly of the new ISC took place in Paris in July 2018; IUHPST was represented by Catherine Jami, Benedikt Löwe, and Michael Osborne. IUHPST nominated a candidate for the Governing Board of the newly formed Council, but was not successful in the elections in Paris.

In the last year, the ISC has started its work; in particular, the four statutory committees of ISC were formed: the Committee for Finance and Fundraising, the Committee for Freedom and Responsibility in Science (CFRS), the Committee for Science Planning, and the Committee for Outreach and Engagement. IUHPST nominated candidates for three of these four committees and ISC appointed our nominee Craig Callender to the CFRS.

IUHPST is one of the participating Unions in the ICSU projects *Gender Gap in Science. A Global Approach to the Gender Gap in Mathematical, Computing, and Natural Sciences: How to Measure It, How to Reduce It?* coordinated by the International Mathematical Union (IMU), the International Union for Pure and Applied Chemistry (IUPAC), and the International Union for Pure and Applied Physics (IUPAP) as well as *TROP-ICSU: Educational Resources for Teachers to Integrate Climate Topics across the Curriculum.* The Gender Gap project has been presented in a special symposium at CLMPST XVI and Benedikt Löwe and Amita Chatterjee represented DLMPST at Gender Gap workshops in Paris (May 2017) and Taipei (November 2017), respectively.

Relationship with CIPSH

The Division is a member of the *Conseil International de Philosophie et des Sciences Humaines* (CIPSH); in this council, DLMPST and DHST are separately members (as opposed to the ISC where the Union is a member). CIPSH is undergoing a process of re-vitalisation and re-organisation. The General Assembly 2015 had given the Executive Committee the mandate to decide whether the Division should remain a member of CIPSH "based on further information concerning the re-organisation of CIPSH and the success of WHC".

The Division was represented by Menachem Magidor at the CIPSH General Assembly in Beijing in December 2015 and by Benedikt Löwe at the CIPSH General Assembly in Liège in August 2017. Benedikt Löwe was a member of the programme committee of the *World Humanities Conference* (WHC) in Liège (Belgium) in August 2017. CIPSH is involved in the project *Global History of Humanity* and the *World Humanities Report*, and has been co-ordinating programs of *UNESCO-CIPSH Chairs* and *CIPSH Chairs* for specially designated chairs in the Humanities.

At the WHC, the Division organised a *Symposium on Logic, Methodology, and Philosophy of Science and Technology* (coordinated by Benedikt Löwe; speakers: Mieke Boon, Stefania Centrone, Inkeri Koskinen, Pierluigi Minari) in order to encourage the exchange of ideas between the humanities-centred disciplines and those that straddle the humanities-science divide. There is the desire in the governance of CIPSH to strengthen this dialogue, but it is unclear how much support this has among the other CIPSH members.

The Division nominated Benedikt Löwe to serve on the CIPSH Executive Committee and he was elected by the CIPSH General Assembly in Liège. He attended the first meeting of the CIPSH Executive Committee in Xiamen (China) in April 2018.

Other activities

In addition to the mentioned symposium on *Logic, Methodology, and Philosophy of Science and Technology* at the World Humanities Conference in Liège in August 2017 and the workshop *Global Perspectives on Reasoning and Scientific Method* organised in combination with the Council meeting in Salzburg in November and December 2017, the Division joined the *Sociedade Brasileira de Lógica* (SBL) in organising a logic satellite meeting (in Niterói) to the *International Congress of Mathematicians* (ICM) organised by the IMU in Rio de Janeiro in August 2018.

IUHPST became one of the sponsors of the *International Year of the Periodic Table* (IYPT 2019); a special symposium marked this involvement at CLMPST XVI.

In 2019, the arrest of Tuna Altınel (Université de Lyon) in Turkey for signing a peace petition got international media attention and many learned societies and international organisations joined the protests against his incarceration. IUHPST sent an open letter to the President of the Republic of Turkey, Recep Tayyip Erdoğan on 1 June 2019. Tuna Altınel was released from prison on 30 July 2019 after 81 days in prison.

CLMPS XV proceedings volume

The proceedings of CLMPST XV in Helsinki (Finland) were published as

> Hannes Leitgeb, Ilkka Niiniluoto, Päivi Seppälä, and Elliot Sober (*eds.*), Logic, Methodology and Philosophy of Science, Proceedings of the Fifteenth International Congress, College Publications (2017).

CLMPST XVI in Prague

After the re-naming of our Division from DLMPS to DLMPST, the sixteenth congress is the first to be held under the new name of *Congress on Logic, Methodology and Philosophy of Science and Technology* (CLMPST). The Congress was excellently hosted by the Institute of Philosophy of the Czech Academy of Sciences under the auspices of the Division; the chair of the Local Organising Committee was Tomáš Marvan; the chair of the Programme Committee was Hanne Andersen (Denmark). Andersen, Marvan, the

Secretary General, and the JC chair formed the *Core Team* within the Programme Committee that dealt with executive decisions. Upon proposal of the Core Team, the Council decided that the theme of CLMPST XVI should be

Bridging across academic cultures,

a theme at the heart of many logicians and philosophers of science who are dealing with mediating between the disciplinary cultures of the humanities and the sciences in their academic daily lives. The Programme Committee implemented the theme of the congress by considering the programme structure and its subdivisions as fluid and open in order to allow people from different disciplinary cultures to interact and talk to each other. The Programme Committee was gender balanced and had representatives from all continents except for Africa with good representation of the traditionally under-represented continents Asia (6 members), Australasia (2 members), and South America (3 members). The Programme Committee produced a list of three plenary and 22 invited speakers that was also gender balanced and included names from all continents.

In addition to the regular sessions and symposia, the Congress had a number of special features, including a jointly organised symposium with the *International Science Council* (ISC) on the topic *Denial of facts: Instrumentation of science, criticism, and fake news* dealing with the important interface between philosophy of science and science policy; a symposium reporting on progress of the *Gender Gap* project of which IUHPST is a party (see above); a celebration of the *International Year of the Periodic Table* 2019 (IYPT) where philosophers of science discussed the relevance of the periodic table for their work; and a *journal panel* where editors of prominent journals discussed issues facing scholarly publishing.

All in all, CLMPST XVI had almost 800 registered participants, over 600 presentations, and forty symposia. The Division plans the publication of the proceedings of CLMPST XVI in Prague (Czech Republic) as

> Hanne Andersen, Tomáš Marvan, Benedikt Löwe, Hasok Chang, Ivo Pezlar (*eds.*), Logic, Methodology and Philosophy of Science and Technology, Proceedings of the Sixteenth International Congress, Bridging Across Academic Cultures, College Publications (to appear).

The Secretary General used this opportunity to invite Hanne Andersen to report to the Assembly on the programme of CLMPST 2019.

Small conference grants

In the years 2017 and 2018, the Division invited applications from members to support conferences and meetings. The following meetings were supported:

2017.

1. Workshop "Consequence and Paradox between Truth and Proof", in Tübingen, Germany, 2–3 March 2017. DLMPST support: USD 1,000.
2. SILFS 2017: Triennial International Conference of the Italian Society for Logic and Philosophy of Science, in Bologna, Italy, 20–23 July 2017. DLMPST support: USD 1,000.
3. Humboldt-Kolleg "Proof Theory as Mathesis Universalis", in Como, Italy, 24–28 July 2017. DLMPST support: USD 1,000.

2018.

1. LSFA 2018: 13th Workshop on Logical and Semantic Frameworks with Applications, in Fortaleza, Brazil, 26–28 September 2018. DLMPST support. USD 1,000.
2. IX SLMFCE: IX Congress of the Spanish Society of Logic, Methodology and Philosophy of Science, in Madrid, Spain, 13–16 November 2018. DLMPST support: USD 1,000.
3. 29th Novembertagung: "History of Mathematical Concepts and Conceptual History of Mathematics", in Seville, Spain, 28–30 November 2018. DLMPST support: USD 1,000.
4. PTS3: Proof-Theoretic Semantics: Assessment and Future Perspectives. Third Tübingen Conference on Proof-Theoretic Semantics, in Tübingen, Germany. 27–30 March 2019. DLMPST support: USD 1,000.
5. Workshop "Responsible Research and Innovation: An HPS/STS Agenda", in Canberra, Australia, 17–18 April 2019. DLMPST support: USD 1,000.
6. CiE 2019: Computability in Europe 2019, in Durham, England, 15–19 July 2019. DLMPST support: USD 1,000.

3c. Treasurer's Report

The Treasurer gave a short summary of his financial report that had been made available to to delegates prior to the General Assembly. The following includes the complete text of the report.

The financial details are listed in Appendix A. Formally the report was given for the years 2015 to 2018, but the comments are based on the four-year periods 2012 to 2015 and 2016 to 2019, as these are the terms of office of the respective Executive Committees. The Executive Committee usually prepares its four-year budget at the beginning of its term of office, that is, at the beginning of the year after the Congress. Moreover, since the Division's financial contribution to the Congress is sometimes split between the year of the congress and the year before, the figures for the four-year period starting with a congress year and ending with the year ahead of the next congress, could be unreliable. The figures for 2012 to 2014 are part of the previous reporting period. The figures for 2019 are based on a (reliable) estimate.

As mentioned in past reports, most of our assets come from the time when the Division received funding from UNESCO via ICSU (now ISC), of which the IUHPST is a member. For more than two decades now (and with no expected changes for the foreseeable future), the only income of the Division has been membership fees. Over the past years this has been roughly USD 16,000 per annum. It varies between years as sometimes members pay with a delay, so that the fee is booked only in the next accounting year. In the period of this report we managed to secure almost all fee payments which were due, sometimes requiring considerable effort in reminding our members of their payment obligations.

The previous Executive Committee (2012 to 2015) initiated a gradual reduction of our assets by around USD 4,000 per year and invested this amount in productive activities, in particular the award of small conference grants and financial support of our commissions. In the three congress-free years, grants of up to USD 1,000 were offered for the organisation of conferences, the application for which had to be submitted through our member committees. Furthermore, our Commissions were able to apply for financial support of their work. Of course, our main activity, also in financial respect, is the Congress, which we supported with USD 45,000 in 2015 (Helsinki). The income during the period 2012 to 2015 was roughly USD 62,000, and our expenses were USD 77,000. This reduced our total assets by USD 15,000 over four years (from USD 89,000 to USD 74,000), which was roughly as planned.

This policy of asset reduction was continued by the current Executive Committee (2016 to 2019). The Treasurer outlined the overall four-year budget for the current term of the Executive Committee formally accepted by the Executive Committee by unanimous vote on 16 November 2016 with expected expenses of USD 80,000 and an expected income of USD 64,000.

DLMPST Four-Year Budget 2016–2019.	
Fixed costs	USD 12,000
Travel expenses	USD 12,000
Congress support	USD 23,500
Commission support	USD 15,000
Support for the Joint Commission	USD 3,500
Council meeting 2017	USD 4,000
Small conference grants	USD 10,000
Total	USD 80,000

Fixed costs are the audit fee, bank charges, and membership fees to ICSU/ISC and CIPSH. In 2018, this budget was extended by USD 5,000 to include *DLMPST/IUHPST Travel Stipends for Researchers from Developing Countries* for CLMPST XVI. This programme was supplemented with USD 2,500 and waivers of registration fees by the Congress organisers: in order to apply for a stipend, reseachers had to be based in a country on the ISI (International Statistical Institute) list; stipends for researchers from European countries on the list were USD 350, and stipends for researchers from outside of Europe were USD 700.

The contribution to the Congress was reduced from the USD 45,000 given to Helsinki 2015 to USD 23,500 (or USD 28,500 including the mentioned travel stipends). Overall, the Congress in Prague costs the Division roughly USD 30,000 (direct contributions, travel stipends, and additional travel expenses for the Congress and its preparation, as well as social expenses at the Congress). The funds saved by the reduction in our congress contribution were spent on an extension of the small conference grants programme and the support of our Commissions. The Joint Commission received a total budget of USD 3,500, and each of the other Commissions could apply for USD 1,000 for each of the years 2017, 2018 and 2019.

Another focus point of the Executive Committee's work was to intensify our contacts with our sister Division DHST (meaning in particular presence at their Congress and their Council meetings), but also the active participation in activities of ICSU/ISC and CIPSH. These activities required increased travel expenses compared to previous years. Due to the higher-than-expected income, the loss (= planned reduction of assets) was only USD 19,000 rather than the intended USD 21,000 (viz. from USD 74,000 to USD 55,000). The higher income resulted from new memberships as well as from the (voluntary) reassignment of existing members to higher membership categories.

At the end of 2019, the income for the entire four-year term will be approximately USD 66,000, and the expenses approximately USD 85,000 as planned.

It was emphasised that the policy of reducing global assets is not a long-term commitment. As the small conference grants programme can be halted immediately if desired or necessary, and contributions to our committees can be reduced at relatively short notice, the current way of spending our assets does not firmly bind any future Executive Committee in their policies. In any case, if we give up the small conference grants programme, our income from membership fees suffices to balance our costs, as long as our total contribution to the Congress does not exceed USD 30,000. Our rate of reducing assets is very moderate, after all, and can run at the present rate for a further three four-year periods, if desired.

All this was made possible as Council members often use financial means available to them from their host universities or from grants to finance trips and meetings on behalf of the Division, as this work provides a service to the scientific community. E.g., our 2017 Council meeting was organised and to a great extent also funded by Charlotte Werndl and the *Universität Salzburg*, as it was combined with a public scientific meeting at which Council members gave talks on their research topics.

The Treasurer reported that it had been a pleasure to serve in the capacity of Treasurer for nearly eight years, first for almost four years as Acting Treasurer in combination to the office of Secretary General, and then for four years as Treasurer. He thanked his secretary, Marine Gaudefroy-Bergmann, without whose professional assistance fulfilling his task as Treasurer would not have been possible.

Agenda item 4. Membership Issues

The Secretary General reported that two of the Ordinary Members upgraded their category: the Czech Republic moved from category A to category B, and Italy moved from category A to category D. These changes were preliminarily approved by the Executive Committee; the Executive Committee requested approval from the Assembly according to Article 6 of the Statutes. The Assembly approved these changes unanimously.

The Secretary General reported that there were seven new membership applications:

1. The *Académie Internationale de Philosophie des Sciences* applied on 3 August 2017 as International Member in category A (annual fee: USD 150; 2 votes).
2. The Association for the Philosophy of Mathematical Practice applied on 22 August 2017 as International Member in category 0 (annual fee: USD 75; 1 vote).

3. The Russian Federation applied on 16 July 2018 as Ordinary Member in category C.
4. The *Gesellschaft für Wissenschaftsphilosophie* applied on 31 July 2018 as International Member in category A (annual fee: USD 150; 2 votes).
5. Mexico applied on 31 January 2019 as Ordinary Member in category B.
6. Moldova applied on 27 February 2019 as Ordinary Member in category A.
7. The Scandinavian Logic Society applied on 22 July 2019 as International Member in category 0 (annual fee: USD 75; 1 vote).

The Executive Committee recommended acceptance of all seven new members with their proposed categories to the General Assembly. The Assembly unanimously voted to accept the membership applications of the seven candidate members with the proposed categories by acclamation.

After agenda item 4, the voting power of the delegations of the Ordinary Members Czech Republic and Italy increases to 2 votes (from 1 vote) and 4 votes (from 1 vote), respectively. The delegations of the seven new members have a combined additional voting power of 12. As a consequence, the total voting power of the Assembly increases from 88 to 104 votes.

Agenda item 5. Commission Issues

The Division introduced commissions in the 2011 Assembly in Nancy and added two new commissions in the 2015 Assembly in Helsinki. It did not have a funding system for its commissions; this was introduced at the beginning of the budget year 2017 based on the model of the commission funding system of DHST.

The Secretary General reported that the Division required its Commissions to submit an annual report on its activities and spending as well as an activity plan for the following year and apply for up to USD 1,000 per annum. Due to its special nature, the Joint Commission was given a separate four-year budget of USD 3,500 for which it did not have to apply annually. The reports are published on the Division's website.

The Secretary General reported that not all commissions have organised annual events, but that the inactive commissions have not applied for their annual funding.

Agenda item 6. Election of the next Council

According to Article 16 of the Statutes, the Council consists of "the 6 members of the Executive Committee and at least 6 assessors." According to Article 17 of the Statutes, the Executive Committee consists of "the President, the First Vice-President, the Second Vice-President, the Secretary, the Treasurer, and the (immediate) Past President of the Division."

Following a suggestion of the French National Committee at the 2015 General Assembly, the Nominations Committee actively involved the members in the nominations process for the new Council. The Nominations Committee requested recommendations for members of the Council from all members and commissions of the Division until 31 January 2019. Five recommendations were received and all recommended candidates are being nominated for a seat on the Council. The Nominations Committee proposed to elect the following new Council; all named individuals had previously accepted the nomination and are willing to serve:

Executive Committee 2020–2023. President: Nancy Cartwright (U.K. & U.S.A.). First Vice-President: Kim Sterelny (Australia). Second Vice-President: Verónica Becher (Argentina). Secretary-General: Benedikt Löwe (The Netherlands, Germany, & U.K.). Treasurer: Pierre Edouard Bour (France). Past President: Menachem Magidor (Israel).
Assessors 2020–2023. Hanne Andersen (Denmark), Rachel Ankeny (Australia), Valeria de Paiva (U.S.A.), Gerhard Heinzmann (France), Concha Martinez Vidal (Spain), Tomáš Marvan (Czech Republic), Dhruv Raina (India), Cheng Sumei (China), Alasdair Urquhart (Canada), and Andres Villaveces (Colombia).

There is no request for a vote with ballots; the slate of proposed officers and Council members is unanimously elected by the Assembly.

Agenda item 7. Hosting the next Congress in 2023

The Executive Committee had received bids to host CLMPST XVII in 2023 from Buenos Aires (Argentina) and Kobe (Japan); the bids had been made available to the delegates before the Assembly. The Secretary General reminded the Assembly that twelve out of the sixteen past congresses had been held in Europe; as a consequence, the Executive Committee welcomed the fact that both bids came from cities outside of Europe, thereby guaranteeing that CLMPST XVII would not take place in Europe. Both proposed hosts gave presentations of their cities and conference facilities.

Discussion. The Secretary General pointed out that the two bids were not based on the same financial assumptions: the Buenos Aires bid assumed DLMPST support of USD 40,000 whereas the Kobe bid assumed DLMPST support of USD 20,000. The Assembly had a long and engaged discussion of the financial details of the two bids; in the end, the representatives of the Buenos Aires bid confirmed that they would be able to re-draft the budget based on DLMPST support of USD 20,000 if only the expenses of those invited speakers who are unable to pay them from other sources have to be covered (rather than the expenses of all invited speakers).

The vote was by secret ballot. There were 104 votes cast, of which 56 votes were in favour of Buenos Aires and 48 votes were in favour of Kobe.

The President thanked the representatives of both potential hosts for the enormous effort they had put into their bids. He congratulated the representatives of Buenos Aires on their success and encouraged the representatives of Kobe to consider submitting a bid for CLMPST XVIII in 2027.

Agenda item 8. Any other business

The Ordinary Member France and the International Member *Societé de Philosophie des Sciences* had submitted a proposal to Council concerning a possible change of Statute 18 concerning the procedure of nominations for the Council. This proposal had been discussed by Council at their meeting on 7 August 2019 in Prague.

Marco Panza read the proposal to the Assembly and the President reported that Council had decided to ask the next Council (after 1 January 2020) to install a committee consisting of members of Council and non-members of Council that should look at the suggestion, possibly discuss with members, and then propose a document to the next General Assembly in 2023 for discussion and decision. The Assembly expresses agreement with the Council's decision.

The President thanked the the members of the Executive Committee and Council for the effort and dedication they put into their work and the delegates for attending the Assembly. The Assembly was adjourned at 8:27pm.

Respectfully submitted,

Benedikt Löwe
Secretary General DLMPST/IUHPST

APPENDIX A: IUHPST/DLMPST ACCOUNTS
Prepared by Peter Schroeder-Heister, Treasurer

Fees from members	2015	2016	2017	2018
Ordinary members				
Category A				
Austria	USD 149.85	USD 148.50	USD 158.20	USD 147.50
Belgium	USD 149.85	USD 148.50		USD 177.00
Croatia (new member since 2016)		USD 148.50	USD 158.20	USD 153.40
Czech Republic (category B from 2017)	USD 149.85	USD 148.50		
Estonia	USD 149.85	USD 148.50	USD 158.20	USD 147.50
Hungary	USD 150.00	USD 150.00	USD 150.00	USD 150.00
India	USD 300.00		USD 288.00	USD 126.41
Iran				
Ireland	USD 149.85	USD 148.50	USD 158.20	USD 147.50
Israel	USD 140.16		USD 300.00	USD 138.47
Italy (category D from 2018)	USD 149.85	USD 148.50	USD 158.20	
Norway	USD 150.00	USD 148.50	USD 158.20	USD 147.50
Romania	USD 300.00	USD 150.00	USD 158.20	USD 138.36
Serbia	USD 150.00	USD 148.50		USD 295.00
South Africa	USD 300.00	USD 150.00	USD 150.00	USD 177.00
Switzerland	USD 150.00	USD 148.50	USD 158.20	USD 147.50
Category B				
Argentina (new member since 2016)				
Brazil (new member since 2016)		USD 274.43	USD 268.94	USD 275.51
Czech Republic (category B since 2017)			USD 316.40	USD 295.00
Denmark	USD 290.00	USD 297.00	USD 316.40	USD 278.44
Finland	USD 299.70	USD 297.00	USD 316.40	USD 295.00
The Netherlands	USD 299.70	USD 297.00	USD 316.40	USD 295.00
New Zealand (new member since 2016)				
Poland	USD 279.00	USD 300.00	USD 300.00	USD 295.00
South Korea	USD 300.00	USD 300.00	USD 300.00	USD 288.61
Spain	USD 299.70	USD 297.00	USD 316.40	USD 295.00
Taiwan	USD 300.00	USD 300.00	USD 300.00	USD 288.27
Category C				
Australia	USD 573.20	USD 575.00	USD 575.00	USD 573.48
Canada		USD 1,200.00	USD 600.00	USD 600.00
China	USD 600.00	USD 590.00	USD 600.00	USD 600.00
Sweden	USD 600.00	USD 600.00	USD 600.00	USD 590.00

Category D				
France		USD 1,069.20	USD 1,224.92	USD 2,251.44
Germany	USD 1,200.00	USD 1,200.00	USD 1,200.00	USD 1,200.00
Italy (category D since 2018)				USD 1,180.00
Japan	USD 1,200.00	USD 1,200.00	USD 1,200.00	USD 1,200.00
United Kingdom	USD 1,201.72	USD 1,149.81	USD 1,200.00	USD 1,163.36
Category E				
United States of America	USD 2,400.00	USD 2,400.00	USD 2,370.00	USD 2,376.54
International members				
Association Computability in Europe		USD 74.25	USD 79.10	USD 73.75
Charles Peirce Society	USD 75.00	USD 225.00		USD 213.24
Polskie Towarzystwo Logiki i Filozofii Nauki	USD 75.00	USD 148.50	USD 82.49	USD 76.70
European Philosophy of Science Association	USD 149.85	USD 148.50	USD 158.20	USD 295.00
FoLLI	USD 149.85	USD 148.50	USD 158.20	USD 147.50
Institut Wiener Kreis		USD 297.00	USD 158.20	USD 147.50
Société de Philosophie des Sciences	USD 299.70	USD 297.00	USD 316.40	USD 295.00
Association for Symbolic Logic	USD 600.00	USD 600.00	USD 570.00	USD 588.24
Total	USD 13,731.68	USD 16,220.19	USD 15,997.05	USD 18,270.72

Expenses

2015	ICSU Subscription	USD 1,334.22
	CIPSH	USD 832.50
	Travel expenses	USD 676.74
	Audit fee	USD 610.50
	Bank fees	USD 268.78
	Domain fees	USD 139.96
	Translation	USD 72.65
	Postage	USD 40.79
	Total	**USD 3,976.14**
2016	ICSU Subscription	—
	CIPSH	USD 825.00
	Travel expenses	USD 4,607.86
	Audit fee	USD 605.00
	Bank fees	USD 386.20
	Domain fees	USD 160.00
	Translation and Notary Public	USD 150.54
	Postage, logos, stamp	USD 382.00
	Total	**USD 7,116.60**
2017	ICSU Subscription	—
	CIPSH	USD 847.50
	Conference support	USD 3,220.50
	Commission funding	USD 2,073.50
	Travel expenses	USD 7,492.92
	Audit fee	USD 678.00
	Bank fees	USD 346.76
	Domain fees	USD 179.60
	Postage	USD 87.69
	Publication costs	USD 678.00
	Total	**USD 15,604.47**
2018	ICSU subscription	—
	CIPSH subscription	USD 590.00
	CLMPST 2019 (prefinancing)	USD 14,160.00
	Conference support	USD 9,560.81
	Commission funding	USD 2,935.20
	Travel expenses	USD 3,813.49
	Audit fee	USD 708.00
	Bank fees	USD 184.64
	Domain fees	USD 162.27
	Postage	USD 52.33
	Total	**USD 32,166.74**

Total assets on December 31

1999	USD 77,418.83	2004	USD 105,138.99	2009	USD 118,854.23	2014	USD 65,502.45
2000	USD 90,768.87	2005	USD 115,061.20	2010	USD 127,909.44	2015	USD 74,162.11
2001	USD 110,776.63	2006	USD 118,918.12	2011	USD 89,509.92	2016	USD 83,186.83
2002	USD 113,071.31	2007	USD 93,332.02	2012	USD 104,098.00	2017	USD 83,514.95
2003	USD 95,691.32	2008	USD 104,185.56	2013	USD 108,625.18	2018	USD 69,786.80

Overview (in USD)

	Total assets on 31 December	Total annual income	Ordinary member dues	International members dues	Other sources of income	Expenses
2015	74,162.11	13,731.68	12,382.28	1,349.40	0.00	3,976.14
2016	83,186.83	16,220.19	14,281.44	1,938.75	0.00	7,116.60
2017	83,514.95	15,997.05	14,474.46	1,522.59	0.00	15,604.47
2018	69,786.80	18,270.72	16,433.79	1,836.93	0.00	32,166.74
2019	c. 55,000.–	c. 16,000.–				c. 30,500.–

www.ingramcontent.com/pod-product-compliance
Lightning Source LLC
Chambersburg PA
CBHW070718160426
43192CB00009B/1236